Annual Conference and Expo of the Association of Avian Veterinarians (AAV 2022)

Held at ExoticsCon 2022

Denver, Colorado, USA
14 – 18 August 2022

ISBN: 978-1-7138-7085-2

Printed from e-media with permission by:

Curran Associates, Inc.
57 Morehouse Lane
Red Hook, NY 12571

Some format issues inherent in the e-media version may also appear in this print version.

Printed with permission by Curran Associates, Inc. (2023)

For permission requests, please contact Association of Avian Veterinarians
at the address below.

Association of Avian Veterinarians
PO Box 9
Teaneck, NJ 07666
USA

Phone: 720-458-4111
Fax: 720-398-3496

office@aav.org

Additional copies of this publication are available from:

Curran Associates, Inc.
57 Morehouse Lane
Red Hook, NY 12571 USA
Phone: 845-758-0400
Fax: 845-758-2633
Email: curran@proceedings.com
Web: www.proceedings.com

Introduction

We invite you to enjoy the 2022 AAV Proceedings from ExoticsCon 2022. Within the proceedings, you will find a wealth of information related to the variety of lectures presented at this year's conference.

AAV recognizes and thanks the following committees and staff for their time and hard work in assuring a quality program and publication:

AAV Executive and Conference Committees

Byron De La Navarre, DVM, *President, Conference Committee Member*
Anna Osofsky, DVM, DABVP (Avian Practice), *President-Elect, Conference Committee Member*
Ashley Zehnder, DVM, PhD, *Immediate Past President, Conference Committee Chair*
Stephanie Lamb, DVM, DABVP (Avian Practice), *Education Committee Co-Chair*
Salina Locke, DVM, *Education Committee Co-Chair*
Susan Tyson, MS, VMD, *AAV Hands-On Workshop Coordinator*
Christina Knight, DVM, *AAV Hands-On Workshop Coordinator*
Len Donato, VMD, DABVP (Avian Practice), *Treasurer*
Brynn McCleery, DVM, DABVP (Avian Practice), *Advisory Council Representative to the AAV Executive Board*
Robert Groskin, DVM, *Executive Director*

Veterinary Review and Editing Committee

Natalie Antinoff
Gunther Antonissen
Heather Barron
Sue Chen
John Chitty
Bob Dahlhausen
Byron de la Navarre
Bob Doneley
M. Scott Echols
Jesse Fallon
Jennifer Graham
Cheryl Greenacre
David Guzman
Rudiger Korbel
Stephanie Lamb
Isabelle Langlois
Elizabeth Mackey
An Martel
Brynn McCleery
Mark Mitchell
Deborah Monks
Bianca Murphy
Olivia Petritz
Christal Pollock
Patrick Redig
Jaime Samour
Nico Schoemaker
Brian Speer
Marcy Souza
Tom Tully
Yvonne van Zeeland
Ken Welle
Paul Welch
Shangzhe Xie

Technical Editor

Charlotte Kirk Baer, MS

Proceedings Layout and Design — BrightLizard, Inc.

Elizabeth Smith

Disclaimer

The material appearing in this publication comes exclusively from the authors and contributors identified in each manuscript. The techniques and procedures discussed reflect the personal knowledge and experience of the authors and contributors, and demonstrate their views of the methods that may be used for these medical procedures. The procedures demonstrated do not incorporate all known techniques, are not exclusive, and other techniques and technology may also be available. Any questions or requests for additional information concerning any of the manuscripts should be addressed directly to the authors.

AAV, does not research, review, or otherwise verify any of the information contained in this publication. Opinions expressed in this publication are those of the authors and contributors and not necessarily those of AAV. AAV is not responsible for errors or for opinions expressed in this publication.

AAV expressly disclaims any warranties or guarantees, expressed or implied, and shall not be liable for damages of any kind in connection with the material, information, techniques, or procedures set forth in this publication.

Publication Terms

Conference proceedings manuscripts are accepted as full manuscripts or as scientific abstracts. Scientific abstracts are accepted if the author(s) express(es) intent to submit their work for consideration for future publication in a peer-reviewed scientific journal. Several distinguished veterinary journals, including the *Journal of the American Veterinary Medical Association* and the *American Journal of Veterinary Research* may reject work that has been previously published of 250 words or more and included additional content such as figures, tables, illustrations, and references. Scientific abstracts published in the AAV Proceedings from ExoticsCon are still required to contain data and meaningful information to the reader.

For permission to reproduce or order copies of this publication, contact the AAV Office, www.aav.org, PO Box 9, Teaneck, New Jersey 07666, USA, e-mail pubs@aav.org.

For information on ExoticsCon educational conferences, visit https://www.exoticscon.org. For information on membership and other association programs visit www.aav.org.

Table of Contents

Association of Avian Veterinarians

Section 1: AAV Plenary Session

Section 2: Pharmacokinetics

Section 3: Analgesia and Anesthesia 1

Section 4: Imaging

Section 5: Wildlife 1

Section 6: AAV Masterclasses

Section 7: Diagnostics 1

Section 8: Surgery

Section 9: Case Reports

Section 10: Infectious Disease

Section 11: Wildlife 2

Section 12: Diagnostics 2

Section 13: Potpourri

Hybrid Sessions

Section 35: Hybrid: Mixed Topics

Section 36: Hybrid: Emergency and Critical Care

Technician Sessions

Section 37: Technical Topics

ExoticsCon Virtual Sessions

Addendum: ExoticsCon Virtual 2022 (AAV Sessions)

Author Index

Section 1

AAV Plenary Session

Byron De La Navarre, DVM
Session Chair

Evaluation of the Thermal Antinociceptive Effects of Hydromorphone Hydrochloride in Great Horned Owls (*Bubo virginianus*)

Marissa Rae Monopoli,
David Sanchez-Migallon Guzman, LV, MS, DipECZM (Avian), DipECZM (Small Mammal), Dipl ACZM,
Joanne Paul-Murphy, DVM, Dipl ACZM, Dipl ACAW,
Hugues Beaufrère, Dr med vet, PhD, DABVP (Avian Practice), DipECZM (Avian), Dipl ACZM,
Michelle G. Hawkins, VMD, DABVP (Avian Practice)

Session #2555

Affiliation: From School of Veterinary Medicine (Monopoli), and the Department of Medicine and Epidemiology, School of Veterinary Medicine (Guzman, Paul-Murphy, Beaufrère, Hawkins), University of California Davis, Davis, CA 95616, USA.

Note: Abstract format selected by author(s). See Front Matter for description of conference proceedings manuscripts submitted as scientific abstracts.

Abstract: Opioid drugs are the cornerstone therapy for moderate to severe pain management and studies have shown mu opioid agonists to be effective in some raptor species.[1-4] This is the first pharmacodynamic study to evaluate hydromorphone in great horned owls (*Bubo virginianus*). Six adult great-horned owls (five females and one male) received IM hydromorphone (0.3 and 0.6 mg/kg) or saline (0.9% NaCl) in a masked, within-subjects, complete crossover study. Agitation-sedation scores were assessed 1-2 minutes prior to each thermal challenge. Foot withdrawal response to increasing thermal stimulus was determined prior to treatment administration and 0.5, 1.5, 3, and 6 hours post-administration. Statistical analysis was performed with linear mixed modeling on withdrawal temperature as outcome variable using time, treatments, treatment order, baseline values, and all interactions as fixed effects, and birds as a random effect. Intramuscular administration of 0.3 and 0.6 mg/kg hydromorphone induced thermal antinociception and agitation. Mean withdrawal temperatures increased between 0.5-3 hours and 0.5-1.5 hours, following administration of 0.3 and 0.6 mg/kg hydromorphone, respectively. These results support the use of hydromorphone for pain management in this species. Further studies are needed to fully determine the analgesic and adverse effects of hydromorphone in great horned owls.

This study was supported by the Wild Animal Health Fund from the American Association of Zoo Veterinarians.

References

1. Guzman DS, Drazenovich TL, Olsen GH, et al. Evaluation of thermal antinociceptive effects after intramuscular administration of hydromorphone hydrochloride to American kestrels (*Falco sparverius*). *Am J Vet Res*. 2013;74:817-822.

2. Ceulemans SM, Guzman DS, Olsen GH, et al. Evaluation of thermal antinociceptive effects after intramuscular administration of buprenorphine hydrochloride to American kestrels (*Falco sparverius*). *Am J Vet Res*. 2014;75:705-710.

3. Guzman DS, Drazenovich TL, Olsen GH, et al. Evaluation of thermal antinociceptive effects after oral administration of tramadol hydrochloride to American kestrels (*Falco sparverius*). *Am J Vet Res*. 2014;75:117-123.

4. Guzman DS, Ceulemans S, Olsen GH, et al. Evaluation of the thermal antinociceptive effects of a sustained-release buprenorphine formulation after intramuscular administration to American kestrels (*Falco sparverius). J Avian Med Surg*. 2018;32(1):1-7.

Grey Parrot (*Psittacus erithacus*) Head Vasculature Imaging and Anatomy

Nick Kirk, DVM, and
M. Scott Echols, DVM, DABVP (Avian Practice)

Session #2615

Affiliation: From the Parrish Creek Veterinary Hospital and Diagnostic Center, Centerville, UT 84014, USA (Kirk, Echols), Kirk Enterprises, Nashville, TN 37214, USA (Kirk), Echols Veterinary Services, Salt Lake City, UT 84054, USA (Echols), Medical Center for Birds, Oakley, CA 94581, USA (Echols).

Note: Abstract format selected by author(s). See Front Matter for description of conference proceedings manuscripts submitted as scientific abstracts.

Abstract: Detailed vascular anatomy is poorly documented in most avian species. Additionally, most publications depict anatomy in 2-dimensional picture formats that do not accurately reflect the 3-dimensional orientation of the vasculature. Three deceased grey parrots (*Psittacus erithacus*) were submitted to the Comparative Anatomy Research Project (formerly the Grey Parrot Anatomy Project) for computed tomography (CT) imaging with BriteVu® post-mortem vascular contrast. Individual segmentation of micro-CT data sets are being reviewed to define the anatomy of the vascular system of the grey parrot head. Advanced 3-dimensional modeling using Avizo 7.1 will be used to segment the vasculature to better understand their relationship with the skeletal and soft tissue structures present in the head and beak. The major vasculature of the skull and beak is identified based on the *Handbook of Avian Anatomy: Nomina Anatomica Avium.*[1] These results allow improved diagnostic evaluation and surgical planning for diseases of the head and beak of grey parrots. At the time of writing, three grey parrot models were included in the study; however, additional samples are expected to minimize the effect of individual variation. The information gained from this study can serve as both a model for anticipated anatomy in grey parrots and other psittacine species and to perform similar studies in other animal models.

References

1. Baumel J. *Handbook of Avian Anatomy: Nomina Anatomica Avium.* 2nd Ed. Cambridge, MA: Harvard Univ Nuttall Ornithological; 1993:407-475.

Section 2

Pharmacokinetics

Trent Van Zanten, BSc (Hons), DVM;
Ashley Zehnder, DVM, PhD
Session Chairs

Pharmacokinetics of Cannabinoids and Metabolites in Orange-Winged Amazon Parrots (*Amazona amazonica*) Following Administration of a Hemp Extract

Mariana Sosa-Higareda, MVZ,
David Sanchez-Migallon Guzman, LV, MS, DipECZM (Avian), DipECZM (Small Mammal), Dipl ACZM,
Heather Knych, DVM, PhD, DACVCP,
Alexander V. Lyubimov, PhD, DABT,
Beatriz Gomez, BS,
Alexander Zakharov, PhD,
Hugues Beaufrère, Dr med vet, PhD, DABVP (Avian Practice), DipECZM (Avian), Dipl ACZM

Session #2595

Affiliation: From the William T. Pritchard Veterinary Medical Teaching Hospital (Sosa-Higareda), Department of Medicine and Epidemiology, School of Veterinary Medicine (Guzman, Beaufrère) and K. L. Maddy Equine Analytical Pharmacology Laboratory (Knych), School of Veterinary Medicine, University of California-Davis, Davis, CA 95616, USA, Department of Pharmacology, Toxicology Research Laboratory (Lyubimov, Gomez, Zakharov) University of Illinois Medical College, Chicago, Illinois, USA.

Note: Abstract format selected by author(s). See Front Matter for description of conference proceedings manuscripts submitted as scientific abstracts.

Abstract: Cannabinoids have been studied for their antinociceptive, anti-inflammatory, antipsychotic and antineoplastic effects in mammals,[1-4] making them attractive therapeutic options in birds with chronic pain. The goal of this study was to determine the oral single and multi-dose pharmacokinetic parameters of cannabinoids and metabolites in a commercial hemp (*Cannabis sativa*) extract (Ellevet, South Portland, ME) that contained cannabidiol (CBD), cannabidiolic acid (CBDA), Δ -tetrahydrocannabinol 9 (THC), tetrahydrocannabinolic acid, cannabigerol, cannabigerolic acid and cannabichromene (45.2, 49, 1.86, 0.56, 0.69. 1.21, 1.53 mg/mL respectively) in orange-winged Amazon parrots. Based on pilot data, a dose of 30/32.5 mg/kg of CBD/CBDA was elected. Eight fasted parrots were used in a single dose pharmacokinetic study with blood samples collected at collected at 0, 0.5, 1, 2, 4, 6, 8, 10, 12, 24 hours after administration. After a 4-week washout period, multiple doses of the hemp extract were administered to seven birds at the previous dose q12h for 7 days, and blood samples were collected at the previous time points. The plasma cannabinoids and specific metabolites (7-Hydroxycannabidiol (7-OH-CBD), 7-Carboxy cannabidiol (7-COOH-CBD), 11-hydroxy-9-tetrahydrocannabinol (11-OH-THC), 11-Nor-9-carboxy-Δ^9-tetrahydrocannabinol (COOH-THC) and tetrahydrocannabinol-9-carboxylic acid glucuronide (COOH-THC-Glu)) were measured by liquid chromatography-tandem/mass-spectrometry and pharmacokinetic parameters using non-compartmental analysis were calculated. Plasma biochemistry and lipoproteins were also obtained during the multiple dose study. For the single-dose CBD/CBDA, mean Cmax was 105/602.1 ng/ml with a Tmax of 2.5/1 hours and a terminal half-life of 2.92/1.72 hours. No adverse effects were detected during the multidose study.

References

1. Go YY, Kim SR, Kim DY, et al. Cannabidiol enhances cytotoxicity of anti-cancer drugs in human head and neck squamous cell carcinoma. *Sci Rep*. 2020;10(1). https://doi.org/10.1038/s41598-020-77674-y.

2. Hartsel JA, Boyar K, Pham A, et al. Cannabis in veterinary medicine: cannabinoid therapies for animals. In: Gupta RC, Srivastava A, Lall R, eds. *Nutraceuticals in Veterinary Medicine*. Cham, Switzerland: Springer International Publishing; 2019:121-155.

3. Ryan D, McKemie DS, Kass PH, et al. Pharmacokinetics and effects on arachidonic acid metabolism of low doses of cannabidiol following oral administration to horses. *Drug Test Anal*. 2021;13(7):1305-1317.

4. Rock EM, Limebeer CL, Parker LA. Effect of cannabidiolic acid and Δ9-tetrahydrocannabinol on carrageenan-induced hyperalgesia and edema in a rodent model of inflammatory pain. *Psychopharmacology (Berl)*. 2018;235(11):3259-3271.

Antimicrobial Treatment of Intraocular Infections: Pharmacokinetics of Enrofloxacin in the Aqueous Humor of Chicken Eyes

Ruediger T. Korbel, Prof Dr med vet habil, DipECZM (avian),
Katrin Fuchs, med vet,
Richard Dietrich, Dr med vet,
Erwin Märtlbauer, Prof Dr med vet, Dr med vet,
Monika Rinder, Prof Dr med vet

Session #2646

Affiliation: From Center for Clinical Veterinary Medicine, Clinic for Birds, Small Mammals, Reptiles and Ornamental Fish, LMU University of Munich, Oberschleissheim, Germany.

Note: Abstract format selected by author(s). See Front Matter for description of conference proceedings manuscripts submitted as scientific abstracts.

Abstract: Ocular diseases of birds frequently result from systemic infections. Among others, pathogens like *Salmonella* spp., *Mycoplasma* spp., *E. coli* or *Staphylococcus* spp. frequently cause intraocular infections leading to severe ophthalmological alterations, including uveitis, chorioretinitis and pectenitis.[1,2] Because of the blood-ocular barrier system, a natural mechanism of protection, the antibiotic treatment of intraocular bacterial infections in general is difficult. To date there are almost no scientifically based data on antimicrobial pharmacokinetics and treatment protocols of intraocular infections in birds available. Transduction studies using enrofloxacin in mammals indicating that this fluoroquinolone is appropriate to treat bacterial eye infections are of limited use in birds because of the substantial species-specific differences including the ocular blood supply.[3-7]

Using a specific avian animal model, 40 adult roosters (*Gallus gallus,* Lohman selected leghorn) received a single IM application of enrofloxacin at a dosage of 25 mg/kg BW. Serial blood samples and aqueous humor samples of the anterior eye chamber (12 samples collected from t = 0 to 32 hours post application) were taken. The concentrations of fluoroquinolones in serum and aqueous humor were determined using a competitive enzyme immunoassay (EIA) with a highly sensitive generic monoclonal antibody for fluoroquinolones.[8]

Concentrations of fluoroquinolones in serum peaked at t = 30 minutes at levels of 14.24 µg/ml. Within aqueous humor they peaked after 7.5 hours at 4.52 µg/ml. Concentrations remained at > 1 µg/ml for more than 16 hours in both serum and aqueous humor. The comparison of the measured aqueous humor concentrations with minimal inhibitory concentrations of relevant sensitive bacteria[9-11] suggests a suitability of enrofloxacin for treatment of intraocular bacterial infections of the chicken eye. Results may be transduced to other bird species and are thus of relevance for treatment of infectious intraocular diseases in birds, especially in the field of pet bird and raptor ophthalmology.

References

1. Korbel R, Schulze C, Sigg R, Rinder M. Comparative investigations in the healthy and diseased avian fovea region using ophthalmoscopy and optical coherence tomography (OCT). *Proc ICARE.* 2019;128.

2. Ammon JA. Augenerkrankungen bei Ziervögeln. Dissertation thesis. University of Munich, Germany. 2013.

3. Reese S, Korbel R, Liebich H-G. The eye (organum visus). In: König HE, Korbel R, Liebich H-G, eds. *Avian Anatomy Textbook and Colour Atlas.* 2nd Ed. Essex, England: 5M Publishing Ltd; 2016:216-242.

4. Cunha-Vaz J, Bernardes R, Lobo C. Blood-retinal barrier. *Eur J Ophthalmol.* 2011;21(S 6):S3-9. doi:10.5301/ejo.2010.6049.

5. Cunha-Vaz JG, Shakib M, Ashton N. Studies on the permeability of the blood-retinal barrier. I. On the existence, development, and site of a blood-retinal barrier. *Brit J Ophthalmol.* 1966;50(8):441.

6. Popp MK. *Enrofloxacin im Glaskörper an Equiner rezidivierender Uveitis erkrankter Pferde.* Dissertation thesis. University of Munich, Germany. 2011.

7. Krastev Z, Haritova A, Lashev L, Hubenov H. Enrofloxacin disposition in aqueous humour after subcutaneous administration in dogs. *Bulgarian J Vet Med.* 2011;14(4):221-230.

8. Acaroz U, Dietrich R, Knauer M, Märtlbauer E. Development of a generic enzyme-immunoassay for the detection of fluoro (quinolone)-residues in foodstuffs based on a highly sensitive monoclonal antibody. *Food Analyt Method.* 2020;13(3):780-792.

9. Kang J, Hossain MA, Park HC, Kim Y, et al. Pharmacokinetic and pharmacodynamic integration of enrofloxacin against *Salmonella enteritidis* after administering to broiler chicken by per-oral and intravenous routes. *J Vet Sci.* 2019;20(2):e15. doi:10.4142/jvs.2019.20.e15.

10. Kaspar H, Steinacker U, Römer A, et al. Bericht zur Resistenzmonitoringstudie 2018- Resistenzsituation bei klinisch wichtigen tierpathogenen Bakterien. Bundesamt für Verbraucherschutz und Lebensmittelsicherheit. Berlin, Germany. 2020.

11. Kronthaler FX. Untersuchungen zur in-vitro-Antibiotikaempfindlichkeit aerober Bakterien beim Wirtschaftsgeflügel mittels Agardiffusionstest und Bouillon-Mikrodilutionsmethode. Dissertation thesis. University of Munich, Germany. 2009.

Determining the Pharmacokinetic Properties of Two Different Doses of Meloxicam in Barred Owls (*Strix varia*)

Zoey Lex, BS,
Laken Russell, DVM,
Corinne Mayer, DVM,
Jordan Padlo, CVT,
Kimberley L. Boykin, DVM, MS,
Mark Papich, DVM, MS, DACVCP,
Mark Mitchell, DVM, MS, PhD, DipECZM (Herpetology)

Session #2660

Affiliation: From Department of Veterinary Clinical Sciences, School of Veterinary Medicine, Louisiana State University, Skip Bertman Drive, Baton Rouge, LA 70803, USA (Lex, Russell, Mayer, Padlo, Boykin, Mitchell), Clinical Pharmacology Laboratory, College of Veterinary Medicine, North Carolina State University, Raleigh, NC 27607, USA (Papich).

Note: Abstract format selected by author(s). See Front Matter for description of conference proceedings manuscripts submitted as scientific abstracts.

Abstract: Anthropogenic activities have negatively affected the lives of many birds, including owls. These interactions have led to a rise in cases admitted to wildlife rehabilitation facilities, especially with traumatic injuries. The Wildlife Hospital of Louisiana has had a more than two-fold increase in barred owl (*Strix varia*) cases presented over the past 5 years (2020, 95; 2014, 42). Because most of these animals present with traumatic injuries, it is essential that analgesics are included in their treatment plan. To date, no study has evaluated an analgesic in barred owls. The goals of this study were to determine the harmonic means, times to maximum concentration, and elimination half-lives for single intramuscular doses (1.0 mg/kg, 2.0 mg/kg) of meloxicam. Our hypotheses were that both doses would be safe and provide adequate plasma concentrations, but that there would be significant differences between doses. Twelve barred owls (1 mg/kg, n = 6; 2 mg/kg, n = 6) admitted to the Wildlife Hospital of Louisiana and determined to be clinically normal based on examination and complete blood count were used in this study. Meloxicam (Ostilox, 5mg/ml, VetOne) was administered intramuscularly in the right pectoralis muscle and blood samples were collected from the jugular vein prior to injection and at 30 minutes, 45 minutes, and 1, 2, 4, 6, and 12 hours post-administration. Samples were placed in lithium heparin tubes and plasma removed following centrifugation. Samples were stored at -80 ^{o}C until being analyzed. Meloxicam concentrations were measured using high performance liquid chromatography. No complications were identified with meloxicam dosing or blood sampling.

Determining a Plasma Retinol Reference Interval in the Quaker Parrot (*Myiopsitta monachus*)

Brandon J. Louth, BS, DVM,
J. Harrison Wellford, DVM,
J. Jill Heatley, DVM, MS, DABVP (Avian Practice), DABVP (Reptile and Amphibian Practice), Dipl ACZM

Session #2699

Affiliation: From Research Boulevard Pet and Bird Hospital, Austin, TX 78759, USA (Louth, Wellford), Department of Small Animal Clinical Sciences, College of Veterinary Medicine & Biomedical Sciences, Texas A&M University, College Station, Texas 77843-4474, USA (Heatley).

Note: Abstract format selected by author(s). See Front Matter for description of conference proceedings manuscripts submitted as scientific abstracts.

Abstract: The quaker parrot (*Myiopsitta monachus*) is a species of bird native to Argentina and other South American countries and it is a popular pet in the exotic pet community around the world. There is minimal knowledge on plasma concentrations of retinol in this or closely related species. Since this is a commonly kept bird and since complications due to poor husbandry and diet are often seen in this species, establishment of concentrations of the essential vitamins is important for appropriate diagnosis of medical conditions. This study evaluated and characterized an interval for plasma retinol concentrations in the quaker parrot. Blood (0.1-0.3 ml) of 20 clinically healthy subjects was drawn from the right jugular vein using a 25-gauge needle on a 1-cc syringe and placed in a green top lithium heparin tube. Blood was spun down using a centrifuge to obtain plasma samples then placed in a retinol standard and analyzed using LCMS analysis for retinol concentrations. With the selected sample size there was a statistically significant range analyzed for the plasma retinol values sampled to help determine a reliable reference interval for quaker parrots to allow clinicians the ability to interpret plasma retinol concentrations from collected blood samples. The novel procedure used in this study can be used to evaluate plasma retinol values with a sample volume as low as 50 μl.

References

1. Ueda K, Akashi F, Kawasaki M, et al. Effects of feeding on plasma concentrations of vitamin A in captive African penguins (*Spheniscus demersus*). *J Vet Med Sci.* 2019;81(11):1580-1585. doi.org/10.1292/jvms.19-0316.

Section 3

Analgesia and Anesthesia 1

Trent Van Zanten, BSc (Hons), DVM;
Ashley Zehnder, DVM, PhD
Session Chairs

Evaluation of Weight-Bearing, Locomotion, Thermal Antinociception, and Footpad Size with Carrageenan-Induced Inflammation in Cockatiels (*Nymphicus hollandicus*)

Nicole Mikoni, DVM,
David Sanchez-Migallon Guzman, LV, MS, DipECZM (Avian), DipECZM (Small Mammal), Dipl ACZM,
Hugues Beaufrère, Dr med vet, PhD, DABVP (Avian Practice), DipECZM (Avian), Dipl ACZM,
Joanne Paul-Murphy, DVM, Dipl ACZM, Dipl ACAW

Session #2537

2022 AAV House Officer Manuscript Competition Winner

Affiliation: From the Veterinary Medical Teaching Hospital (Mikoni), and the Department of Medicine and Epidemiology, School of Veterinary Medicine (Guzman, Beaufrère, Paul-Murphy), University of California Davis, Davis, CA 95616, USA.

Note: Abstract format selected by author(s). See Front Matter for description of conference proceedings manuscripts submitted as scientific abstracts.

Abstract: This study evaluated a novel carrageenan-induced inflammatory model using the left foot pad of the cockatiel (*Nymphicus hollandicus*), not previously described in the literature.[1-8] Sixteen healthy cockatiels were randomly assigned into two groups, either treatment (carrageenan injection, n = 8) or control (handling only, n = 8). Treatment of cockatiels involved unilateral subcutaneous injection of 0.05 ml of 1% lambda carrageenan solution into the left footpad. Control birds were handled in a similar manner without an injection. Following treatment or control procedures, measurements included weight-bearing perch load (for up to 336 hours), locomotive abilities when placed on a rotating perch (for up to 96 hours), thermal threshold withdrawal (for a 24- to 30-hour period), and both vertical and horizontal left footpad dimensions and degree of inflammation (up to 84 days). Statistical analysis was performed with linear mixed modeling and $p < 0.05$ was considered significant. Treatment cockatiels had a significant decrease in left foot weight-bearing load, increase in left footpad dimensions, and subjective gross evaluation of visible inflammation compared to control cockatiels. Rotational perch locomotion and thermal threshold withdrawal did not differ significantly between groups. Cockatiels injected with carrageenan returned to normal weight-bearing within 2 weeks; however, left footpad dimensions remained significantly increased up to 3 weeks post injection. These findings highlight that carrageenan footpad injection prompts a measurable and grossly visible inflammatory response and induces alterations in weight-bearing distribution in injected cockatiels. This model provides a novel method to evaluate inflammation and lameness in cockatiels.

This study was supported by the UC Davis Center for Companion Animal Health and the Richard M. Schubot Parrot Wellness and Welfare Program.

References

1. Patil KR, Mahajan UB, Unger BS, et al. Animal models of inflammation for screening of anti-inflammatory drugs: implications for the discovery and development of phytopharmaceuticals. *Int J Mol Sci.* 2019;20(18):4367.

2. Morris CJ. Carrageenan-induced paw edema in the rat and mouse. *Methods Mol Biol.* 2003;115-121.

3. Möller KÄ, Berge OG, Hamers FPT. Using the CatWalk method to assess weight-bearing and pain behaviour in walking rats with ankle joint monoarthritis induced by carrageenan: effects of morphine and rofecoxib. *J Neurosci Methods.* 2008;174(1):1-9.

4. Ito NMK, Noronha AMB, Bohm GM. Carrageenan-induced acute inflammatory response in chicks. *Res Vet Sci.* 1989;46(2):192-195.

5. Roach JT, Sufka KJ. Characterization of the chick carrageenan response. *Brain Res.* 2003;994(2):216-225.

6. Fehrenbacher JC, Vasko MR, Duarte DB. Models of inflammation: carrageenan- or complete Freund's adjuvant (CFA)-induced edema and hypersensitivity in the rat. *Curr Protoc Pharmacol.* 2012;56(1):5.

7. Paul-Murphy JR, Krugner-Higby LA, Tourdot RL, et al. Evaluation of liposome-encapsulated butorphanol tartrate for alleviation of experimentally induced arthritic pain in green-cheeked conures (*Pyrrhura molinae*). *Am J Vet Res.* 2009;70:1211-1219.

8. Guzman DSM, Douglas JM, Beaufrere H, Paul-Murphy JR. Evaluation of the thermal antinociceptive effects of hydromorphone hydrochloride after intramuscular administration to orange-winged Amazon parrots (*Amazona amazonica*). *Am J Vet Res.* 2020;81(10):775-782.

Behaviors Associated with a Carrageenan-Induced Model of Inflammation in Cockatiels (*Nymphicus hollandicus*)

Nicole Mikoni, DVM,
David Sanchez-Migallon Guzman, LV, MS, DipECZM (Avian), DipECZM (Small Mammal), Dipl ACZM,
Hugues Beaufrère, Dr med vet, PhD, DABVP (Avian Practice), DipECZM (Avian), Dipl ACZM,
Joanne Paul-Murphy, DVM, Dipl ACZM, Dipl ACAW

Session #2538

Affiliation: From the Veterinary Medical Teaching Hospital (Mikoni), and the Department of Medicine and Epidemiology, School of Veterinary Medicine (Guzman, Beaufrère, Paul-Murphy), University of California Davis, Davis, CA 95616, USA.

Note: Abstract format selected by author(s). See Front Matter for description of conference proceedings manuscripts submitted as scientific abstracts.

Abstract: This study evaluated behavioral changes associated with a novel method of inducing inflammation via carrageenan injection in the foot pad of the cockatiel (*Nymphicus hollandicus*), not previously described in the literature.[1-9] Sixteen healthy cockatiels were randomly assigned as either treatment (carrageenan injection, n = 8) or control (handling only, n = 8). Cockatiels in the treatment group received a subcutaneous injection of 0.05 ml of a 1% lambda carrageenan solution (TCI America) into the left footpad. Following treatment or control procedures, all cockatiels were video recorded individually for 9.5 hours. Ten minutes of video at each of 11 time points post injection and/or handling were evaluated. Three observers assessed 25 behaviors in addition to monitoring the crest position (i.e., erect, lowered, or flat) and mentation (i.e., alert, quiet, depressed, or non-vigilant) of each cockatiel. Differences of individual behavior scores between treatment and control cockatiels were assessed using serial Wilcoxon sum rank tests with a false discovery rate (FDR) of 5%. A P and q (after FDR adjustment) of 0.05 were used for statistical significance. A volcano plot was used to identify variables of interest. Interobserver reliability was assessed using an intraclass correlation coefficient for a two-way design for consistency among multiple raters. Individual behaviors that differed significantly between the two groups included focal preening ($q = 0.023$), burst preening ($q = 0.036$), and upright stance ($q = 0.036$). The level of agreement between observers varied between the individual behaviors. These findings demonstrate that carrageenan-induced inflammation leads to behavioral differences as noted between injected and unaltered cockatiels and may be valuable in evaluating inflammatory pain in small psittacine species.

This study was supported by the UC Davis Center for Companion Animal Health and the Richard M. Schubot Parrot Wellness and Welfare Program.

References

1. Anil SS, Anil L, Deen J. Challenges of pain assessment in domestic animals. *J Am Vet Med Assoc.* 2002;220(3):313-319.

2. Bufalari A, Adami C, Angeli G, Short CE. Pain assessment in animals. *Vet Res Commun.* 2007;31.1:55-58.

3. Paul-Murphy JR, Hawkins MG. Bird specific considerations: recognizing pain behaviour in pet birds. In: Gaynor JS, Muir WW, eds.). *Handbook of Veterinary Pain Management.* 3rd Ed. St. Louis, MO: Elsevier Mosby. 2015;536-554.

4. Turpen KK, Welle KR, Trail JL, Patel SD, Allender MC. Establishing stress behaviors in response to manual restraint in cockatiels (*Nymphicus hollandicus*). *J Avian Med Surg.* 2019;33(1):38-45.

5. Patil KR, Mahajan UB, Unger BS, et al. Animal models of inflammation for screening of anti-inflammatory drugs: implications for the discovery and development of phytopharmaceuticals. *Int J Mol Sci.* 2019;20(18):4367.

6. Morris CJ. Carrageenan-induced paw edema in the rat and mouse. *Methods Mol Biol.* 2003;115-121.

7. Roach JT, Sufka KJ. Characterization of the chick carrageenan response. *Brain Res.* 2003;994.2:216-225.

8. Fehrenbacher JC, Vasko MR, Duarte DB. Models of inflammation: carrageenan- or complete Freund's adjuvant (CFA)-induced edema and hypersensitivity in the rat. *Curr Protoc Pharmacol.* 2012;56(1):5.

9. Bateson P, Martin P. *Measuring Behaviour: An Introductory Guide.* 3rd Ed. Cambridge UK: Cambridge University Press; 2007:48-61.

Evaluation of a Point-of-Care Viscoelastic Coagulation Device in Hispaniolan Amazon Parrots (*Amazona ventralis*)

Laken Russell, DVM,
Thomas N. Tully, Jr., DVM, MS, DABVP (Avian Practice), DipECZM (Avian),
M. Ryan Smith, DVM, DACVECC

Session #2726

Affiliation: From the Department of Veterinary Clinical Sciences, Louisiana State University, School of Veterinary Medicine, Baton Rouge, LA 70803, USA.

Note: Abstract format selected by author(s). See Front Matter for description of conference proceedings manuscripts submitted as scientific abstracts.

Abstract: Patient-side viscoelastic coagulation monitors are easy to use, portable, and provide rapid results. Viscoelastic testing evaluates the formation and lysis of a clot over time, providing information to better assess a patient's true clotting potential.[1] Traditional tests, prothrombin time (pT) and partial thromboplastin time (pTT) measure only coagulation factor function.[2] Viscoelastic testing requires a small blood sample and is useful in diagnosing and treating coagulopathies. Currently, there is no standardized coagulation testing method available for birds.[2,3] In this cross-sectional study, a viscoelastic coagulation device, the Entegrion Viscoelastic Coagulation Monitor, Vet© (VCM-Vet©) was evaluated. Blood samples were obtained from 26 Hispaniolan Amazon parrots (HAP) via manual restraint. Results were recorded on the device as a graphical output and quantitative viscoelastic measurements. Results were reported using standard rotational thromboelastometry terminology: clotting time (CT), clot formation time (CFT), alpha angle (AA), maximum clot firmness (MCF), clot firmness amplitude at 10 (A10) and 20 (A20) minutes after clot formation, and clot lysis at 30 (LY30) and 45 (LY45) minutes.[2,3] Statistical analysis found that all parameters were normally distributed aside from clotting time in seconds. Clot lysis at 30 and 40 minutes were all 100%. There was no appreciable breakdown of the clot during the 60-minute device runtime, and there was no significant difference in any parameter based on sex. Resulting data were consistent across this particular population of HAP, enabling creation of reference ranges for this avian species. Based on our findings the VCM can be used to assess clot potential in HAP and possibly other avian species.

References

1. McMichael MA, Smith SA. Viscoelastic coagulation testing: technology, applications, and limitations. *Vet Clin Pathol.* 2011;40(2):140-153.

2. Strindberg S, Nielsen TW, Ribiero AM, et. al. Thromboelastography in selected avian species. *J Avian Med Surg.* 2015;29(4):282-289.

3. Keller KA, Guzman DSM, Aciern, MJ, et al. Thromboelastography values in Hispaniolan Amazon parrots (*Amazona ventralis*): a pilot study. *J Avian Med Surg.* 2015;29(3):174-180.

Outcomes and Complications Associated with Caudal-Thoracic and Abdominal Air-Sac Cannulation in 68 Birds

Dominic Byron-Chance, BVMedSci, BVM, BVS, GPCert(ExAP), MRCVS,
Lucía Gomez, DVM, MRCVS,
Ashton J. Hollwarth, BSc, BVMS, CertAVP (Zoo Med), MRCVS,
Thomas A.G. Dutton, BVM&S, CertAVP (Zoo Med), DipECZM (Avian), MRCVS

Session #2558

Affiliation: From Great Western Exotics, Unit 10, Berkshire House, County Business Park, Shrivenham Rd, Swindon SN1 2NR, UK.

Note: Abstract format selected by author(s). See Front Matter for description of conference proceedings manuscripts submitted as scientific abstracts.

Abstract: Air-sac cannulation is used as an emergency procedure in avian patients with severe upper respiratory compromise, as well as a means of routine ventilation for surgery of the head and neck. The objective of this retrospective study was to describe and quantify the complications associated with air-sac cannulation in birds. Medical records were retrieved for all patients that underwent caudal-thoracic or abdominal air-sac cannulation at a single centre between August 2004 and October 2020. Patient signalment, primary reason for air-sac cannulation, location of air-sac cannula placement, occurrence and category of complications encountered, and survival data were recorded. Eighty-four air-sac cannulas were placed in 68 birds across six orders. Eighty of 84 cases (95.2%) survived general anaesthesia for initial air-sac cannula placement. Forty-three of 80 (53.8%) cases survived to air-sac cannula removal. Complications were observed in 26 of 80 (32.5%) cases. Of these 26 cases, 3 (11.6%) cases died as a direct result of the complication. The most common reported air-sac complication was loss of patency (n = 19/26 [73.1%]). Right-sided placement was associated with a significantly reduced risk of complications ($P = 0.003$). Increased likelihoods of survival to air-sac cannula removal were seen with peri-interventional antibiotic use ($P = 0.007$), and in left-sided ($P = 0.03$) and intercostal ($P = 0.04$) placements. This study demonstrates a moderate risk of complication with a guarded overall prognosis for survival to air-sac cannula removal. Inter-costal and left-sided air-sac cannula placement, and the use of peri-interventional antibiosis appear to improve patient outcomes.

Relationship Between Arterial and Intraosseous Pressure in Pigeons (*Columba rupestris*) in a Hypovolemic Model

Po-Hsiang Wang, DVM, and
Pin-Huan Yu, DVM, PhD

Session #2637

Affiliation: From Institute of Veterinary Clinical Sciences, School of Veterinary Medicine, National Taiwan University, No.1, Sec. 4, Roosevelt Road, Taipei, 10617, Taiwan R.O.C.

Note: Abstract format selected by author(s). See Front Matter for description of conference proceedings manuscripts submitted as scientific abstracts.

Abstract: Blood pressure is an important vital sign in veterinary medicine; however, both direct and indirect blood pressure measurement methods are not practical due to invasiveness and inaccuracy, respectively.[1] Studies in African grey parrots showed that the inflow arterioles likely account for the retained arterial waveform radiating into the intraosseous space.[2] To further investigate if the routinely used intraosseous route could serve as a blood pressure assessing method, the aim of this research is to find out the relationship between intraosseous pressure (IOP) and arterial pressure (AP) in pigeons (*Columba rupestris*) using a hypovolemic model and whether needle placement sites affect the results. This research was approved by the Institutional Animal Care and Use Committee (IACUC) of National Taiwan University. Thirty healthy pigeons were included in the study and divided into two groups randomly. The intraosseous route was set up in the distal ulna in one group and in proximal tibiotarsus in another using a 21-gauge 1.25-inch hypodermic needle. At the same time, a 24-gauge 0.75-inch catheter was placed into the deep radial artery. AP and IOP were recorded simultaneously under four stages: 0%, 20%, 30% and > 40% of blood loss. Pressure data are descriptively summarized, with the percent of change of IOP at each stage compared to change of AP using a multilevel mixed effects linear model. Data showed a significant difference between each stage for AP and IOP, and both are positively correlated at each stage. Results of this study provide a potential alternative option for blood pressure assessment in hypovolemic pigeons, but more studies should be conducted before practical use.

References

1. Johnston MS, Hill AE. Precision of repeated, Doppler-derived indirect blood pressure measurements in conscious psittacine birds. *J Avian Med Surg*. 2011;25(2):83-90.

2. Peng Y-W, Chi C-H, Yu P-H. Relationship between arterial and intraosseous pressure in African grey parrots (*Psittacus erithacus*). *J Avian Med Surg*. 2022;35(4):402-411.

Section 4

Imaging

Dana Franzen-Klein, DVM, MS;
Emily Hsieh, DVM
Session Chairs

Comparison of Contrast Computed Tomography Renal Lesions and Uric Acid Values in Avian Cases

Nick Kirk, DVM, and
M. Scott Echols, DVM, DABVP (Avian Practice)

Session #2614

Affiliation: From Parrish Creek Veterinary Hospital and Diagnostic Center, 86 N 70 W St, Centerville, UT 84014, USA (Kirk, Echols), Kirk Enterprises, Nashville, TN 37014, USA (Kirk), Echols Veterinary Services, Salt Lake City, UT 84121, USA (Echols), Medical Center for Birds, 3805 Main St, Oakley, CA 94561, USA (Echols).

Abstract: Uric acid is commonly used for evaluation of kidney function in avian patients and may become significantly elevated with advanced renal disease, dehydration, or postprandially in carnivorous and piscivorous birds.[1] Advanced imaging such as computed tomography (CT) has become notable in avian medicine for increased detail and precise information about patient anatomy and pathology. Avian orders examined included Psittaciformes, Bucerotiformes, Anseriformes, Pelecaniformes, and Accipitriformes. Iodinated intravenous contrast (370 mg/ml) was used at 4-6 ml/kg given over 2 minutes. All patients underwent pre-, during- and post-contrast scans. The 'during' scan was taken 30-60 seconds into contrast administration. A post-contrast scan was started after the completion of contrast delivery and the 'during' scan. Contrast studies allowed for kidney shape, size and position characterization and renal contrast uptake and clearance. Renal lesions identified on contrast CT include renomegaly, atrophy, cysts, hypoperfusion, mass invasion, and avulsion/trauma. At the time of submission, 47 avian cases met case inclusion criteria which included the CT protocol described above identifying renal lesions and uric acid values collected within 6 months prior to imaging. While the published normal values for uric acid range between species, we designated values greater than 12.0 mg/dl as hyperuricemic.[2] Six of 47 cases had hyperuricemia. Of the 47 cases, 8/8 submitted for histopathology had confirmed renal disease with 2/8 of those cases having hyperuricemia. This study demonstrates poor correlation between hyperuricemia and renal lesions identified on contrast-based CT scans.

References

1. Vergneau-Grosset, C, Beaufrère H, Ammersbach M. Advances in clinical pathology and diagnostic medicine. In: Speer B, ed. *Current Therapy in Avian Medicine and Surgery.* St. Louis, MO: Elsevier; 2016:492-492.

2. Hawkins MG, Guzeman DSM, Beaufrère H, et al. Birds. In: Carpenter JW, ed. *Exotic Animal Formulary.* 5th Ed. St. Louis, MO: Elsevier; 2018:167-375.

Intraosseous Iodinated Contrast Administration is Comparable to Intravenous Administration for Computed Tomography in Birds

Rachel Jania, DVM, Dipl ACVR,
Thomas N. Tully, Jr., DVM, MS, DABVP (Avian Practice), DipECZM (Avian),
L. Abbigail Granger, DVM, Dipl ACVR

Session #2509

Affiliation: From the Department of Veterinary Clinical Sciences, Louisiana State University, School of Veterinary Medicine, Baton Rouge, LA 70803, USA.

Note: Abstract format selected by author(s). See Front Matter for description of conference proceedings manuscripts submitted as scientific abstracts.

Abstract: Establishing and maintaining intravenous access for contrast medium during computed tomographic imaging can be difficult in birds due to their small size and thin-walled vessels.[1] Intraosseous (IO) catheters are an alternative to intravenous catheters and are often used for fluid or medication administration in birds.[2] To determine the efficacy of intraosseous iodinated contrast enhancement for computed tomography in birds, ten adult Hispaniolan Amazon parrots (*Amazona ventralis*) weighing 260 to 325 g, underwent pre and post contrast computed tomographic (GE Lightspeed 16-slice scanner; GE Medical Systems, Milwaukee, WI, USA) scans using both routes of contrast administration with a wash out period of at least 1 week between the two methods. Scans were evaluated subjectively for diagnostic quality, and objectively with Hounsfield units measured over three organs: the brain, right kidney, and liver. All scans were diagnostic and there was no measurable difference in contrast enhancement between the two methods in any of the three organs. Subjectively I0 catheters were technically more difficult to place. Once in place, the IO catheters were easier to manipulate for the imaging procedure and provided no complications upon removal. Minimal adverse side effects were noted from the 10 catheters. Intraosseous iodinated contrast administration is safe and comparable in enhancement characteristics to intravenous administration for computed tomographic imaging in avian species. The confirmation that 10 administration of an iodinated contrast agent is as effective as IV placement will lend confidence to those that perform this procedure. :

References

1. van Zeeland Y, Schoemaker NJ, Hsu EW. Advances in diagnostic imaging. In: Speer BL, ed. *Current Therapy in Avian Medicine and Surgery*. St. Louis, MO: Elsevier; 2016:531-549.

2. Dubé C, Dubois I, Struthers J. Intravenous and intraosseous fluid therapy in critically ill birds of prey. *J Exot Pet Med*. 2011;20(1):21-26.

Validation of a Portable X-ray Fluorescence Analyzer for Quantification of Bone Lead in Trumpeter Swans

Miranda Torkelson, DVM, CWR, CertAqV, and
Aaron Specht, PhD

Session #2609

Affiliation: From the Wildlife Rehabilitation Center of Minnesota, 2530 Dale St N, Roseville, MN 55113, USA (Torkelson), School of Health Sciences, Purdue University, 550 Stadium Mall Drive, West Lafayette, IN 47907, USA (Specht).

Note: Abstract format selected by author(s). See Front Matter for description of conference proceedings manuscripts submitted as scientific abstracts.

Abstract: Lead toxicosis is a common cause of morbidity and mortality in wildlife. Diagnosis is typically achieved through blood lead testing; however, this value does not necessarily correlate with total body burden of lead.[1] Blood lead level also does not differentiate between acute and chronic lead exposure.[2] Portable x-ray fluorescent (XRF) analyzers are used in industry to noninvasively quantify the concentrations of elements present in various materials. Previous studies have validated the use of portable XRF analyzers to measure bone lead in children, California condors (*Gymnogyps californianus*), common loons (*Gavia immer*), and wedge-tailed eagles (*Aquila audax audax*).[3-6] Bone lead is more reflective of total body burden, as the majority of lead is stored in mineralized tissue.[7] We sought to validate the use of a portable XRF analyzer on trumpeter swans (*Cygnus buccinator*) to understand the degree and chronicity of lead exposure. We performed measurements on tarsometatarsal bones from trumpeter swans using both a commercial portable XRF analyzer (Niton XL3t 950 GOLDD+, Thermo-Fisher, Waltham, MA USA) and inductively coupled plasma mass spectrometry (ICP-MS). XRF measurements were taken on bone through soft tissue, on excised bone, and on soft tissue alone. Significant correlations were found between bone XRF measurements with and without soft tissue. A linear relationship was found between XRF measurements of bone lead and ICP-MS results. These results suggest that a portable XRF analyzer can be used to measure bone lead in trumpeter swans. Future study will determine if these data better predict patient outcomes in rehabilitation.

References

1. Tell I, Somervaille LJ, Nilsson U, et al. Chelated lead and bone lead. *Scand J Work Environ Health.* 1992;18(2):113-119.

2. Barbosa Jr F, Tanus-santos JE, Fernanda Gerlach R, Parsons PJ. A critical review of biomarkers used for monitoring human exposure to lead: advantages, limitations, and future needs. *Environ Health Perspect.* 2005;113:1669-1674.

3. Specht AJ, Lin Y, Weisskopf M, et al. XRF-measured bone lead (Pb) as a biomarker for Pb exposure and toxicity among children diagnosed with Pb poisoning. *Biomarkers*. 2016;21(4):347-352.

4. Specht AJ, Parish AN, Wallens EN, et al. Feasibility of a portable X-ray fluorescence device for bone lead measurements of condor bones. *Sci Total Environ.* 2018;615:398-403.

5. Specht AJ, Kirchner KE, Weisskopf MG, Pokras MA. Lead exposure biomarkers in the common loon. *Sci Total Environ.* 2019;647:639-644.

6. Hampton JO, Specht AJ, Pay JM, et al. Portable X-ray fluorescence for bone lead measurements of Australian eagles, *Sci Total Environ.* 2021;789:147998.

7. Hu H, Rabinowitz M, Smith D. Bone lead as a biological marker in epidemiologic studies of chronic toxicity: conceptual paradigms. *Environ Health Perspect.* 1998;106:1-8.

Validation of Optical Coherence Tomography (OCT) Results in Birds Using Confocal Laser Scanning Microscopy (CLSM) and Bright Light Histology

Ruediger T. Korbel, Prof, Dr med vet habil, DipECZM (Avian),
Maren Kiessling, Dr med vet,
Christoph Schmitz, Prof, Dr med,
Katharina Düfert-Kammerloher, med vet,
Elisabeth Hagen, Dr med vet

Session #2647

Affiliation: From the Center for Clinical Veterinary Medicine, Clinic for Birds, Small Mammals, Reptiles and Ornamental Fish, LMU University of Munich, Oberschleissheim, Germany (Korbel, Düfert-Kammerloher, Hagen) and Faculty of Medicine, Institute of Anatomy, LMU University of Munich, Munich, Germany (Kiessling, Schmitz).

Knowledge of the micromorphology of the anterior and especially posterior eye segment with the retinal and choroidal structures and thickness including the avian foveae and pecten is indispensable for diagnostics and therapy of ocular fundus lesions and disease-related conditions both in mammals[1,2] and in avian medicine. Within the latter special consideration of infectious (i.e., ocular borna virus manifestations)[3] and frequently occurring trauma related lesions[4] is of special interest. Standard reference values of the corneal thickness comparing central (CCT) and peripheral avian corneal thickness (PCT) in diurnal and nocturnal raptors and various poultry species (chickens, turkeys) as well as for the raptor fovea[5-9] had been established in the past for ocular surgery and evaluation of light related conditions and comparison with pathologic conditions earlier using fast domain Fourier anterior segment optical coherence tomography (AS-OCT).

Optical coherence tomography (OCT) is a sophisticated clinical diagnostic method established for human ophthalmology, which allows contact free, non-invasive cross-sectional imaging of anterior and posterior ocular structures at almost histologic resolution in the living individual. It is a highly sensitive interferometric, image-based technique, that measures light, reflected from ocular tissue layers and which is used in human ophthalmology being developed as one of three basic approaches besides diffraction and diffuse optical tomography since the early 1980s. Based on initially case description based publications in raptors[10] and chickens[11] this technique was established for routine use for both the anterior[5] and posterior eye segment[9] in commercial kept poultry for measuring low light induced ocular conditions of animal welfare relevance such as dim light buphthalmos (DLP) and light induced avian glaucoma (LIAG)[12-14] pet birds and raptors.[12,13,15-21] Using specific additional optical devices OCT may be used for anterior segment examination as well (AS-OCT). Clinically OCT resp. AS-OCT allows in vitro examination of the posterior and anterior eye segment with most detailed 2-D and 3-D sections through the various ocular layers such as retina, fovea and choroid as well as the cornea at almost histologic resolution (resolution approximately 10-15 µm). General advances of OCT include contact-free and non-invasive operation. The possibility for continuous recording of clinical stages of diseases and therefore progressive studies is of major value as could be shown for avian borna virus (ABV) related ocular conditions.[12,13,15] Even though the unit used (SPECTRALIS HRA+OCT unit, Heidelberg Engineering, Heidelberg/Germany) is equipped with a so-called active eye tracking function (i.e., retrieving specific fundus areas after patient head/eye movements). We found volume-based scans taking up to 20-30 seconds with the complete OCT scan examination taking several minutes. General anesthesia therefore might be advantageous in easily stressed prey species (e.g., chickens and psittacine birds). OCT examinations in raptor species can be performed without anesthesia resulting in excellent images. Evaluation of the images recorded, demonstrated OCT being a suitable tool even in smaller avian eyes including the advantages of high depth and transversal resolution and a depth resolution, which is decoupled from transverse resolution as well as high probing depth in scattering media. For specific use in avian ophthalmology and easily stressed patients, the contact-free and non-invasive operation is of major value as well as the possibility to create various function dependent image contrasting methods.

Confocal laser scanning microscopy (CLSM) is an advanced optical imaging technique, resulting in high optical resolution and contrast images by means of using a spatial pinhole to block out-of-focus light in image formation and capturing multiple 2-D images at different depths in a sample to allow the reconstruction of 3-D structures within an object.

The aim of the first study was to validate OCT results originating from the posterior eye segment of a variety of wild bird species by means of confocal laser scanning microscopy (CLSM) with fifty healthy eyes representing seven different diurnal raptor species (n = 27 individuals) and four eyes of two nocturnal species, which underwent complete routine ophthalmologic examination procedure including OCT (study 1). The wild birds within the study had to be sacrificed by means of (non-eye-related) non-treatable animal welfare issues allowing the eyes to be enucleated, histologically processed and examined. The retinae were immunohistochemically labelled and visualized using CLSM. The depths of the foveae, individual retinal layer thicknesses and pathologic findings of in vivo OCT images were compared to values and findings found using ex vivo CLSM. As a second study OCT examination results of the eyes of 50 commercial kept broiler chickens kept under a standardized light protocol (50 lux with a consecutive step-down-protocol down to 5 Lux)[8] were examined using bright light histologic techniques and Azan staining of the cornea with the eyes taken after slaughtering at the end of the standard (42-day) fattening period.[14]

As to results concerning study 1, all ocular structures examined using OCT showed low correlation with histology results with increasing correlation towards increasing layer thickness and additive measurement of several layers. There was a high correlation for total retinal thickness and a low correlation and high dispersion of values for foveal depths. Pathologic changes and trends in OCT accurately correlated with histology. The investigated chicken corneae of the second study seem to show a decreasing thickness with increasing fattening days and decreasing light intensity in OCT and brightfield microscopy.

In conclusion, measurements of ocular tissues using OCT should be viewed critically, as within our study we were able to show, that a significant deviation between OCT images and demonstration of ocular layers with histomorphologic images and values resulting from CLSM is given. The nature of problem is complex and relevant not only for avian and mammalian interpretation of OCT images but is given also in human ophthalmology. Potential reasons may include influences of fixation techniques, image resolution and profound technical differences in both techniques. Further studies are necessary to develop a standardized interpretation scale to compare OCT and histologic images and interpretation of pathologic changes using OCT in the living individual image. OCT is a perfect tool for examining and monitoring ocular alterations in the living animal, as it allows a better evaluation of developmental ocular disorders in the living patient and thus is a solid basis for prognosis and monitoring therapeutic processes.

References

1. Kafarnik C, Fritsche J, Reese S. In vivo confocal microscopy in the normal corneas of cats, dogs and birds. *Vet Ophthalmol.* 2007;10(4):222-230.

2. Werther K, Candioto CG, Korbel R. Ocular histomorphometry of free-living common kestrels (*Falco tinnunculus*). *J Avian Med Surg.* 2017;31(4):319-326.

3. Rinder M, Korbel R, Schlottau K, Hoffmann B. Diagnosis of bornavirus infections in birds using a broad-range RT realtime PCR. *Proc ICARE.* 2017;660.

4. Korbel R, Hagen E, Rinder M, Medizinische Versorgung von verletzt aufgefunden heimischen Wildvögeln. *In J. Asmus (Hrsg.) Wildvogelhaltung.* 2021:387-409.

5. Korbel R, Sigg G. Central and peripheral avian corneal thickness measured with optical coherence tomography (OCT) in the normal eye. *Proc ICARE.* 2015.

6. Velasco-Gallego ML. Imaging of physiological retinal structures in various raptor species using optical coherence tomography (OCT). Vet Dissertation. München, Germany. 2015.

7. Schulze C. Zur Morphologie der physiologischen Foveae am Augenhintergrund von Greifvögeln und Eulen mittels Optischer Kohärenztomografie (OCT). Vet Dissertation. München, Germany. 2016.

8. Korbel R, Schneeganss A, Reisenbichler B, Moser K. Comparative evaluation of ocular reference values in growing broiler chicks using ultrasonography and optical coherence tomography. *Proc ICARE.* 2017.

9. Korbel R, Schulze C, Sigg R, Rinder M. Comparative investigations in the healthy and diseased avian fovea region using ophthalmoscopy and optical coherence tomography (OCT). *Proc ICARE.* 2019.

10. Ruggeri M, Major JC Jr, McKeown C, et al. Retinal structure of birds of prey revealed by ultra-high resolution spectral-domain optical coherence tomography. *Invest Ophthalmol Vis Sci.* 2010;51:5789-5795.

11. Moayed AA, Hariri S, Song ES, et al. In vivo volumetric imaging of chicken retina with ultrahigh-resolution spectral domain optical coherence tomography. *Biomed Opt Express.* 2011;2:1268-1274.

12. Korbel R. Investigations in the turkey eye using optical coherence tomography (OCT) as a basis to evaluate light induced alterations in the turkey eye. *Proc Hafez Int Symp Turkey Dis.* 2013.

13. Korbel R, Schulze C, Velasco-Gallego ME, et al. Optical coherence tomography (OCT) investigations in the filtration angle and other ocular structures of the turkey eye. *Proc Hafez Int Symp Turkey Dis.* 2016.

14. Düfert-Kammerloher K. Lichtmikroskopische Untersuchung der kornea am masthähnchenauge unter diskontinuierlichen beleuchtungsbedingungen. Vet Dissertation. Munich, Germany. In prep.

15. Korbel R, Rinder M. Ophthalmological findings in birds affected with PDD. *Proc Annu Conf Assoc Avian Vet.* 2011;215.

16. Korbel R, Barwig S, Rinder M. Vergleichende untersuchungen zur ophthalmoskopie, digitalen scanning ophthalmoskopie (SDO) und optischen kohärenztomographie (OCT) unter besonderer berücksichtigung bornavirus-assoziierter ophthalmologischer befunde am vogelauge. *Proc DVG.* 2012;43-46.

17. Korbel R. Ocular 2D and 3D ultrasonography and ocular coherence tomography (OCT). *Proc Annu Conf Assoc Avian Vet.* 2012;27.

18. Velasco-Gallego ML, Rinder R, Schulze C, et al. Imaging of physiological retinal structures in various raptor species using optical coherence tomography (OCT). *J Vet Ophthalmol.* In prep.

19. Hagen E. Validierung von OCT-befunden am hinteren augensegment von greifvögeln mittels konfokalmikroskopie. Vet Dissertation. Munich, Germany. 2020.

20. Hagen E, Kiessling M, Schmitz C, Korbel R. Optische kohärenztomographie bei greifvögeln: validierung von befunden mittels konfokalmikroskopie. 4. *Jahrestagung der Fachgruppe Zier-, Zoo- und Wildvögel, Reptilien, Amphibien und Fische.* 2021;113-115.

21. Hagen E, Korbel R, Schmitz C. Validation of optical coherence tomography (OCT) in avian ophthalmology using confocal laser scanning microscopy (CLSM). *Vet Ophthalmol.* In prep.

Section 5

Wildlife 1

Dana Franzen-Klein, DVM, MS;
Emily Hsieh, DVM
Session Chairs

Utilization of Welfare and Health Monitoring Tools in Geriatric Raptors

Victoria Hall, DVM, MS, Dipl ACVPM,
Gail Buhl, BS,
Michelle Willette, DVM, MPH, Dipl ACVPM,
Dana Franzen-Klein, DVM, MS

Session #2685

Affiliation: From The Raptor Center, the University of Minnesota, 1920 Fitch Ave, St Paul, MN 55108, USA.

Note: Abstract format selected by author(s). See Front Matter for description of conference proceedings manuscripts submitted as scientific abstracts.

Abstract: In the zoo and nature center world, raptors serve as incredibly effective ambassadors for conservation. Coming often from wildlife rehabilitation settings, these birds have permanent disabilities that need to be managed their entire lives. Even captive bred raptors will have unique challenges as they age. Living substantially longer than they would in the wild, these captive individuals' present unique circumstances and needs as they age. Approaching the bird's care with a multifaceted program helps to ensure excellent welfare throughout their life. At The Raptor Center, this approach includes geriatric health monitoring and utilization of quality-of-life measurement tools including medical exams, ethograms, activity budgets, overall behavior tracking and positive reinforcement training. Approaching the bird's care as a team helps to ensure a more complete picture of their needs. By establishing baseline values and metrics for individuals prior to the development of chronic conditions, animal care providers can ensure that impacts on the welfare of each individual bird are measured in both a quantitative and qualitative way, and thus better inform and drive medical and management decisions for the life of that animal.

Post-Release Survival of California Brown Pelicans (*Pelecanus occidentalis californicus*) Rehabilitated 2009-2019

Rebecca S. Duerr, DVM, MPVM, PhD

Session #2705

Affiliation: From International Bird Rescue, 4369 Cordelia Rd., Fairfield, CA 94534, USA.

Abstract: During the period from 2009 to 2019 International Bird Rescue applied easily read blue plastic auxiliary leg bands to 1418 rehabilitated California brown pelicans prior to release, in addition to metal U.S. Geological Suvey federal bands. All birds were of independent age at admission, categorized as after-third-year (ATY) mature adults, sub-adults (SA) and hatch-year (HY) birds by plumage. As of April 4, 2022, 707/1418 birds (49.6%) were resighted at least once; the most-recent resighting for each bird was tallied with 561/707 (79.3%) last observed alive and 146/707 (20.7%) reported as dead. Care records were used to categorize medical problems during rehabilitation. These data were compared to post-release survival duration to ascertain whether specific injuries or conditions showed a relationship to duration of post-release survival. We hypothesized that all birds would have similar post-release survival duration by problem or injury type because each bird had to meet similar pre-release criteria but that age may affect survival. Medical problems considered against survival included severe emaciation (intake weight < 70% release weight), severe intake anemia (PCV < 20%), severe intake hypothermia (cloacal temperature < 100 °F/37.8 °C), general debilitation, fishing gear injuries, all wing injuries, patagium injuries, wing constriction injuries, all fractures, wing fractures, toe fractures, toe amputations, large bite wounds, pouch lacerations, and petroleum oiling. Overall, mean (SD) minimum survival time for all reported birds was 2.95 (3.1) years, median = 2.0 years. For birds last reported alive: longevity n = 561, mean (SD) = 1244 (1165) days; median = 898 days; range 0-4442 days (12.2 years). For those reported dead: n = 146, mean (SD) = 439 (671) days; median = 155 days; 1-3324 days (9.1 years). Longevity by major medical problems is shown in Table 1.

Table 1. Post-release survival by major categories of medical problems during rehabilitation.

	Total		Alive at most recent re-encounter (days)						Dead at most recent re-encounter (days)					
Medical problems of interest	N	% resighted	n	n(%)	Mean	StDev	Median	Range	n	n(%)	Mean	StDev	Median	Range
Severe emaciation	631	43%	221	81%	1168	1171.4	780	0-4170	53	19%	369	593.5	127	4-3324
Severe anemia	126	46%	50	86%	1243	1423.9	459	11-4307	8	14%	438	365.7	243	123-1074
Severe hypothermia	127	53%	56	84%	1181	1121.9	784	7-3688	11	16%	289	401.7	103	50-1138
General debilitation	510	48%	186	76%	1435	1286	1110	0-4307	58	24%	513	727.2	154	1-3324
Fishing gear injuries (all)	463	48%	185	83%	1188	1106.2	1017	1-4334	38	17%	269	452.6	91	9-2315
Wing injuries (all)	456	51%	185	80%	1208	1125	869	1-4334	47	20%	327	497	216	4-2319
Patagium damage	182	46%	71	85%	1433	1265	1061	5-4334	13	15%	452	656.6	296	9-2315
Constriction injury, wing	115	52%	53	88%	1307	1128.7	1061	28-4334	7	12%	444	835.2	78	9-2315
Wing fracture	72	58%	32	76%	1099	1046.7	778	11-3973	10	24%	288	244.9	252.5	14-860
Toe fracture	56	52%	26	90%	668	752.8	571	11-3258	3	10%	114	53.7	89	78-176
Any fracture	160	53%	66	79%	919	902.4	675	9-3973	18	21%	279	283.7	214	14-1078
Toe amputation	53	51%	23	85%	1071	1033.7	1087	5-4170	4	15%	201	89.5	210	89-296
CSL bite	110	50%	44	80%	1421	1214.1	1154	6-4170	11	20%	236	412.6	87	7-1441
Pouch laceration	116	41%	38	81%	1059	1066.6	773	22-3741	9	19%	396	554.5	216	12-1705
Petroleum oiled	21	57%	11	92%	1568	1357.6	1188	69-4238	1	8%	425			

Program MARK was used to estimate annual survival, by age, sex, and major medical problems, incorporating the likelihood of detection into the modeling process.[1] In brief, age had a significant effect on longevity post-release with ATY birds showing an approximate 10% increase in annual survival than birds stranding as HY post-fledges (76.5% vs 65.7%). Data from survival modeling shows reduced annual survival probabilities in birds with a history of wing fractures or toe fractures, although these findings were not significant in each age class. Birds that stranded due to petroleum oiling showed higher than average annual survival across all age classes. Analyses of these data are ongoing.

References

1. White GC. Analysis of Data from Marked Individuals. 2014. Available at: https://sites.warnercnr.colostate.edu/gwhite/program-mark/. Accessed May 21, 2022.

The Migrating Songbird: Injuries and Ailments

Renée Schott, DVM, CWR

Session #2523

Affiliation: From The Wildlife Rehabilitation Center of Minnesota, 2530 Dale St. N, Roseville, MN 55113, USA.

Abstract: Wild passerine birds often present to veterinary clinics during migration. They primarily present with trauma, but a variety of other illnesses often are seen as well. This interactive master class dives into the most common presentations of songbirds during migration. It covers how to diagnose, how to decide whether to treat or euthanize, and all the considerations that go into treating various species. Data from the largest wildlife rehabilitation center in the country is used to inform prognosis.

Introduction

Private practitioners are often regarded as experts across all taxa, regardless of the focus of their practice. Regular clients and the public occasionally will call with questions about injured or orphaned wildlife, or simply show up in the lobby with the animal. It behooves the clinic to be prepared to deal with these scenarios so that the finders have resources and options. It is important that clinics are prepared to handle and refer these calls appropriately for the sake of efficiency, the welfare of the wild animal, and the safety of the patients in the hospital. In addition, if there are no permitted rehabilitators in the area that treat wildlife, it is important for private practitioners to be armed with basic triage and euthanasia skills to humanely end the suffering of wildlife that cannot recover. This paper discusses generic background information regarding wildlife in a private practice setting and then goes into more detail about songbird-specific conditions. For the purposes of this paper, songbird is defined as any species within the order Passeriformes.

Data from The Wildlife Rehabilitation Center of Minnesota (WRCMN) also are presented. The WRCM admits >19,000 sick and injured wild animals each year. Asterisks below are used to indicate older WRCMN data. Updated data are presented during the master class. It is difficult to compare data from one rehabilitation center to another due to differences in location, resources, capacity, diseases, among other things. For data presented in this paper, "all" means every one of that condition that was admitted and evaluated, and "treated" means "all" except those animals that were dead on arrival or euthanized on arrival. For more details about how the data were compiled, please contact renee@wrcmn.org.

Preparation

Before wildlife is in the lobby or calls start coming in, clinics should prepare by following the steps below.

Find a wildlife rehabilitator or facility to work with

It is illegal to possess native wildlife without the proper permits. Veterinarians must be working with a permitted rehabilitator to possess wildlife they are treating. Conversely, most states require rehabilitators to have a licensed veterinarian to work with to keep their permits active. Rehabilitators know the natural history and you know veterinary medicine; working together ensures the best outcome for the animal. In addition, rehabilitators are knowledgeable about in which scenarios animals need to be rescued and when they should be left alone. Calls are best referred to rehabilitators or facilities set up to field these questions, as questions are complicated, and answers aren't always straight forward.

Compile resources

Where can you refer phone calls or injured patients? Do you know the addresses or hours of facilities in the area? Who is a contact at your local state fish and game department and at US Fish and Wildlife? Making a resources sheet and/or SOPs for your reception will limit miscommunications.

Know the laws

Typically, each state regulates who can rehabilitate wildlife and how. Permits are required, and federal regulations exist for migratory birds as well. While it is usually legal for veterinarians to treat wildlife, they cannot possess it for longer than 24-48 hours if they aren't already working with a permitted rehabilitator. There often are additional guidelines and regulations surrounding threated or endangered species. Keep in mind, species that have a hunting season in your area must be treated as food animals; withdrawal times and the list of prohibited medications apply.[1] More details about legal responsibilities and restrictions have been published.[2]

Euthanasia

For a wild animal to be caught, it is usually very debilitated. For an animal to be releasable, it needs to recover to near 100% and function as a truly wild animal; therefore, animals with disabilities (e.g., limb amputations, cognitive impairment, habituation) are unable to be released. Release proportions vary according to species, age and injuries, but range from 24%-85%.[3-6] Euthanasia is common, necessary and humane. In most areas, the number of animals that need rehabilitation are too numerous for the available resources. When considering the below cases, it is important to consider these resources, predicted demand, rehabilitator and facility limits, and welfare. The ability of the rehabilitator, facility or clinic to accept animals for rehabilitation will be dependent on the above factors. If resources are low or burnout is high, the overall ability of the caretakers will diminish, which will lower welfare for everyone, and decrease release proportions.

What types of conditions should be euthanized?

Schott (2018) provides more in-depth discussion of the conditions below.[7] Consider your ability, the ability of the wildlife rehabilitator, and the species. Below are generalizations that apply to most taxa unless specified otherwise.

- Necrotic or old, open fractures
- Traumatic joint injuries
- Permanent vision loss in >50% of one eye
- Spinal trauma with no deep pain
- Any condition that is causing chronic pain or may cause chronic pain in the future
- Shoulder fractures in aerial insectivores such as swallow, swift and nightjar species

Treatable Conditions

Mycoplasma gallisepticum conjunctivitis in house finches

Mycoplasma gallisepticum (MG) originally spread from the production poultry industry to house finches (*Haemorhous mexicanus*) in the 1990s[8] and is now endemic in the house finch population.[9] When it first appeared, the poultry industry and biologists were afraid of MG-recovered house finches becoming latent carriers and some antibiotics being ineffective;[10,11] however, it is difficult for wild house finches to actually transmit *M. gallisepticum* to domestic poultry[12,13] and there is a treatment that when given appropriately can eliminate the carrier state.[14] Euthanizing one or even several hundred house finches because they have this disease will not affect the population or disease status. MG is difficult to diagnose so any house finch with classic symptoms is presumed to have MG.

Treatment: Tylosin tartrate 1 mg/ml as sole source of drinking water for 21 days; ciprofloxacin ophthalmic drops OU q12h for 5-7 days, or until conjunctiva appears normal; meloxicam 0.5 mg/ml PO q12h for 5-7 days, or until conjunctiva appears normal.

Prognosis: Adult house finches with presumed MG have a great prognosis. From 2011-2018, 41% of all (n = 448) and 73% of treated recovered and were released. Most of the finches who were euthanized on arrival had other injuries that precluded rehabilitation (e.g., open fractures, luxated joints); house finches even with severe clinical signs (eyes swollen and crusted shut) have a great prognosis. For juveniles, 23% of all and 35% of treated recovered and were released; however, the sample size was much smaller (n = 80). Because of the difficulty in diagnosing MG (i.e., culturing is difficult, PCR is expensive), most diagnoses are made presumptively based on clinical signs.

Birds missing digits

Certain species can survive well in the wild with certain digits amputated. The type and number depend on the species and natural history. Recommendations exist in the literature;[15] however, recommendations and judgements change as new natural history and post-release data becomes available. WRCMN will usually amputate up to two digits on one foot in species that do not walk, crawl, creep, or hang vertically. For instance, an American robin (*Turdus migratorius*) may do well with two toes missing on one foot; however, a brown creeper (*Certhia americana*) may not because they climb vertically up trees. The contralateral foot must be monitored for pododermatitis prior to release.

Shoulder fracture in non-aerial insectivores

Simple, mid-shaft fractures of one of the shoulder girdle bones (coracoid, clavicle, scapula) in passerine birds that do not eat on the wing heal very well. Because of the complicated structure of the avian pectoral girdle, a 3-view radiograph series is ideal (VD, lateral and H-view[16]) to determine prognosis.

Treatment: Meloxicam 0.5-1 mg/kg PO q12h, tramadol 15 mg/kg PO q12h, supplemental calcium, wrap wing to body for 7 days, cage rest for 10 days total, then consider flight room for self-PT. The bird should be able to easily gain height and turn both ways in the air. Evaluate flight on slow motion video to ensure wing beats are symmetrical and extension is full.

Prognosis: As long as the shoulder joint is not affected (well-aligned, not dropped, no intra-articular fractures) and the fracture is simple, prognosis is good. From 2011-2018, songbirds with any type of shoulder (coracoid, scapula or clavicle fracture), (n = 645) the release proportion was 27% of all and 43% of treated.

Head trauma

Head trauma is very common in songbird wildlife admissions, especially during migration season when migrating groups and individuals hit windows or other structures. Head trauma in a songbird may present as severe periocular swelling, ataxia, other neurologic deficits or a slight decrease in activity level as compared to healthy individuals of that age or species. Always do a full ophthalmic exam.

Treatment: Meloxicam 1 mg/kg PO q12h, gabapentin 20 mg/kg PO q12h, eye lubricant if ocular trauma or inability to fully sweep third eyelid, oxygen cage if available, decreasing stress and handling is priority when treating head trauma.

Prognosis: From 2011-2018 patients with abnormal neurologic or brain clinical signs (e.g., nystagmus, ataxia, head tilt), which may or may not have been caused by trauma specifically, (n = 3159) had release proportions of 15% for all and 28% for those treated.

Cat exposure

Any animal found near (within a few feet) of a cat should be considered a cat-bite victim and should be treated as such. Cat's small teeth cause deep puncture wounds; the skin of the victim often seals by the time of presentation, making wounds difficult to see. Cat bites often damage internal structures and have secondary bacterial infections so all cat exposure patients should be put on prophylactic antibiotics and analgesics. Take the mentation of the patient into consideration; if the patient is fluffed, lethargic and not moving much, it likely has severe internal injuries.

Treatment: Clavamox (Zoetis, Parsippany, NJ) 125 mg/kg PO q12h for 7+ days (7 days if no injury seen, otherwise continue until any wounds healed), meloxicam 0.5-1 mg/kg PO q12h, tramadol 15 mg/kg PO q12h.

Prognosis: From 2011-2018, the release proportion of all cat exposure songbirds (n = 1529) was 21% and 38% for treated birds.

Dog attack

Dogs crush and shake their prey so generally injuries are a little more obvious. Injuries tend to be physical and less secondary bacterial infection, so antibiotics are only warranted if there is a visible injury that warrants it.

Treatment: Meloxicam 0.5-1 mg/kg PO q12h, tramadol 15 mg/kg PO q12h.

Prognosis: From 2011-2018, the release proportion of all dog attack songbirds (n = 807) was 21% and 40% for treated birds.

Subcutaneous emphysema

Punctures into an air sac often present as pockets of air under the skin called subcutaneous emphysema (Fig 1). While the instinct is to puncture the emphysema with a needle, it always refills with air soon after (e.g, in minutes or hours) and this also introduces new bacteria into the body. Puncturing the pockets does not speed healing either. WRCMN will leave the emphysema unless it is causing a problem with ambulation, eating, or other activities in which case the emphysema will be punctured, and a tape vest will be applied to keep it from refilling.

Treatment: Trimethoprim-sulfa 50 mg/kg PO q12h until emphysema resolved, meloxicam 0.5-1 mg/kg PO q12h, tramadol 15 mg/kg PO q12h. Do not administer SC fluid therapy or medications until the emphysema is resolved.

Prognosis: From 2011-2018, the release proportion of all songbirds admitted with subcutaneous emphysema (n = 267) was 17% and 27% for treated birds.

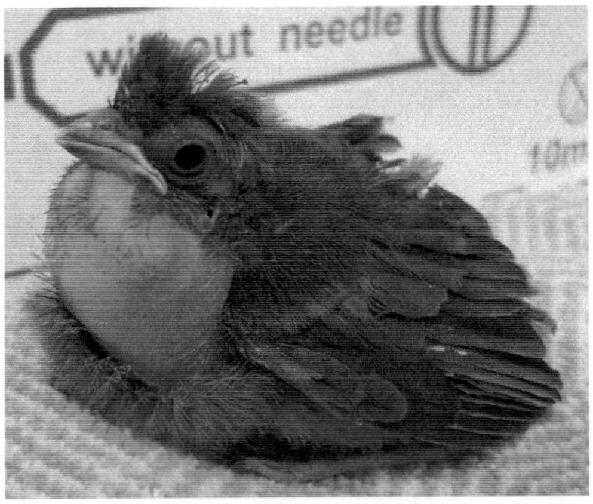

Figure 1. Young Northern cardinal chick with severe subcutaneous emphysema in the cervical region.

Severe emaciation

Severe emaciation often is secondary to another injury/illness or secondary to the end of a long migration, or at the start of first-year birds' migration. Look for an underlying cause; if none is found, consider the season and species when making decisions on these patients.

Treatment: Heat support (i.e., incubator), fluid support (10-20% BW [kg] SC fluids q12h), nutritional support if not eating (e.g., gavage feed Emeraid omnivore/carnivore 3% BW q12h), analgesics.

Prognosis: Prognosis is poor if severely emaciated and slightly better if very thin. Adult American robins from 2011-03/2019 with emaciated body condition (n = 51) had a release prognosis of 6% and 12% of the treated ones.

Lead (Pb) toxicity

While uncommon overall, songbirds from industrialized areas,[17] shooting ranges,[18] or other areas with lead contamination may be prone to lead toxicity. Due to the small size and high stress of passerine birds, blood lead levels (BLL) are not routinely checked. WRCMN has found severe lead toxicity in adults of species presenting with neurologic signs and no other indication of trauma in species that are not routinely admitted (i.e., adult blue jays, common grackles). It is recommended to check the BLL of passerines with odd presentations or persistent neurologic signs when safe. Likely, more testing will reveal more positives. WRCMN treats all BLL over 10 µg/dl.

How to Euthanize Songbirds

The techniques described below are used most frequently at WRCMN (Table 1). There are other technique options; however, the ones below have proven reliable and have the least amount of apparent stress induced to the animal in the author's experience. Euthanasia via isoflurane overdose is in the AVMA 2020 Euthanasia Guidelines.[19]

Table 1. Euthanasia techniques.

Species or group	Age	First choice	Second choice	Notes
Songbird species	Any	Isoflurane box alone	Isoflurane via facemask then pentobarbital IV or IC	Birds have such efficient gas exchange in their lungs, they become anesthetized in two breaths in the isoflurane box

Select euthanasia techniques for wild songbirds

Isoflurane (or sevoflurane) via face mask: Restrain the patient and place head in mask through diaphragm or hook up vaporizer to an induction box with the animal inside. Start at maximal percentage until patient is fully anesthetized. Then administer appropriate amount of pentobarbital IV or IC.

Isoflurane box:

- Liquid isoflurane is poured into a clear Rubbermaid-type container with a lid. Either do this in a vent hood or ensure container is on the ground/floor and you remain above to prevent accidental inhalation of isoflurane, which is heavier than room air. Ideally, a respirator would be worn to avoid inhalation of all fumes when not in the hood.
- Use 1 tablespoon of liquid isoflurane for every 4 gallons container volume.
- The box should be mostly sealed (but not completely sealed as the gas expands).
- After isoflurane is poured into the container, the animal (comfortably wrapped in a small towel) can be transferred to the container and the lid replaced.
- Monitor the patient to ensure it quickly loses consciousness.
- Leave patient in the tub for at least 10 minutes.
- Rock tub before opening to ensure no movement.
- Open tub outside (tub on ground, ensure you are above tub or are wearing a respirator) or in a vent hood.
- Use a stethoscope to ensure heartbeat of the patient has stopped.
- Leave tub outside or in vent hood for at least 10 minutes.

Basic Wildlife Stabilization

During all of the steps described below, decreasing stress is the most important thing we can do for wildlife. Stress will kill wildlife and often a hands-off protocol is preferable over an invasive one.

First. Do visual triage. Is there an obvious injury that would render the animal non-releasable? If so, euthanize immediately. Immediately treat any breathing difficulties (place in oxygen chamber) or active hemorrhage (apply bandage to stop bleeding).

Second. Treat thermoregulatory deficits. If the animal is hypothermic, warm it up. If the animal is hyperthermic, cool it down.

Third. Leave the animal alone! Leave the animal in a quiet, dark room so it can destress and relax. This also gives the animal time to adjust internal body temperature.

Fourth. Give fluid therapy. Assume at least 5% dehydration in all animals admitted. Subcutaneous fluids are fast to administer and a great choice in almost all cases. If gaping, use oral fluids for baby songbirds instead of subcutaneous. Songbirds have a much higher metabolic rate compared to larger birds; assume a maintenance daily fluid requirement of 100-120 ml/kg/d.

Fifth. Give pain medications. If there is any injury, appropriate pain management is paramount.

References

1. Schott R. Extra-label drug use in wildlife rehabilitation medicine. *Wildl Rehabil Bull.* 2017;35.

2. Duerr R, Whittington JK. Legal responsibilities and restrictions on veterinarians working with wildlife. *Wildl Rehabil Bull.* 2017;35:25-37.

3. Kelly A, Bland, M. Admissions, diagnoses, and outcomes for Eurasian sparrowhawks (*Accipiter nisus*) brought to a wildlife rehabilitation center in England. *J Raptor Res.* 2006;40:231-235.

4. Molina-López RA, Casal J, Darwich L. Final disposition and quality auditing of the rehabilitation process in wild raptors admitted to a wildlife rehabilitation centre in Catalonia, Spain, during a twelve year period (1995-2007). *PLoS ONE.* 2013;8:1-8.

5. Fix AS, Barrows SZ. Raptors rehabilitated in Iowa during 1986 and 1987: a retrospective study. *J Wildl Dis.* 1990;26:18-21.

6. Orós J, Montesdeoca N, Camacho M, et al. Causes of stranding and mortality, and final disposition of loggerhead sea turtles (*Caretta caretta*) admitted to a wildlife rehabilitation center in Gran Canaria Island, Spain (1998-2014): a long-term retrospective study. *PLoS ONE.* 2016;11:1-14.

7. Schott R. Basic wildlife rehabilitation triage. *Proc ExoticsCon.* 2018;167-176.

8. Citti C, Blanchard A. Mycoplasmas and their host: emerging and re-emerging minimal pathogens. *Trends Microbiol.* 2013;21:196-203.

9. Dhondt AA, DeCoste JC, Ley DH, Hochachka WM. Diverse wild bird host range of *Mycoplasma gallisepticum* in Eastern North America. *PLoS ONE.* 2014;l 9.

10. Williams ES, Yuill T, Artois M, et al. Emerging infectious diseases in wildlife. *Rev Sci Tech.* 2002;21(1): 139-57.

11. Wellehan JF, Zens MS, Calsamiglia M, et al. Diagnosis and treatment of conjunctivitis in house finches associated with mycoplasmosis in Minnesota. *J Wildl Dis.* 2001;37:245-251.

12. Stokholm NM, Permin AA, Bisgaard BM, Christensenac JP. Causes of mortality in commercial organic layers in Denmark. *Avian Dis.* 2010;54(4):1241-1250.

13. Luttrell MP, Stallknecht DE, Kleven SH, et al. *Mycoplasma gallisepticum* in house finches (*Carpodacus mexicanus*) and other wild birds associated with poultry production facilities. *Avian Dis.* 2001;45:321.

14. Mashima TY, Ley DH, Stoskopf MK, et al. Evaluation of treatment of conjunctivitis associated with *Mycoplasma gallisepticum* in house finches (*Carpodacus mexicanus*). *J Avian Med Surg.* 1997;11:20-24.

15. Duerr RS, Purdin GJ. *Topics in Wildlife Medicine: Orthopedics.* Vol 4. Duerr RS, Purdin GJ, eds. Bloomington, MN: National Wildlife Rehabilitators Association; 2017.

16. Visser M, Hespel A-M, de Swarte M, Bellah JR. Use of a caudoventral-craniodorsal oblique radiographic view made at 45° to the frontal plane to evaluate the pectoral girdle in raptors. *J Am Vet Med Assoc.* 2015;247(9):1037-1041.

17. Zahran S, Laidlaw MAS, McElmurry SP, et al. Linking source and effect: resuspended soil lead, air lead, and children's blood lead levels in Detroit, Michigan. *Environ Sci Toxicol.* 2013;47(6):2839-2845.

18. Vyas NB, Spann JW, Heinz GH, et al. Lead poisoning of passerines at a trap and skeet range. *Environ Pollut.* 2000;107:159-166.

19. American Veterinary Medical Association. *AVMA Guidelines for the Euthanasia of Animals: 2020 Edition.* Schaumburg, IL: American Veterinary Medical Association; 2020.

Altered Mercury:Selenium Ratios in Florida Osprey with Neurologic Signs: Wildlife and One Health Implications

Laura Kellow, MVsc, DVM,
Heather W. Barron, DVM, DABVP (Avian Practice), CertAqV,
Julia S. Lankton, DVM, DACVP,
Barbara L. Bodenstein, BS,
Laura Perkins, DVM, PhD, DACVP,
Mark Cunningham, DVM, MS, Dipl ACVPM

Session #2669

Affiliation: From Clinic for Rehabilitation of Wildlife (CROW), 3883 Sanibel Captiva Rd, Sanibel, FL 33957, USA (Kellow, Barron, Perkins), U.S. Geological Survey National Wildlife Health Center, 6006 Schroeder Road, Madison, WI 53711, USA (Lankton, Bodenstein), Florida Fish and Wildlife Research Institute, Florida Fish and Wildlife Conservation Commission, 1105 SW Williston Rd, Gainesville, FL 32601, USA (Cunningham).

Note: Abstract format selected by author(s). See Front Matter for description of conference proceedings manuscripts submitted as scientific abstracts.

Abstract: Eight western osprey (*Pandion haliaetus*) presented to the Clinic of Rehabilitation of Wildlife (CROW, Sanibel, Florida) over a 2-month period with neurologic signs. Clinical signs included generalized ataxia, head tremors, and depressed mentation. Diagnostics performed included complete blood count, radiographs, and a blood gas point-of-care blood analyzer. Patients were treated supportively with antibiotics and parenteral fluids; however, all patients were euthanized due to declining neurologic status. Three patients were necropsied and fresh frozen tissues (brain, liver, and kidney) were submitted for heavy metal analysis, infectious disease testing, and histopathology through the U.S. Geological Survey National Wildlife Health Center. Tissues from all three of these patients demonstrated elevated mercury:selenium ratios > 1 (1.14, 1.29, 1.37, 1.35, 1.35), suggestive of mercury toxicosis. Brevotoxicosis and infectious disease testing, including for avian influenza, were negative, and tissue concentrations of other heavy metals were within normal limits. Control samples were obtained from three apparently healthy adult osprey that presented to CROW with severe trauma and were humanely euthanized. These tissue samples demonstrated normal mercury:selenium ratios (< 1) in the liver, brain, and kidney, which are supportive of protective levels of selenium. Research has previously demonstrated bioamplification of mercury in fledgling and adult healthy osprey, with the species playing an important sentinel role in water and ecosystem health.[1-3] Recent literature has demonstrated that mercury toxicity produces a selenium deficient state and directly inhibits essential oxidative mechanisms of selenium.[4,5] We postulate that clinical signs in these neurological osprey resulted from selenium deficiency secondary to mercury toxicity, which broadens horizons for potential treatments and has One Health implications for water ecosystems in Florida.

References

1. Burger J, Gochfeld M. Marine birds as sentinels of environmental pollution. *EcoHealth.* 2004;1(3):263-274. doi.org/10.1007/s10393-004-0096-4.

2. Ogden JC, Baldwin JD, Bass OL, et al. Waterbirds as indicators of ecosystem health in the coastal marine habitats of southern Florida: selection and justification for a suite of indicator species. *Ecol Indic.* 2014;44:148-163. doi.org/10.1016/j.ecolind.2014.03.007.

3. Grove RA, Henny CJ, Kaiser JL. Osprey: worldwide sentinel species for assessing and monitoring environmental contamination in rivers, lakes, reservoirs, and estuaries. *J Toxicol Environ Health B.* 2009;12:1:25-44. doi.org:10.1080/10937400802545078.

4. Spiller HA. Rethinking mercury: the role of selenium in the pathophysiology of mercury toxicity. *Clin Toxicol.* 2017;56(5):313-326. doi.org/10.1080/15563650.2017.1400555.

5. Berry MJ, Ralston NV. Mercury toxicity and the mitigating role of selenium. *EcoHealth.* 2008;5: 456-459. doi.org/10.1007/s10393-008-0204-y.

Serologic Responses to West Nile Virus Vaccination in Captive California Condors (*Gymnogyps californianus*)

Michelle G. Hawkins, VMD, DABVP (Avian Practice),
Shoshanna J. Lee, BS,
Terra R. Kelly, DVM, Dipl ACZM, PhD,
Pranav Pandit, BVSc & AH, MPVM, PhD,
Jeff Zuba, DVM,
Joanne Paul-Murphy, DVM, Dipl ACAW, Dipl ACZM

Session #2511

Affiliation: From the Department of Medicine and Epidemiology (Hawkins ,Paul-Murphy), William T. Pritchard Veterinary Medical Teaching Hospital (Lee), Karen C. Drayer Wildlife Health Center (Kelly), EpiCenter for Disease Dynamics and One Health Institute (Pandit), School of Veterinary Medicine, University of California, Davis, Davis, CA 95616, USA, Department of Veterinary Services, San Diego Zoo Safari Park, San Diego Global, Escondido, CA 92027, USA (Zuba).

Note: Abstract format selected by author(s). See Front Matter for description of conference proceedings manuscripts submitted as scientific abstracts.

Abstract: The critically endangered California condor (*Gymnogyps californianus*, CA condor) is the largest North American land bird. California condors are vulnerable to West Nile virus (WNV). [1] The objectives of this study were to evaluate associations between WNV antibody titers in CA condor chicks and WNV vaccination schedule, type of vaccine administered, maternal antibody titers (if known), location, age and season. Plasma/serum samples were collected from CA condors between 2010-2020 by three captive propagation sites that support and contribute to the U.S. Fish and Wildlife Service's CA Condor Recovery Program: San Diego Zoo Global, Los Angeles Zoo and Oregon Zoo. Data from medical records including signalment, pedigree, history, physical exam findings, type of WNV vaccine used, intramuscular verses subcutaneous injection technique, and vaccination history of each bird was recorded. Serological titers were analyzed using a plaque reduction neutralization test. Mixed-effects generalized additive models (GAMMs) with a negative binomial distribution and mixed effects logistic regression models were used to understand associations between WNV antibody titer and vaccination-related factors in the chicks. Samples (n = 373) from 117 CA condor chicks and 23 maternal adults were utilized for analysis. Condor dam WNV titers were found to be positively associated with chick WNV titers. Model results suggest that condor chick WNV titers decreased with increasing numbers of vaccinations administered to the chicks, with a significant reduction in WNV titer for each additional vaccination a chick received. Although an association between humoral antibody induction and protection has been suggested in other bird species,[2,3] as no challenge was conducted in this study this could not be confirmed.

References

1. Beaty B, Calisher C, Shope R. Arboviruses, 1995. In: Lennette E, Lennette D, eds. *Diagnostic Procedures for Viral, Rickettsial and Chlamydial Infections.* 7th Ed. Washington DC: American Public Health Association; 189-212.

2. Jiménez de Oya N, Escribano-Romero E, Blázquez A-B, Martín-Acebes MA, Saiz J-C. Current progress of avian vaccines against West Nile virus. *Vaccines*. 2019;7:126. https://doi.org/10.3390/vaccines7040126.

3. Bunning ML, Fox PE, Bowen RA, et al. DNA vaccination of the American crow (*Corvus brachyrhynchos*) provides partial protection against lethal challenge with West Nile virus. *Avian Dis.* 2007;51:573-577.

Section 6

AAV Masterclasses

Richie Burdeaux, DVM, DABVP (Avian Practice);
Len Donato, VMD, DABVP (Avian Practice);
Karen Holmes, DVM;
Kaitlyn Nikirk, BS, DVM Candidate, Class of 2024;
Jennifer Graham, DVM, DABVP (Avian Practice), DABVP (Exotic Companion Mammal Practice), DACZM;
Bianca Murphy, DVM, DABVP (Avian Practice);
Anna Osofsky, DVM, DABVP (Avian Practice);
Andrea Winkel, DVM, MPH
Session Chairs

Confiscated Birds and Bird Markets: Addressing the Welfare Implications

Anthony A. Pilny, DVM, DABVP (Avian Practice), and
Pat Latas, DVM

Session #2549

Affiliation: From Arizona Exotic Animal Hospital, 2340 E Beardsley Rd. Ste 100, Phoenix AZ 85024, USA (Pilny), Wild Parrot Coalition, Tucson, AZ 85715, USA (Latas).

Abstract: Although birds may have some protections under the Wild Bird Conservation Act (WBCA), Endangered Species Act (ESA), and the Convention on International Trade in Endangered Species of Wild Fauna and Flora (CITES), millions of wild birds are still smuggled illegally and traded or sold to meet the demand for the pet trade. This continues to create serious welfare implications; however, avian veterinarians can play a positive role in changing the outcomes for these birds.

Capturing live wild birds for the pet trade is an escalating crisis of biodiversity loss and horrendous cruelty. The trade deeply affects avian welfare and conservation and can lead to serious repercussions in both captive and pet bird populations. Authorities confiscate thousands of birds every year, but there is little veterinary involvement, from planning through implementation of post confiscatory care and minimal welfare assessment. These birds are often viewed solely as evidence and little consideration is given to their health, welfare and well-being during and following confiscation. The disposition of specimens, in other words, what happens to the unfortunate birds after they are released as evidence, is rarely a topic of welfare discussion and only has limited involvement with avian/wildlife-experienced veterinarians. The fate of confiscations is not well known within the avian veterinary community nor the public.

Birds can suffer from injuries and illness prior to, during, and following confiscation. This lecture will discuss several scenarios to illustrate the extent of suffering brought on by illegal and legal trade in wild birds and promote the need for experienced veterinary involvement at every level. Professional partnering with law enforcement can diminish potential introduction of invasive species, potential zoonotic and pandemic pathogens, protect and conserve biodiversity, and help assist a pathway for re-introduction to the wild.

This lecture discusses the negative impacts on avian welfare observed in birds confiscated from wildlife trafficking and legal wildlife trade, and how informed and pro-active avian veterinarians can make a difference in transformative change, and make a world of difference for each individual bird which has suffered.

The Intersection of Aviculture, Conservation and Welfare of Birds: The Bird Markets of the USA and the World

This lecture discusses the effects bird markets, exhibitions, and expos have on the welfare and conservation of wild birds and potential negative impacts on captive and pet birds. The sale of wild birds via bird markets has negative impacts on health, welfare, and well-being of the birds for sale and on exhibit, whether wild-origin or captive-bred. These impacts include the following:

- crisis of biodiversity loss and habitat destruction around the globe;
- declining and endangered populations of popular and rare species;
- potential introduction of invasive species and epizootic/pandemic pathogens; and
- threats from lax biosecurity, economic impact, and organized crime.

Avian veterinarians need to play a major part in the discussion on all levels, but particularly regarding health, contagion, and biosecurity. We offer some guidelines for evaluation and observation of typical bird marts and what conditions, illnesses and injuries to expect.

AAV Position Statement: *The illegal international wildlife trade, also known as wildlife trafficking, is devastating to wild animals, plants and human communities around the world. Wildlife trafficking poses an immediate threat to already critically endangered species, increasing the likelihood of their extinction, and is directly linked to some of the most important underlying causes of biodiversity loss. Wildlife trafficking furthermore poses significant health and safety risks for both humans and animals, by contributing to the spread of infectious, zoonotic, and food-borne diseases. With a multitude of human infectious diseases having their origins in wild animals, putting an end to the illegal trade in wildlife is not only critical for species' conservation, but also to human health.*

Recommended Reading

1. Animal Welfare Institute. Taking care of confiscated animals: a CITES solution? Available at: https://awionline.org/awi-quarterly/winter-2017/taking-care-confiscated-animals-cites-solution. Accessed May 21, 2022.
2. BirdLife International. Tackling the illegal trade and trafficking of birds. Available at: https://www.birdlife.org/projects/tackling-illegal-trade-inc-songbirds/. Accessed May 21, 2022.
3. BirdLife International. Illegal wild bird trade. Available at: https://flightforsurvival.org/threat/illegal-trade/. Accessed May 21, 2022.
4. Audubon. The illegal parrot trade. Available at: https://www.audubon.org/news/the-illegal-parrot-trade. Accessed May 21, 2022.
5. Berman C. Smithsonian Magazine. Wildlife Trafficking. Available at: https://www.smithsonianmag.com/travel/wildlife-trafficking-149079896/#:~:text=Birds%20are%20the%20most%20common,well%20as%20mammals%20and%20insects. Accessed May 21, 2022.
6. Hruby D. He once trafficked in rare birds. Now, he tells how it's done. New York Times. Available at: https://www.nytimes.com/2020/11/27/world/europe/rare-birds-macaws-trafficking-brazil.html. Accessed May 21, 2022.
7. Vélez C. Bird trafficking in the Amazon. Amazon Cooperation Treaty Organization. Available at: http://otca.org/en/bird-trafficking-in-the-amazon/. Accessed May 21, 2022.

Studying the Behavior of Captive Parrots: Respecting Their Cognitive Abilities Creates Mutual Respect and Trust

Irene Pepperberg, PhD, and
Jan Hooimeijer, DVM, RPBC

Session #2541

Affiliation: From The Alex Foundation, 30 Curry Circle, Swampscott, MA 01907, USA (Pepperberg), Consultancy Practice for Birds, Schoolstraat 7 A 7941 CA Meppel, The Netherlands (Hooimeijer).

Part I. Grey Parrot Communication and Cognition

Irene M. Pepperberg, PhD

Manuscript adapted from references cited below.

Deciphering nonhuman communication—particularly nonhuman vocal communication—has been a longstanding human quest. We are, for example, fascinated by the songs of birds and whales, the grunts of apes, the barks of dogs, and the croaks of frogs; we wonder about their potential meaning and their relationship to human language. Do these utterances express little more than emotional states, or do they convey actual bits and bytes of concrete information? Humans' numerous attempts to decipher nonhuman systems have, however, progressed slowly. We still wonder why only a small number of species are capable of vocal learning, a trait that, because it allows for innovation and adaptation, would seem to be a prerequisite for most language-like abilities.[1] Humans have also attempted to teach nonhumans elements of our system, using both vocal and nonvocal systems.[2-10] The rationale for such training is that the extent of success in instilling symbolic reference provides some evidence for, at the very least, the cognitive underpinnings of parallels between human and nonhuman communication systems.

In the 1980s, Macphail claimed language was crucial for all advanced forms of problem-solving, placing humans above all other creatures.[11] Premack argued that chimpanzees could achieve some measure of symbolic reference and that successful apes then processed information differently from those that failed;[7,12] he also argued that symbolic reference acquisition was limited to primates. Notably, symbolic reference is only a subset of actual language, and even separating acquisition of reference from simple object-label association is not a simple matter, as reference begins with such associations, and the point at which true reference emerges is not always obvious. I begin by discussing these points and questions, predominantly from the viewpoint of someone studying avian abilities. I end by examining the question posed by Premack: do nonhumans that have achieved some level of symbolic reference then process information differently from those that have not? I suggest he was correct about the importance of symbolic representation for complex cognitive processing, but incorrect about limiting it to primates; I give examples from my research on grey parrots as discussed in Pepperberg.[13,14]

References

1. Beecher MD. Why are no animal communication systems simple languages? *Front Psychol.* 2021;12:602635. doi:10.3389/fpsyg.2021.602635.

2. Gardner RA, Gardner BT. Teaching sign language to a chimpanzee. *Science.* 1969;165:664-672.

3. Herman L, Richards DG, Wolz JP. Comprehension of sentences by bottlenosed dolphins. *Cognition.* 1984;16:129-219.

4. Miles HL. Language acquisition in apes and children. In: Peng FCC, ed. *Sign Language Acquisition in Man and Ape.* Boulder, CO: Westview Press; 1978:108-120.

5. Patterson F. The gestures of a gorilla: language acquisition in another pongid. *Brain Lang.* 1978;5:72-97.

6. Pepperberg IM. *The Alex Studies.* Cambridge, MA: Harvard University Press; 1999.

7. Premack D. On the assessment of language-competence in the chimpanzee. In: Schrier AM, Stollnitz F, eds. *Behavior of Nonhuman Primates.* Vol. 4. New York, NY: Academic Press; 1971:186-228.

8. Reiss DR, McCowan B. Spontaneous vocal mimicry and production by bottlenosed dolphins (*Tursiops truncates*): evidence for vocal learning. *J Comp Psychol.* 1993;107:301-312.

9. Rumbaugh DM. Language *Learning by a Chimpanzee.* New York, NY: Academic Press; 1977.

10. Schusterman RJ, Krieger K. Artificial language comprehension and size transposition by a California sea lion (*Zalophus californianus*). *J Comp Psychol.* 1986;100:348-355.

11. Macphail E. The comparative psychology of intelligence. *Behav Brain Sci.* 1987;10:645-695.

12. Premack D. The codes of man and beast. *Behav Brain Sci.* 1983;6:125-167.

13. Pepperberg IM. The comparative psychology of intelligence: some thirty years later. *Front Psychol.* 2020;11:973. doi: 10.3389/fpsyg.2020.00973.

14. Pepperberg IM. Nonhuman communication: some issues and questions. *Front Psychol.* 2021;12:647841. doi: 10.3389/fpsyg.2021.647841.

Part II. Behavioral Interventions

Jan Hooimeijer, DVM, RPBC

Manuscript adapted from Hooimeijer J. Welfare and behavior in parrots. Proc Annual Conf Assoc Avian Vet. 2013:289-297. See also (1).

Introduction

In general, there is still a lack of awareness that parrots are nondomesticated prey animals and highly intelligent animals with remarkable cognitive abilities that are being kept in captivity under unnatural circumstances. If we do not understand the normal behavior and underestimate the high intelligence and cognitive abilities of parrots, a wide variety of behavioral problems are apt to occur depending upon the individual parrot, species, and circumstances surrounding the bird. To understand behavior, we must understand and learn about the principles of ethology. Looking at the world through the eyes of parrots, we must consider that our body language and attitude determines their body language and behavior.

Parrots are built and behave as prey animals

Parrots are prey animals in nature. The fear of being killed as a prey animal determines a major part of normal behavior in nature. Fear of getting killed as a prey animal also can be expected as normal behavior within captivity. A typical anatomic feature of parrots as prey animals is the positioning of their eyes. The eyes of humans are positioned as in predators such as dogs, cats, owls and birds of prey, which also have binocular vision. Humans that are talking and smiling to the parrot, showing their teeth, can be perceived by a parrot as intimidation by a predator and be a reason for insecure and defensive behavior.

Parrots are intelligent birds with amazing cognitive abilities. From studies demonstrating the intelligence and cognitive abilities of parrots, we can infer that they understand human intentions, posture and attitude.

Parrots learn by observing other birds, other animals and humans as part of their environment. Parrots evaluate information, make inferences, and behave accordingly.

Allowing myself (JH) some anthropomorphism, from the viewpoint of parrots, it must be a frustrating experience not to be appreciated for their intelligence and talents but just to be considered as beautiful, funny, and cute.

Definitions of behavior

Normal behavior

Normal behavior is species-typical individual behavior that is not meant to manipulate the caretaker. Examples are eating, drinking, grooming, playing, sitting in the cage, and perching on the hand. It is the experience of the author (JH) that the outcome of rewarding normal behavior will cause the bird to develop additional behavior patterns that are then used to manipulate the caretaker to get attention. These additional patterns can be considered undesired behavior.

Undesired behavior

Undesired behavior is intended to manipulate the behavior of the caretaker. This may include screaming, talking, biting, feather picking. A human response of fear, frustration, anger, concern and laughter might be considered a reward by the bird for such behavior and thus as encouragement to act in this manner.

Consequently, demonstrating that a bird's behavior is problematic may actually increase its occurrence.

Desired behavior

The definition, developed within the Consultancy Practice for Birds, for desired behavior is that which is created and manipulated by the caretaker. It is behavior that the caretaker would like to see repeated in the future.

- Desired behavior is not behavior that is the result of a command or a request.
- Desired behavior also is not behavior the bird is showing to get a treat.
- Desired behavior involves (but is not limited to) actions such as the bird demonstrating an interest in novel toys or actions.

Rewarding desired behavior, including rewarding intelligence, is a way to show respect for the bird and can aid in creating self-esteem, mutual respect, and might prevent undesired behavior.

Enforced behavior

Enforced behavior is the result of a command, any kind of force, or behavior motivated by a treat or food. Enforced behavior also is behavior that occurs out of fear of punishment.

Displacement behavior

Tinbergen, one of the founders of ethology, has described displacement behavior in gulls during territorial behavior, when a gull generally either defends its territory or cedes to another individual (e.g., fight or flight).[2,3] Displacement behavior can be considered the ultimate alternative behavior when fight or flight responses are not the best choices or even impossible. Instead of a fight or flight response, the gulls started to preen themselves, started to pull at grass as if they were collecting materials to improve the nest, and looked away from their competitors. Note that a freeze response also might occur.[4,5] In a conflict situation in nature as well as in captivity, parrots also have the choice between a fight or flight response and displacement behavior. Displacement behavior seems to be completely out-of-order but has the effect that the other parrot or the human caretaker becomes distracted and a new situation is created. Typical examples of displacement behavior are nail biting, scratching, grooming, flapping the wings, shaking the tail, shaking the head, and looking up with the head tilted sideways. Grooming and preening have been associated with normal behavior that birds perform when they are comfortable but can also be a form of displacement behavior that is displayed in a conflict situation.[6] When humans respond to displacement behavior in a way that is rewarding for the bird, this behavior, which is normal in a conflict situation, becomes undesired behavior.

When dealing with parrots, we can expect to see displacement behavior in different conflict situations. The most common conflict situation is when parrots as prey animals are approached by humans in an intimidating manner; that is, face forward with the typical characteristics of a predator, especially when the parrot is in or on top of its cage. Under those circumstances it is important to know how an avian species analyzes and combines its perceptual experiences to understand what is happening in the surrounding world.[7] There are reasons to hypothesize that parrots utilize extensive information processing (e.g., at least to some degree, they do not simply react mindlessly to environmental stimuli) but they actually process the stimuli and choose to react in a certain ways.[8]

When displacement behavior is not recognized or not understood and caretakers respond to displacement behavior in a way that is considered as rewarding, displacement behavior can easily result in abnormal repetitive behavior (also known as maladaptive behavior) without the primary trigger that was causing the conflict situation to begin with.[9] It can be one of the explanations when we are dealing with complex behavior problems like feather destructive behavior. It can be assumed that the type of displacement behavior has a genetic component and that the type of displacement may therefore differ between species. It can help explain why some species are more prone to feather destructive behavior.[10] It can be postulated that the way displacement behavior is expressed also will be influenced by environmental factors and the observational learning ability of the parrots. For that reason, it has been argued that hand feeding baby parrots is contraindicated because birds need to learn by observing normal behavior of their parents. Such observation includes normal displacement behavior in conflict situations.[11]

How to deal with displacement behavior: It is important to recognize and acknowledge displacement behavior and to consider that the parrot is showing this behavior because it perceives itself to be in a conflict situation. Any direct response to the bird at that moment can potentially be rewarding to the animal and can create maladaptive behavior. By improperly reacting to displacement behavior when in the role of caretaker or veterinarian, we are interrupting normal behavior; however, displacement behavior can be an important tool for humans to use when dealing with parrots.

5-step behavior protocol to create a relationship based on mutual respect and mutual trust

The most important aspect of dealing with parrots is the attitude of the person or persons caring for the parrot. A basic protocol for veterinarians and caregivers of parrots can be summarized in five steps. This protocol also can be used with any parrot showing any behavior problem.

Step 1: Show the parrot that you have no negative intentions by acknowledging the fact that parrots are prey animals. For example, avoid staring directly at the bird. Allow the bird to observe you and your actions from a short distance, as though it was in the audience of a play. Relax your body, show that you are not (overtly) attending to the bird, and playfully interact with objects in your environment. For example, toss and catch a parrot toy, shake the toy, fiddle with your pen, tear up pieces of paper, rub a towel on the side of your face. In that way, be a positive role model. Engage in these actions without any interaction with the parrot; don't look directly at the bird. Note via peripheral vision whether the bird is attending and try different playful actions until it does. Parrots are more likely to feel comfortable with people who are comfortable with parrots and with themselves. Your actions (e.g., positive, confident, or nervous, insecure) around the parrot will provide the bird with the information it needs to decide how to react. People often have to act and play a role around the parrot. What we create by doing so is desired behavior and what we create is the experience of showing respect for the bird.

Step 2: This step can be considered rewarding the behavior that has been created and manipulated by your attitude in performing step 1. You have created an environment that encourages the bird to exhibit desired behavior—interest in your actions—and you reward this interest to by telling the parrot that it's the most beautiful creature on this planet and how precious the bird is to you.

Step 3: Establish yourself as the best teacher the parrot can wish for. You do this by showing respect for the bird's intelligence by describing, in detail, what there is to see around it (e.g., birds in the trees, toys on the table, their colors, shapes, sizes, materials). You also can describe the color of the bird's tail or beak. Parrots truly enjoy observing and learning. Pepperberg has shown that birds will request labels for novel items in their environment.[12] Reward the bird for its intelligence in a nonintimidating way (e.g., be aware of the direction of your gaze).

Step 4: After rewarding the intelligence, allow the bird to touch your pen, toy, paper or towel and praise the bird enthusiastically. Only let them touch items when you tell it is okay. When they reach out for anything, just take it away without a word and present it later after telling that now it's okay to touch or even bite into the item. Then they engage, in general, in very gentle touching with their tongue. Ask the bird to participate in the learning process by offering it new, small, unthreatening objects. Tell the bird that it may touch and feel the objects and allow it to investigate. Praise each positive move. You also can present your hand and tell the bird that it may step up, and praise the bird for doing so. This is the opposite of giving a command or making the bird step up to get a treat.

Step 5: Because, in the previous steps, you have shown the parrot that you are safe to be around, you can now extend the experience to new situations. When the parrot shows fear, do not reassure the bird. Reassurance can easily exacerbate the bird's sense of fear and insecurity as it can function as a reward for the behavior. It is the experience of the author (JH) that after steps 1-4, a situation of mutual respect and mutual trust is established. When the parrot fears a place or object like the towel, simply start over again with step 1 of the protocol. The bird will eventually realize that there is no reason for fear. Then tell the bird how smart and wonderful it is. This will help to stimulate the self-esteem that all parrots need in order to display normal parrot behavior. The outcome is to create a situation in which a parrot can accept novel situations like an examination, taking blood, grooming, wing clipping, gently towelling and others.[1,13]

Dealing with behavior problems of parrots

Dealing with behavior problems involves looking for a diagnosis.[11] It is important to consider the circumstances and to understand the reason for the behavior and its consequences. In general, unwanted behavior of parrots in captivity can be considered behavior that started as normal behavior under unnatural circumstances. When dealing with unwanted behavior, we must consider that unwanted behavior will become more and more a problem when the consequence of engaging in that behavior results in a reward for the parrot. Any response, whether positive, negative, emotional or even subconscious, to the (unwanted) behavior, including ignoring a behavior, could be a reward for the (undesired) behavior.

Preventing undesired behavior

Preventing undesired behavior requires understanding the reason for the undesired behavior: looking at the circumstances, the consequences, the natural behavior, and body language of parrots as a social, intelligent prey animal. Preventing undesired behavior involves optimizing health and welfare by making sure that birds benefit from certified organic pelleted food, sunlight, fresh air, and exercise. Preventing undesired behavior is accomplished by creating an enriched environment in which the parrot is allowed to express its intelligence and skills, providing toys and food in a way that stimulates parrots to express normal foraging behavior. Preventing undesired behavior is enabled with social interactions with other birds, other animals and humans. Taking a bird outside for a walk, a bike ride, a picnic or a family visit (with appropriate safety measures) will prevent undesired behavior. Preventing undesired behavior also includes understanding that the cage of a bird can be considered by the bird as its nesting site. In nature, birds spend their time during the day and during the night at the same location when they are breeding as part of their reproductive cycle. Being at what it sees as its nesting site, the bird can feel threatened by interlopers and try to defend its nest site by displaying territorial behavior. To prevent undesired behavior, an owner can create a more natural day-night routine in which the bird gets 10-12 hours of sleep in a special small sleeping cage that is located in another room of the house, away from the location the bird spends the daylight hours. The underlying reason for this is that birds in nature do not sleep at night where they spend their daytime hours, foraging and drinking. Preventing undesired behavior is about understanding that birds prefer to sit in a high position when they feel insecure. For that reason, allowing the bird to sit on the shoulder is not encouraged. This point will be discussed in further detail during the talk.

Redirecting undesired behavior, displacement behavior

We have to consider that every response to undesired behavior can be perceived as a reward. Instead of responding to the behavior of the parrot and instead of ignoring the behavior, the advice is to act in a way that signals to the bird that there is no perceived problem with a demonstration of displacement behavior. Part of the displacement is looking away (i.e., using only peripheral vision) and distracting the bird with your interest in objects with which you wish the bird eventually to engage. The result of demonstrating displacement behavior is that insecurity and aggression are replaced by interest in these objects. Over the years, the author (JH) has observed this behavior both as a birdwatcher in nature and as an avian veterinarian dealing with parrots in captivity. In a situation of undesired behavior, step 1 of the 5-step behavior protocol can be considered as showing displacement behavior, resulting in desired behavior that can be rewarded as step 2. Using this protocol in this order prevents undesired behavior from being rewarded. Having the bird watch us when we are not looking at the bird, acknowledging and rewarding their intelligence and rewarding the behavior by allowing the bird to touch, feel and even bite a specific object can all be done within 30 seconds.

Using this protocol, undesired behavior is redirected into desired behavior, for which the bird is rewarded.

Wing clipping

In the wild, flying is essential for the survival of parrots, with the typical exception of the kakapo (*Strigops habroptilus*). Flying enables parrots to find food, water, and a safe place to sleep. Flying is a way to escape from predators; flying in flocks also offers protection from predators. For parrots in captivity that are kept as companion birds, flying is no longer essential for their survival. Within the Consultation Practice for Birds, clipping can be recommended for safety reasons but is mainly recommended because it makes it possible to provide the parrot more freedom and social interactions both inside the house and during the daily outside activities with the caretaker; the bird has a wider range to become more a part of the family (flock). In the experience of the author (JH), wing clipping done with a parrot-friendly attitude helps prevent behavior problems and prevents parrots from ending up in parrot rescue facilities.

Hand-rearing baby parrots

Basically, the necessity for hand rearing baby parrots in the past was the result of management failures concerning housing, nutrition and care complemented with a lack of knowledge about the natural breeding behavior of parrots. Over the years hand-rearing baby parrots became a commercial activity because females produced more eggs.

The Dutch Parrot Foundation, supported by the Consultancy Practice for Birds has been successfully campaigning in the Netherlands against the hand-rearing of baby parrots and the trade in unweaned baby parrots. The outcome has been that since July 1 2014, it is unlawful to separate a baby parrot from the parents until the baby parrot is fully weaned (i.e., able to eat completely on its own). The new legislation puts parrots on the existing list with dogs, cats, rabbits, chimpanzees and other species for which the timing of separation of youngsters from the parents has been regulated for many years. The new regulation is based on the data that have become available during the past 15 years showing the negative consequences of separating baby parrots from their parents. A range of behavior problems and welfare issues have been recognized as consequences of separating youngsters from the parents, particularly before or during the important first imprinting phase. Some of the behavior problems can occur after months or even after years, especially when the bird becomes reproductively active. Insecure behavior, phobic behavior, biting, feather picking and self-mutilating can become serious consequences of separating a baby parrot from the parents.[14,15] Some researchers have shown that enrichment at an early age is important in maintaining good interactions later in life.[16]

Conclusions

We have to learn how to interact with parrots by looking at the world through their eyes. Parrots belong to the most charismatic, intelligent and social animals on the planet, and can have a very long life expectancy when provided with the appropriate living environment. Consider and acknowledge parrots as prey animals. By adjusting our body language, we show respect for parrots instead of showing an intimidating posture, creating insecure and defensive behavior. By showing respect for their social skills and intelligence, we demonstrate respect for parrots. The natural learning process involves observing the behavior and skills of their parents, other members of the flock, and humans in captivity. Showing that we have a problem with a parrot's behavior can create insecurity and behavior problems. Instead, using the 5-step behavior protocol can help create a relationship based on mutual trust and respect. It is vital to reward desired behavior, *but in the appropriate manner*. It is vital to reward the intelligence and acknowledge the cognitive abilities in parrots, but, again, *in the appropriate manner*. It is vital *not* to reward normal behavior. It is vital *not* to reward undesired behavior. It is important to recognize displacement behavior and to use displacement behavior when we are in a conflict situation. Taking care of the health, welfare, well-being, and conservation of parrots in the world is our primary responsibility as avian veterinarians. Parrots' behavior patterns show whether you are doing a good job, just as children show whether their teachers are doing a good job. In that way parrots have become my best teachers because they correct me when I'm wrong. I'm grateful for the parrots in my life because they force me to be a better avian veterinarian.

If Not Us, Who? If Not Now, When?

Acknowledgments: Authors would like to thank Harrison's Bird Foods, The Netherlands for sponsoring the parrot events in the Netherlands. Finally, we thank the staff of the Consultancy Practice for Birds for their support and understanding. Without them, nothing could have been accomplished over the years.

References

1. Hooimeijer J. Dealing with behavior problems of parrots. *Proc Annu Conf Deutsche Veterinärmedizinische Gesellschaft.* 2008.

2. Tinbergen N. *Spieden en Speuren in de Vrije Natuur.* Amsterdam, The Netherlands: Uitgeverij Ploegsma; 1959.

3. Tinbergen N. *Curious Naturalists.* London,UK: Country Life; 1958.

4. Hagenaars MA, Oitzl M, Roelofs K. Updating freeze: aligning animal and human research. *Neurosci Biobehav Rev.* 2014;47:165-176.

5. Bracha HS. Freeze, flight, fight, fright, faint: adaptationist perspectives on the acute stress response spectrum. *CNS Spectrums.* 2004;9:679-685.

6. Spruijt BM, van Hooff JARAM, Gispen WH. Ethology and neurobiology of grooming behavior. *Physiol Rev.* 1992;72:825-852.

7. Zucca P. Mind of the avian patient: cognition and welfare. *Proc Annu Conf Euro Assoc Avian Vet.* 2007; 357-365.

8. Pepperberg IM. The African grey: how cognitive processing might affect allospecific vocal learning. In: Balda RP, Pepperberg IM, Kamil AC, eds. *Animal Cognition in Nature.* London, UK: Academic Press; 1998:381-410.

9. Garner JP. Perseveration and stereotype-systems-level insights from clinical psychology. In: Rushen J, Mason G, eds. *Stereotypic Animal Behavio: Fundamentals and Applications to Welfare.* Wallingford England, UK: CABI; 2006:121-152.

10. Seibert LM. Feather-picking disorder in pet birds. In: Luescher AU, ed. *Manual of Parrot Behavior.* Oxford, England: Blackwell; 2006:255-265.

11. Orosz S. Diagnostic workup of suspected behavioral problems. In: Luescher AU, ed. *Manual of Parrot Behavior.* Oxford, England: Blackwell; 2006:233-245.

12. Pepperberg IM. *The Alex Studies.* Cambridge, MA: Harvard University Press; 1999.

13. Hooimeijer JA. Practical behavioural protocol for dealing with parrots. *Proc Annu Conf Assoc Avian Vet.* 2003;177-181.

14. Engebretson M. The welfare and suitability of parrots as companion animals: a review. *Anim Welf.* 2006;15:263-276.

15. Schmidt R. The influence of the breeding method on the behaviour of adult African grey parrots. Inaugural Dissertation zur Erlangung der Doktorwurde der Vetsuisse-Falkultat der Universitat Bern; 2004.

16. Fox RA, Millam JR. The effect of early environment on neophobia in orange-winged Amazon parrots (*Amazona amazonica*). *Appl Anim Behav Sci.* 2004;89:117-129.

Veterinary Care of Breeding Flocks of Birds of Prey and Psittacine Species: Tips and Tricks to Help Your Client

Michael Lierz, DZooMed, DipECZM (Wildlife Population Health), DipECPVS

Session #2536

Affiliation: From Clinic for Birds, Reptiles, Amphibians and Fish, Faculty of Veterinary Medicine, Justus Liebig University Giessen, Frankfurter Str. 91-93, D-35392 Giessen.

Abstract: Captive breeding has contributed to the successful restoration of many species of birds of prey or psittacine birds. Avicultural breeding techniques include double clutching, direct fostering, cross fostering, hatch and switch, hacking, imprinting and assisted reproduction techniques such as semen collection and artificial insemination; however, reproductive failure is a common occurrence. Failures are mainly related to management problems, including hygiene measures, food quality issues, breeding flock structure, or to individual health issues of breeding birds. These may result in non-egg laying females, low quality eggs or infertile eggs due to male infertility. Therefore, veterinary care of breeding collections is extremely important. This master class provides an overview of veterinary involvement in birds of prey and psittacine bird breeding projects. Insights into the most common management problems are provided, basic aviculture techniques explained, and state of the art as well as cutting-edge assisted reproduction techniques in avian reproduction are described. Information from this veterinary field allows the veterinarian to expand expertise and service to clients, including prophylactic health measures, preparation of birds and cages, semen collection, evaluation and cryopreservation techniques, artificial insemination and ex-ovo fertilization as the latest development.

Introduction

Many and raptor and psittacine species are endangered or threatened and extinction and species conservation programs exist. In addition, commercial breeding is common. Reproduction attempts with birds of prey and psittacine birds in captivity regularly fail or the reproduction rate is low. The reason for this is multifaceted. Management-related problems are the most common reason, followed by diseases and infertility. To solve problems in a breeding flock, a detailed investigation is necessary. The veterinarian plays the most important role in this process as s/he has the capability to perform all necessary detailed investigations. A veterinarian in this role needs to understand how to manage a breeding flock and which aspects are of special importance. This master class provides a basic introduction. For more detail the reader is directed to relevant literature.[1-16]

Basics of Bird of Prey Reproduction

Birds of prey are monogamic and reach their sexual maturity usually between 2-6 years. Pairs are very territorial and the bond to their nesting site is strong. The male is around one-third smaller than the female and provides the food for the brooding female. Males do brood to a much lesser extent compared to the female. Birds of prey usually have a sibling sexual imprint starting around day 8 after hatch. Purely human hand-reared single birds of prey are unable to reproduce naturally but can be stimulated by man and artificially inseminated. This is usually used to produce species hybrids. Birds of prey hand-reared in groups may be used for natural breeding and sexual stimulation by humans, but parent-reared birds are best for natural reproduction.

Birds of prey usually lay 2-4 eggs in one clutch. By removing the eggs before the clutch is completed, the female may be stimulated to lay more eggs (egg pulling). If the eggs are removed within 2-3 weeks after clutching, a second clutch will be laid. Breeding birds should be kept in flying cages with a nesting site. If assisted reproduction techniques are used, imprinted semen donors can be kept by falconry methods.

Basics of Psittacine Bird Reproduction

Compared to birds of prey, psittacine birds have multiple reproductive strategies, from strongly monogamic to group nesting and polygamic. Imprinting birds for reproduction purposes is not used in this avian group as imprinting usually causes behavioral problems preventing natural breeding or even nesting. Also, artificial insemination is not routinely used because semen collection is much more difficult in this group compared to birds of prey. Breeders usually use double clutching to increase the number of offspring. Psittacine bird reproduction usually relies on natural pair breeding.

Assisted Reproduction Techniques

In falconry outside Germany hybrid falcons are commonly used. Those hybrids are mixed breeds of two different falcon species, which under natural circumstances, do not reproduce together. Therefore, artificial insemination techniques are used. Breeding birds are imprinted to humans and when they reach sexual maturity the breeder can stimulate the bird. Semen is collected from the males by massage methods or by voluntary semen donation. This semen is transferred into the cloaca or the oviduct of a female immediately prior to egg laying.[12,16]

In psittacine species artificial insemination techniques are only recently introduced, as semen collection is more difficult compared to birds of prey. These techniques are usually used in species conservation programs, but commercial breeders have begun to include them in management. More and more parrot owners ask for species hybrids to produce uncommon colours, but there is really no argument to support this change.

In several psittacine or bird of prey breeding facilities eggs are removed and placed in an incubator. Natural incubation in captivity can be risky, especially with inexperienced females because they might leave the nest or damage the eggs. Artificial incubation of eggs is not easy, and each egg needs to be treated individually. Apart from temperature (37.2-37.4 °C), humidity is an important key factor. Each egg should lose between 14-16% of its weight during the complete incubation period. Therefore, the egg needs to be weighed and humidity adjusted to reach those loss rates. If alteration is insufficient, the eggshell needs to be manipulated. In addition, vitality of the embryo should be controlled, either by candling or by a device that is able to monitor the heart rate (Buddy ® Digital Egg monitor).[11]

Chicks break the eggshell from inside (pipping) at the end of the species-specific incubation period. The total incubation period is highly variable between species. After pipping it still takes 48-50 hours until hatch is completed and during this time no manipulations should be done. Chicks are then fed at the earliest 6 hours after hatch and usually are placed back with the parents between day 6-10 of life (birds of prey), or purely hand reared, which must be done in groups in psittacine birds to reduce risk of behavioral problems later in life.

Veterinary Considerations of Bird of Prey and Psittacine Bird Breeding

The process of breeding birds includes many risk factors that need to be considered. Pairs are in large cages with no direct control of the individual. Especially during the reproduction period disturbances can interfere with breeding activity. In imprinted birds of prey single breeding birds are easier to control from a health point-of-view; however, repeated inseminations do carry an increased risk of diseases of the reproductive organs. Finally, reproductive behavior and egg laying is an increased stress to the birds. In Europe, psittacine and raptorial birds are usually kept in outside aviaries where those birds have contact to the free-ranging populations of various birds. Also, partners are usually placed together without an intensive health check. Breeders regularly place eggs from other breeders in their incubator or place chicks of different sources together, even with the same pair. All this significantly increases the risk of infectious disease.

Problems in Reproduction

Infertile clutches

The most common problem in bird of prey and parrot breeding facilities are infertile eggs. Pair behavior is not always ideal and despite the owner not seeing a problem, copulation does not occur successfully. Psittacine birds are highly social birds with a complex partner selection system. Force pairing might regularly led to insufficient partner acceptance. In birds of prey, the male as the smaller partner must approach the larger female and must fly onto her back. Females might interpret this as aggression and hurt the male. Compared to nature, males cannot escape aggressive behaviors because they are in an aviary. Therefore, males must approach the female carefully, offer her food and stimulate her for copulation. Even after an attack from the female, males need to be brave enough to try copulations again and again and again. Many males fail this, offer food, demonstrate reproductive behavior, but do not try to copulate again after one poor experience. Therefore, the male represents the key factor in natural reproducing pairs. As food-offer behavior from the male to the female is important, males must be supported by the breeder. During the reproduction period, the pair should be fed several times during the day with high-quality food.

In psittacine birds as well as birds of prey, infertility of the male is a common problem. Infections (e.g., *Salmonella* sp., *Chlamydia* sp., *Mycoplasma* sp.) involving the reproductive tract, environmental toxins (e.g., wood protecting paint, heavy metals, pesticides, mycotoxins), medications (e.g., coccidiostats given in the feed of raptors) or a vitamin deficiency (e.g., vitamin E, A) are possible reasons. In cases where infertility of a male is suspected, a detailed spermatologic investigation (e.g., motility, PH, live/death rate, sperm morphology) should be performed. The semen of falcons contains so-called round bodies (RB), which are a normal structure, but not found in all avian species. In falcons the number of RB can be high despite that their function is presently unknown.[5] Reference values are available for the different falcon species,[4] but not for all birds of prey species and are very rare in psittacine birds.[2] If reference values are not available, the comparison of ejaculate parameters from the investigated male with proven breeding males is useful. To date spermatologic investigations can be supported by computer assisted semen analysis (CASA) which provides additional very useful information.[7] Finally, functional spermatologic tests, such as the perivitelline membrane penetration test[9] are very helpful in investigating infertile males which demonstrates motile semen. Electron microscopic investigations of individual spermatozoa[15] assist in the determination of the reason for infertility. Infertile clutches occur after artificial insemination with contaminated semen. Especially when urate or fecal contamination decreases spermatozoa vitality enormously.

In addition, a detailed endoscopic investigation of the reproductive tract is recommended. It should include a testicular biopsy in case the spermatologic values or the endoscopic appearance of the testicle are altered or abnormal.

Reduced egg quality

Alterations of the eggshell occur if the female faces stressful events. These stresses might come from outside or from disturbed pair behavior. Calcium deficiencies also lead to thin eggshells. Causes include low calcium content or imbalanced calcium:phosphorus ratio in the food, vitamin D deficiency and intestinal or kidney disorders. Diseases of the reproductive tract, such as oophoritis, salpingitis or cloacitis need to be considered.

Reduced hatching rate

A reduced hatching rate is caused by incubation management failure. Incorrect temperatures (check the thermometer!) and not controlling the weight loss of each egg. In addition, poor incubation hygiene and malpositioning of embryos in the eggs also reduce hatching rate.

An egg is a closed system. Once the egg is laid, nothing can be added or removed. Therefore, the feeding management of the parents, especially the female, is of major importance to obtain a satisfying hatching rate. The egg follicle starts growing approximately 6 weeks prior to ovulation and the egg yolk contains most of the nutritional ingredients for the growing embryo. High-quality, vitamin-enriched (E, B_2, D_3) food must be supplied well before the first reproductive activity is visible. This is in many cases not considered by the breeder.

Vertically transmitted pathogens need to be considered. Regularly infected females are clinically healthy, but carriers and infected embryos die. The same occurs with environmental toxins (pesticides, mycotoxins), which may not affect the female clinically, but are transmitted via the egg yolk and affect the embryo severely.

The investigation of a decreased hatching rate involves a detailed judgement of the breeding and feeding management and a postmortem examination of the egg including evaluation of the time of death of the embryo. Common reasons for early embryonic death are egg transport, egg storage prior to incubation, severely inappropriate incubation temperature, microbiologic egg contamination or environmental toxins. Late embryonic death usually occurs after malnutrition of the parents, vertically transmitted pathogens, failure in incubation management such as deficient egg turning, slightly inappropriate incubation temperature, insufficient or increased egg weight loss.

The incubator needs to be controlled, all instruments must work correctly, and swabs for microbiologic investigations should be taken to control incubation hygiene. Investigation of the egg and embryo should involve a microbiologic examination of the egg yolk. The egg yolk membrane and the trachea of the embryo should be taken to investigate for mycoplasmas, the egg shell for detection of *Salmonella*, liver and kidney of the embryo should be stored for virological and molecular biological examination and the complete embryo should then be placed in formalin and sent for histopathologic examination.

Veterinary Management of Breeding Birds

Breeding birds demand special attention, not only because of their high value, but also due to their increased health risk (see above). On a regular basis (every 3-6 months) a fecal sample should be taken and investigated for parasites and *Salmonella* spp. At least once per year the birds should be evaluated in detail. This involves a swab for detection of *Chlamydia* sp. and a stress test,[3] investigating the function of the respiratory tract.

In cases where the stress test demonstrates alterations, an endoscopic evaluation of the bird should be performed. A blood sample should be taken to investigate the antibody level against paramyxovirus-1 and avian influenza virus H5 and H7, blood chemistry to investigate liver and kidney function, hematology to detect chronic diseases and determination of vitamin and trace element levels, especially if a nutritional imbalance is suspected. In psittacine birds, samples to check for circovirus, polyomavirus, bornavirus, herpesvirus, and adenovirus should be taken, including antibody screening- particularly for antibodies directed against avian bornavirus. Treatments (e.g., against parasites) or prophylactic measures such as vaccinations (e.g., in birds of prey: paramyxovirus-1 [PMV-1], avian influenza [AIV] [H5/H7] and West Nile virus) can be done at that time. In birds of prey, it has been demonstrated in vaccination and challenge studies, that vaccination against H5 using inactivated vaccine protects the birds completely from clinical disease[10] and partial protection against West Nile virus can be achieved.

Conclusions

Breeding birds of prey and psittacine birds in captivity requires complex systems. The basics of management and behavior of the birds are not fully understood by many owners. Reproductive success rates of 40-50% (hatched chicks to laid eggs) are considered successful, which is usually well below natural reproduction rate (> 80%), not to mention pairs that do not lay at all. The role of a veterinarian in a breeding facility is critical. Increasing success rate is usually possible. The veterinarian plays a key role in increasing the success of such facilities and maintaining the health of the breeding birds.

References

1. Bailey T, Lierz M. Veterinary aspects of bird of prey reproduction. *Vet Clin North Am Exot Anim Pract.* 2017;20(2):455-483.

2. Bublat AD, Fischer S, Bruslund H, et al. Seasonal and genera-specific variations in semen availability and semen characteristics in large parrots. *Theriogenology.* 2017;5(91):82-89.

3. Chitty J, Lierz M, eds. *BSAVA Manual of Raptors, Pigeons and Passerine Birds.* Gloucester, UK: British Small Animal Veterinary Association; 2008.

4. Fischer D, Garcia De La Fuente J, Wehrend A, Lierz M. Semen quality and semen characteristics of large falcons with special emphasis on fertility rate after artificial insemination. *Reprod Domest Anim.* 2011;46:14.

5. Fischer D, Garcia De La Fuente J, Wehrend A, Lierz M. Celibate birds needs assistance-spermatology and artificial insemination in falcons. *Proc 4th Int Giessen Grad Centre Life Sci.* 2011;127.

6. Fischer D, Neumann D, Purchase C, et al. The use of semen evaluation and assisted reproduction in Spix´s macaws in terms of species conservation. *Zoo Biol.* 2014;33(3):234-244.

7. Fischer D, Neumann D, Wehrend A, Lierz M. Comparison of conventional and computer-assisted semen analysis in cockatiels (*Nymphicus hollandicus*) and evaluation of different insemination dosages for artificial insemination. *Theriogenology.* 2014;82(4):613-620.

8. Fischer D, Angenvoort J, Ziegler U, et al. DNA vaccines encoding the envelop protein of West Nile virus lineages 1 or 2 administered intramascularly, via electroporation and with recombinant virus protein induce partial protection in large falcons (*Falco* spp.). *Vet Res.* 2015;46:87.

9. Krohn J, Meinecke-Tillmann S, Schneider H, et al. Technical note: investigations on the accomplishment of the avian perivitelline membrane-penetration-assay. *Reprod Domest Anim.* 2012;33.

10. Lierz M, Hafez HM, Klopfleisch R, et al. Protection and virus shedding of falcons vaccinated against highly pathogenic avian influenza A virus (H5N1). *Emerg Infect Dis.* 2007;13(11):1667-1674.

11. Lierz M, Gooss O, Hafez HM. Non-invasive heart rate measurement using a digital egg monitor in chicken and turkey embryos. *J Avian Med Surg.* 2006;20(3):141-146.

12. Lierz M. Raptors: reproductive disease, incubation and artificial insemination. In: Chitty J, Lierz M, eds. *BSAVA Manual of Raptors, Pigeons and Passerine Birds.* Gloucester, UK: British Small Animal Veterinary Association; 2008:235-249.

13. Lierz M, Reinschmidt M, Müller H, et al. A novel method for semen collection and artificial insemination in large parrots (Psittaciformes). *Sci Rep.* 2013;3:2066.

14. Lierz M. Reproduction: advancements in methods for improving reproductive success.

In: Speer B, ed. *Current Therapy in Avian Medicine and Surgery.* St. Louis, MO: Elsevier; 2016:433-445.

15. Schneider H, Fischer D, Bergmann M, et al. Investigations on crypreservation of psittacine semen. *Proc ICARE.* 2013;386-387.

16. Temple SA. Artificial insemination with imprinted birds of prey. *Nature.* 1972;237:287-288.

Practical Management of Iron Storage Disease in Captive Birds

Trent C. Van Zanten, BSc (Hons), DVM,
Shangzhe Xie, BSc/BVMS, MVS (Conservation Medicine), PhD, DABVP (Avian Practice),
Ellen K. Rasidi, BBiomedSc, BSc(Vet) (Hons), BVSc

Session #2625

Affiliation: From Conservation, Research and Veterinary, Mandai Wildlife Group, 80 Mandai Lake Road, Singapore, 729826.

Abstract: Hemochromatosis or iron storage disease (ISD) refers to the excessive accumulation of iron within the liver, resulting in damage to hepatocytes due to excessive dietary iron intake or variations in iron metabolism. The disease is challenging to resolve and understanding its pathophysiology and natural history of affected birds is essential to developing an effective treatment plan. Diagnosis is reliant on hepatic biopsy using Prussian-blue staining, and the severity of the condition can be graded based on the number of affected hepatocytes per slide. Hematology, biochemistry, diagnostic imaging, and iron serum testing generally are inadequate to confirm the diagnosis but may assist in evaluating concurrent pathologies. Clinical management is dependent on the disease severity, and dietary review ensuring iron intake remains below 50 mg/kg of dry matter fed in susceptible species may be sufficient in mild cases. Severe ISD additionally requires treatment with iron chelators or serial phlebotomy, and control of concurrent conditions. Treatment of chronic cases can be challenging. A surveillance program was initiated in 2020 examining coelioscopic hepatic biopsies of the hornbills from Jurong Bird Park to evaluate case prevalence and severity in the collection. 20 species were sampled, comprising a total of 85 birds. 27.1% of birds had no evidence of ISD, 54.1% had mild (Grade 1-2) ISD, and 18.8% had severe (Grade 3-4) ISD with notable species differences in disease occurrence. This presentation reviews the management practices utilized at Jurong Bird Park, the outcomes of treatment and expected prognosis in a range of species.

Introduction

Iron is an essential element in all virtually all orders of life: in animals, it is required by red blood cells for heme formation to allow oxygen transport, in addition to other critical metabolic processes such as DNA synthesis, electron transport, and other oxidation-reduction reactions.[1] Because of its ability to form toxic free-radicals when unbound to transport proteins, the concentration of bodily iron must be tightly regulated to prevent tissue damage when present in excessive amounts. The pathological accumulation of hepatic iron has been documented in birds since the 1970's and is labelled by many synonyms: Iron Storage Disease (ISD), iron overload (IO), iron overload disorder (IOD), or hemochromatosis.[2] The term hemosiderosis denotes hepatic accumulation of stainable iron without evidence histopathologic or biochemical toxicity, while hemochromatosis in a veterinary context refers to IO with associated organ damage. By contrast, in human medicine, the term hemochromatosis is typically reserved for primary hereditary ISD. Hemosiderosis can progress to hemochromatosis, however the accumulation of high iron concentrations within various organs can occur without associated functional or structural damage if the toxic, free iron can be sequestered as hemosiderin through the action of ferritin transporters with sufficient speed to prevent cell death.[3] In captive birds, the disease is typically nutritional in origin with numerous species having evolved to maximize iron absorption on wild diets with low iron availability, and needs to be carefully considered when preparing foods for such species under human care.[4] ISD can also present as a secondary process to other disorders and may necessitate the treatment of concurrent diseases in order to achieve clinical resolution.[2] The condition is challenging to treat and often long term, multimodal management through dietary reassessment, medical interventions, and regular disease screening in endemically affected populations.

Regulation of Iron Absorption and Homeostasis

Significant research has been conducted to elucidate iron transport and regulatory pathways in humans and other mammals. Iron mat be ingested as an insoluble ferric (Fe^{3+}) or soluble ferrous (Fe^{2+}) form; within the stomach gastric acid lowers duodenal pH reducing ferric iron via ferric reductases to Fe^{2+} within the duodenal lumen for enhanced solubility and uptake.[1] Ferrous iron is then transported across the apical membrane of proximal intestinal enterocytes by divalent metal transporter 1, a solute carrying transporter protein and is then transferred from duodenal mucosa across the basolateral membrane to the bloodstream via the Fe^{2+} transporter ferroportin. From here, a circulating transporter protein (transferrin) captures unbound plasma iron, reduces it to the redox-inert state Fe^{3+}, to then be distributed to cells of the body (generally liver, spleen, cardiac/skeletal muscle) for absorption, or to the bone marrow for erythropoiesis.[1] Within organ parenchymal and muscle cells, transferrin binds to the cell surface transferrin receptor TfR1, resulting in endocytosis and cellular uptake of the metal cargo.[5] Once internalized, iron is transported to mitochondria to form heme or iron-sulfur clusters (a component of metalloproteins), and excess iron is stored and detoxified in cytosolic ferritin molecules. Hemosiderin is an additional iron storage complex that minimally releases iron when the body requires it. Together with ferritin, this compound stores insoluble iron within the liver/spleen/bone marrow.[3] Once absorbed and stored as ferritin, no physiological excretion mechanism exists for iron except for blood loss.

From these mechanisms allowing iron flux within the body, regulation of the absorption, transport, storage, and utilization. Hepcidin is a circulating peptide hormone released by hepatocytes, and is the primary regulator of systemic iron homeostasis, coordinating the use and storage of iron with iron acquisition.[1] Hepcidin binds to ferroportin on enterocytes and macrophages, inducing the transport protein's internalization and degradation, preventing iron entry into plasma. Reduced iron in plasma results in lower transferrin saturation, and less iron delivery to erythroblasts.[6] In contrast, reduced hepcidin expression results in greater cell surface ferroportin and increased iron absorption. Tight control of plasma and tissue iron concentration is needed as excess iron can lead to toxicity from the generation of reactive oxygen species.[7] Subsequently, iron homeostasis is regulated at the level of intestinal absorption, as there is no excretory pathway for iron in all animals.[8] Plasma hepcidin concentrations are regulated by various cytokines, plasma iron levels, periods of anemia and hypoxemia.

As the key iron transport and storage molecules and are highly conserved between birds and mammals, these principles are essential to understanding the dysregulation that occurs in avian hemochromatosis, and why this condition can be so specific to certain species.[9]

Etiologies: Why Do Birds Accumulate Iron in their Organs?

ISD is considered a nutritional or metabolic disorder in birds depending on the underlying cause, and typically has a low incidence in most captive housed species, with an incidence rate of the more common incidental condition hemosiderosis being detected in 4.6% of avian post-mortems submitted to a U.S. pathology service over a five-year period.[10] Most commonly, the excessive intake of dietary iron can result in the deposition of this mineral in various organs, which is considered primary IO due to the bird having a constitutional, genetic propensity to accumulate iron. Certain species (often zoological birds, and less frequently those held in private collections) are much more susceptible and a large proportion of captive birds can develop clinical ISD without careful consideration of causal factors and preventative care. Disease susceptibility is higher in frugivorous and insectivorous species than those which are granivorous, omnivorous, or carnivorous birds are considerably more ISD-resistant. Even within the same families of birds, there is great difference in the susceptibility of frugivorous vs. granivorous/omnivorous species. Among the Columbidae, are fruit pigeons and doves considered the only ISD-susceptible group compared to the remaining granivorous genera,[11] and similarly in psittacines birds, most taxa are granivorous and have a low incidence of diseases occurrence, while lories, Pesquet's parrots (*Psittrichas fulgidus*), and eclectus which are nectivores/frugivores and red spectacled amazons (*Amazona pretrei*) feeding primarily on pine nuts (both considered low iron diets) are known to have a greater incidence of clinical ISD compared to generalist granivores.[12,13] The theory behind this difference in susceptibility is that bird species which have evolved consuming foods with low iron content have evolved highly efficient iron absorption pathways which cannot be downregulated when systemic iron stores are at capacity; thus the continued provision of foods with high bioavailable iron to such birds results in a continued state of iron sequestration in the body's organs beyond homeostatic control.[11] This state can be induced in ISD-resistant species which are provided access to inappropriate dietary choices; one report described fatal ISD in two African grey parrots (*Psittacus erithacus*) subsequent to prolonged ingestion of carnivorous pellets provided to a cohabiting Buff-crested bustard (*Lophotis gindiana*) with approximately 6.5× the recommended daily iron intake for most psittacine birds, over a period of 7-8 months.[14] Similarly, 36 thrushes (*Turdidae*), a species previously considered ISD resistant, from three gamebird collections in Italy, presented with acute mortalities subsequent to severe hemochromatosis as a result of being fed a pelleted diet falsely labelled as "low iron" which was found to contain over 3× the recommended passerine bird iron intake using atomic absorption spectrophotometry performed on pellet samples.[15]

Iron overload can also occur secondary to a variety of causes such as excessive enteral or parenteral intake (e.g., post-transfusions), liver disease (intoxications, infections), systemic inflammatory processes (iron sequestration occurs as a means of antibacterial defense), erythropoietic disorders, chronic renal disease, and malnutrition.[2]

Typically, ISD is considered a disease of captivity and has rarely been documented in wild birds. Hemosiderosis is reported in wild Magellanic penguins (*Spheniscus magellanicus*) in significantly higher frequency in females when associated with histopathological diagnosis of *Plasmodium* sp. infection, however the antecedents to IO progression in penguins were considered multifactorial.[16] A comparative screening of serum iron concentration, total iron binding capacity, and percent transferrin saturation between wild and captive kori bustards (*Ardeotis kori*) found no significant difference in between the two populations, except that percentage transferrin was slightly higher in adult over juvenile bustards.[17] Birds living in close association to iron and steel mills are considered at risk of progressive iron accumulation in their tissues, though this assertion has not been confirmed from samples obtained by wild birds under such circumstances to date.[18] A contributing factor as to why such a causal link has not been established is that industrial iron waste leaked to the environment is typically inorganic/ferric, which has a lower bioavailability and absorption across duodenal transporters in most animals unless it is reduced to its ferrous form, a process which is often incomplete within the gastrointestinal tract.[19]

Which Birds can be Affected?

Frugivorous and insectivorous bird Families appear to be the most frequently affected, with clinical primary hemochromatosis reported in Bucerotidae (hornbills, frugivorous species), Cotingadiae (cotingas, capuchin birds), Paradisaeidae (birds of paradise), Pipridae (manakins), Ptylonorhynchidae (bower birds), Ramphastidae (toucans), Sturnidae (starlings) Thraupidae (tanagers), and Trogonidae (trogons and quetzals).[2,8,11,20]

Secondary hemochromatosis on the other hand can be detected in almost any bird from any species which suffers dysregulation in iron sequestration after chronic illness, particularly inflammatory disease. Flamingos are frequently reported to demonstrate hepatocellular and Kupffer cell hemosiderosis with the severity of these changes increasing with age, often associated with hepatitis, hepatic adenocarcinoma, and liver cirrhosis. Disorders which could incite systemic or severe localized inflammation e.g., amyloidosis, pododermatitis, and aspergillosis are present.[2] In the author's experience, greater than 80% of all flamingo post-mortems demonstrate hemosiderosis or hemochromatosis, and all flamingos identified as having systemic amyloidosis on post-mortem will concurrently have hemochromatosis.

In addition to birds, a diverse range of mammal species are affected by iron storage disease including rhinoceros, deer, rodents, fruit bats, dolphins, non-human primates, and humans.[6,9,21]

Pathophysiology and Clinical Presentations

A range of mechanisms can result in the development of hemochromatosis across the various affected vertebrate taxa, but the pathophysiology of ISD in birds is less diverse. In humans, primary hemochromatosis is largely hereditary and related to mutations in any of four genes (TfR2, HFE, HFE2, and HAMP) associated with hepcidin production; the subsequent low hepcidin expression resulting in IO through uncontrolled elevated iron transport into plasma,[5] with subsequent transferrin saturation and excess iron deposition into the liver.[22] While avian primary ISD greatly resembles the genetic disruptions to the hepcidin-ferroportin axis identified in human and rodent models, there is as yet little evidence that this is the mechanism by which iron is excessively absorbed in birds.[11] Rather, susceptible avian species are considered to have evolved on a diet that is low in bioavailable iron, with subsequent very high capabilities for iron absorption when provided, and inadequate down-regulation of iron absorption when provided diets that have a high bioavailable iron content.[11] Susceptible species such as mynahs and toucans have a much higher number of DMT1 enterocyte apical transporters than resistant species, and some birds also have higher ferroportin levels in the duodenum, resulting in very high affinity for iron uptake with little apparent regulation in these susceptible animals.[23] ISD-susceptible birds also have a reduced capacity to down-regulate iron absorption despite the filling of systemic storage pools when compared to ISD-resistant species, as their wild dietary intake would not typically expose them to foods with high iron concentrations.[23]

In addition to inadequate downregulation of intestinal iron uptake, additional means by which IO can occur include the accelerated clearance of red blood cells, and altered iron metabolism as a result of the acute phase response to pathogens, trauma or cancers.[11,24] These mechanisms are more prominent in birds which are normally ISD-resistant. Alterations to iron metabolism with subsequent iron deposition into the liver are consequences of the acute phase response to pathogens or cellular injury in mammals and birds and is subject to the activity of proinflammatory cytokines released from phagocytes to reduce iron availability to infectious microbes, limiting their growth within extracellular and intracellular tissues.[9,11] In some mammals, this results in a diversion of iron from erythropoiesis to reticuloendothelial storage sites, resulting in functional iron deficient anemia if the condition persists, yet is readily reversed once the inflammatory stimulus is removed as erythropoiesis quickly depletes hepatic iron.[25] In birds, the acute phase response differs slightly, and chronic or repeated infections result in greater accumulation of iron in the liver, likely because iron is not the first limiting factor in microbial production in birds as it is in mammals (biotin is first limiting in avian tissues).[11] Additionally, transferrin in birds acts as a positive acute phase protein mediating transport of iron from systemic and intestinal sources to the liver, while in mammals' transferrin behaves as a negative acute phase protein under similar conditions, reducing iron absorption (instead, lactoferrin acts as a mammalian positive acute phase protein).[11,26] As a result, transport of extracellular iron to the liver is not always accompanied by reduced iron absorption or hematopoiesis, and consequently anemia is not a reliable indicator of infection in birds. Iron is first transitioned to phagocytes such as Kupffer cells, then eventually moved to hepatocytes as the condition progresses.[2]

As iron accumulates within hepatic Kupffer cells, macrophages, and hepatocytes, or in further to the heart and spleen - the cytotoxic effects of unbound iron trigger peroxidation of cell membranes, resulting in focal necrosis and disruption of organ function.[7] Marked tissue fibrosis, with subsequent cirrhotic liver failure, ascites, type 2 diabetes, cardiomyopathies, dysrhythmias, and neoplastic progression such as lymphosarcoma or hepatocellular carcinoma may all result depending on the site of accumulation.[2,24] Clinical ISD lacks pathognomonic signs, and the disease presentation is typically quite variable depending on the species, organs affected, concurrent pathology and chronicity of presentation. In many cases, clinical signs will not be evident until the condition is advanced, and the bird is severely compromised.[4] Generalized lethargy or malaise, weight loss, anorexia, hepatomegaly, coelomic distension, cardiomegaly (resulting from secondary hypertension), dyspnea, coughing, and wheezing may all be observed in clinical avian cases.[27] A single report exists in a sulphur crested cockatoo (*Cacatua galerita*) of unilateral joint swelling and lameness resolution associated only with the treatment of underlying ISD, although causal confirmation of this link between pathology and clinical signs was not established.[28]

Diagnosis

At present, liver biopsy remains the only reliable antemortem diagnostic test in avian species.[23] Efforts should be made to stabilize the patient prior to coelioscopic or exploratory biopsy. On H&E staining, histopathological examination of the affected liver will typically demonstrate the accumulation of large amounts of course hemosiderin granules within hepatocytes and Kupffer cells of the periportal region, often with associated with extensive periportal fibrosis and hepatocellular degradation.[9] To demonstrate the extent of iron aggregate within affected cells, the Perl's Prussian blue stain is used to identify ferric iron both within hemosiderin and ferritin, highlighting iron deposits bright blue in color. A semi-quantitative scoring system developed in kori bustards (*Ardeotis kori*) based on area of iron deposition, presence of fibrosis and cell necrosis, to approximate analytic quantification, allowing comparison of iron levels across specimens or post-treatment.[29] Liver iron concentration (LIC) can be evaluated using atomic absorption spectrophotometry (AAS) or inductively coupled plasma mass spectroscopy (ICP-MS), however normal reference ranges are lacking for most species and must be interpreted considering seasonal variations associated with breeding, molting, and migration activities in birds.[2] Samples submitted can be fresh, frozen, or formalin fixed. Of the avian species evaluated, LIC in health is generally higher than those recorded in mammals, ranging from 50-230 μmol Fe/g of wet tissue.[14]

Other less specific methods may be used to indicate the presence of ISD in birds. Radiographs of clinical cases may demonstrate hepatomegaly, cardiomegaly, and increased aortic radiodensity.[30] Magnetic resonance imaging (MRI) in hornbills severely affected by hemochromatosis showed clear correspondence between MRI signal intensity and hepatic iron concentrations, which mirrored earlier trials using MRI in pigeons with experimentally induced hemochromatosis.[31] With respect to clinical pathology, potential findings include non-regenerative microcytic anemia, leukocytosis with toxic heterophilia and/or monocytosis, hypoproteinemia (with or without hyperglobulinemia) and an elevation of liver enzymes (AST, LDH, bile acids) and creatine kinase may all be observed,[32,33] although not all birds will demonstrate these changes on bloodwork, and some will show no significant aberrations to hematology and biochemical parameters at all.

Serum iron analytes, total iron binding capacity, and percentage transferrin saturation are of little diagnostic value as these are not considered consistent across specimens within a species or between similar species; however, they may be of value when monitoring response to treatment.[2] Similarly, serum ferritin, which represents the best non-invasive measurement of body iron status in mammalian species, measurement is not commercially available for avian species. Feathers have also been evaluated for iron levels in Long-billed vultures (*Gyps indicus*) as a means of assessing heavy metal toxicity as a result of environmental contamination, however this technique has not been evaluated in a clinical setting.[34]

Treatment Options: Can it be fixed?

For birds that are receiving a diet too high in iron content, reducing this to acceptable levels will help prevent the condition from worsening; however, reduced iron intake will not be sufficient to rapidly deplete the hemosiderin stores already saturated within the liver and associated organs of clinical cases.[31] Often clinical birds have suffered considerable organ damage and require supportive care in the form of thermal, fluid, and nutritional support, in addition to the control of underlying etiologies or concurrent disorders in the case of secondary hemochromatosis. Severely affected cases are given a guarded prognosis despite extensive treatment, however birds which present stable and self-sufficient (in particular, without the need for oxygen or IV fluid support in the author's experience) have a fair prognosis for recovery.

Two primary means of treatment exist: serial phlebotomy to draw upon the hepatic iron reserves to replace the amount lost as heme is regenerated, and chelation therapy to bind systemic extracellular iron and increase its excretion through urine and feces.[11] Protocols of phlebotomy that have described successful treatment of ISD in starlings (*Sturnus* sp.) and toucans are weekly blood draws of 1% bodyweight in blood volume for 16 weeks or removing 0.3% of bodyweight by blood volume once a day until a PCV just below reference range is achieved, followed by removal of 1% of blood volume weekly for eight weeks.[23,35] The PCV of the bird must be taken before each phlebotomy procedure to ensure the bird is not excessively debilitated prior to starting the procedure, and the bird is ideally sedated or anaesthetized for the process to ensure the animal's safety through preventing accidental injury to the vein, and to preserve unused sites for future blood draw. If the bird's PCV indicates any degree of anemia prior to the phlebotomy procedure, then the treatment should be aborted, and the bird re-evaluated the following week. Phlebotomy is not considered a safe technique when used in anemic or severely compromised patients, and in these cases chelation therapy should be considered as the first treatment choice.[2] Chelation in birds relies on two main drugs: deferiprone and deferoxamine. Reported dose ranges of deferiprone which have successfully reduced hepatic iron concentration vary from 50 mg/kg q24h PO over 6 days for ISD resistant species, to 75 mg/kg q24h PO for 90 days in ISD sensitive birds.[31] Deferoxamine treatment courses have ranged from 50 mg/kg q6h for 14 days provided as eight treatments over 20 months, through to 100 mg/kg SC q24h for 60 days.[35] Treatment with both chelator medications and phlebotomy has been safely combined in several trials to accelerate the speed of recovery. One investigation comparing the different combinations of treatment protocols in mynahs (*Acridotheres tristis*) that chelation therapy need only continue for 30 days for both drugs with and without phlebotomy before effective reduction in hepatic iron concentration occurs, and that combining, deferiprone with phlebotomy can lead to iron deficiency if treatment continues for 60 days.[35] The same investigation found that the administration of iron chelators is not without impact on the metabolism of other metallic ions, significantly reducing copper and zinc – however the clinical relevance of this finding is as yet unknown.[35]

Disease Prevention

Detail and care when curating the nutritional profile (and particularly the dietary iron intake) in ISD susceptible species represents the single most important factor in controlling incidences of hemochromatosis in avian collections. In domestic chickens, as an example of ISD-resistant species, iron requirements for optimal growth and production vary between 70-110 ppm on a dry matter (DM) basis of feed intake.[36] In most psittacines, the recommended iron intake is 100-150 ppm DM feed.[14] Conversely, ISD-sensitive birds have the fewest complications when provided a dietary iron content of 25-50 ppm DM, with 30 ppm often being considered optimal,[31] although commercial diets with such a low iron formulation are challenging to produce and market widely. The addition of domestic and wild-sourced feedstuffs with low iron content (e.g., wild figs, papaya, cantaloupe) are expensive and can be difficult to source, and many low iron fruits (e.g., grapefruit, oranges, mango, papaya) contain high levels of ascorbic acid, which maintains iron in its ferrous (Fe^{2+}) form, maximizing absorption through the intestinal wall.[37-39] As such, food sources containing ascorbic acid, citric acid, and saturated fatty acids–all of which will enhance the absorption of dietary iron–should be avoided.

Additives to the diet, such as phenolic compounds (phenolic acids, tannins, lignin) and certain sources of fiber act to bind up ingested iron, thus preventing intraluminal absorption.[2,40] Phenolics may also mobilize iron from its intracellular storage from ferritin, and reduce iron to the ferrous state, to permit secretion.[41] It has been demonstrated that the natural diets of many wild herbivorous and frugivorous species, including birds, have a higher phenolic and fiber content than both commercial pelleted and domestic produce based diets, and as a consequence captive bird's diets have a higher available iron content than that ingested by their wild counterparts. [41] This suggests that supplementing captive diets with wild food sources may aid in minimizing iron absorption in a readily accepted, naturally beneficial way for birds under managed care. The addition of tea (tannic acid) and inositol to the diets of ISD-sensitive species was found to prevent IOD when birds are provided diets enriched in iron. These compounds can bind many essential trace minerals, theoretically to a state of deficiency – therefore, while potentially of great benefit in managing iron over-supplementation, care should be taken when using phenolics regularly, as "tea" regardless of brand is typically highly variable in composition compared to the tightly regulated tannic acids used in past studies, making replication of historic supplementation schedules virtually impossible. [41]

The above considerations and preventive measures are key components of ISD management in large avian collections. Certain challenges exist in zoological collections that are rarely observed in private facilities. Mixed species exhibits may necessitate dietary options outside of a species' natural intake requirements (particularly mixing carnivorous and frugivorous or insectivorous species).[42] The provision of excess unnecessary or harmful nutrients in large aviaries may predispose to selective feeding by more dominant individuals. The housing of birds in unnatural climates or with aggressive conspecifics may represent stressors which predispose the bird to chronic illness and subsequent secondary hemochromatosis. [42] Finally, sourcing feed items with an appropriate iron content and supplement ratio consistently and affordably for a diverse range of species and requirements represents a significant challenge in preparatory effort. Institutions housing ISD-sensitive species as a major component of the collection can avoid these concerns through carefully curating the species compatibility, environmental compatibility, and enclosure design for the desired collection plan. [42] Seeking the advice of experienced exotic or zoological veterinarians and wildlife nutritionists to compose and appropriate, evidence-based nutritional profile for each species being maintained is the most effective means of ensuring iron levels are controlled in large collections. Where questions arise as to the iron concentration within food items from a given source, quantitative analytic methods such as mass spectrometry should be considered as a means of evaluating the diet's iron content comprehensively, both as a complete diet and per component, to determine food choices which should be removed or replaced to lower daily iron intake. It is important to note that quantitative analyses of food sources may not necessarily represent the true available iron content, as food items may also contain inorganic iron which is incompletely absorbed. Finally, in collections where the incidence of ISD is deemed to be a concern, developing screening or monitoring programs for IOD in at risk species permits earlier detection and resolution of subclinical cases prior to the disease becoming clinical.[43] This can take place in the form of routine coelioscopic hepatic biopsies for H&E and Prussian blue staining as part of the annual health assessment in previously affected specimens or birds with exposure to potential risk factors in disease development, allowing ongoing monitoring of the collection's ISD occurrence and tighter control of disease progression in valuable specimens.

Managing Iron Storage Disease in Bucerotiformes and Ramphastids at Jurong Bird Park

After the loss of three hornbills successively in 2020 with evidence of iron storage disease on post-mortem, the decision was made to investigate the prevalence of hemochromatosis in the hornbill (Bucerotidae) and toucan (Ramphastidae) population of Jurong Bird Park, Singapore. The sample population consisted of 88 hornbills and six toucans of equal sex ratios, and varying ages and durations of care at the bird park. Over a period of one year in 2020, birds older than a year of age (determined as the lower age limit for value in assessing IO status) were sequentially admitted for a complete annual health assessment including bloodwork, radiographs, and were anaesthetized for coelioscopic biopsy of the liver to examine for hemochromatosis, with 85 birds being sampled in total. Hepatic biopsies were then examined via histopathology using H&E and Prussian blue stain and graded 0-4 using an in-house classification system to rank the lesion's severity. All hornbills with high grade (3-4) ISD were treated with deferiprone at 75 mg/kg q24h PO for 90 days, and three birds were additionally treated with phlebotomy comprising 0.3% bodyweight by blood volume being removed daily until the hematocrit fell just below reference range, followed by weekly blood draws of 1% bodyweight by blood volume for eight weeks. The results of this investigation are summarized in Table 1. 27.1% of birds had no ISD on initial screening, 54.1% of birds had low (Grade 1-2) severity ISD lesions, while 18.8% of the population had high (Grade 3-4) severity ISD. *Buceros* and *Tockus* hornbills were the least affected species, whereas Anthracoceros, *Berenicornis,* and *Rhyticeros* were the most severely affected hornbill species. Of the 19 birds which had repeated hepatic biopsies after treatment was completed, 73.7% showed resolution or improvement of ISD scoring, with the remaining 26.3% of birds displaying no histopathologic improvement in ISD status. Birds which were older or geriatric, and which had high periportal fibrotic changes on liver biopsy, showed the least improvement post-treatment. Future recommendations include mass spectrometry of all dietary components to confirm these items appropriately reflect the low iron content they are calculated to contain, and combined treatment of high grade, chronic cases with combined chelation, phlebotomy, anti-fibrotic treatments, and periodic tannin supplementation within drinking water.

Table 1. Assessment of the incidence and severity of iron storage disease in hornbills and toucans at Jurong Bird Park, Singapore.

Genus	Total	Total sampled	Grade 0	Grade 1	Grade 2	Grade 3	Grade 4	Repeat biopsy	Improved	Resolved	Unchanged
Horizocerus	4	4	0	1	2	1	0	0	NA	NA	NA
Berenicornis	6	6	1	0	2	1	2	3	2	0	1
Bycanistes	7	7	1	3	3	0	0	1	1	0	0
Ceratogymna	4	4	0	3	0	0	1	1	0	1	0
Buceros	15	15	10	5	0	0	0	2	0	2	0
Anthracoceros	14	10	3	0	2	2	3	5	1	1	3
Penelopides	5	5	1	1	1	1	1	3	2	1	0
Rhyticeros	13	12	0	5	4	3	0	4	3	0	1
Tockus	20	16	7	8	1	0	0	0	NA	NA	NA
Ramphastos	4	4	0	2	1	1	0	0	NA	NA	NA
Pteroglossus	2	2	0	2	0	0	0	0	NA	NA	NA
Total	94	85	23	30	16	9	7	19	9	5	5

Conclusions

The dysregulation of iron homeostasis has severe and potentially fatal consequences in affected animals. Hemochromatosis is a complex disorder in birds whose etiology has not been completely elucidated. In its primary form, it is suspected to result from an evolutionary requirement for increased dietary iron absorption as an adaptation to low iron intake of frugivore, folivore and insectivore diets, or because of another disease process in the case of secondary hemochromatosis. The condition is difficult to diagnose without hepatic biopsy, and treatment options involve reducing dietary iron intake, chelation therapy, serial phlebotomy and providing iron-binding nutritional supplements in the diet. Management of ISD in large avian collections presents a unique set of challenges which relies on control over the nutritional, husbandry, and health management protocols available to the facility.

Acknowledgments: The authors acknowledge Dr. Chia-Da Hsu and Dr. Shin Min Chong for their assistance in developing and evaluating histopathology slides used in this investigation.

References

1. Abbaspour N, Hurrell R, Kelishadi R. Review on iron and its importance for human health. *J Res Med Sci.* 2014;19:164-174.

2. Lowenstine LJ, Stasiak IM. Update on iron overload in zoologic species. In: Fowler ME, Miller RE, eds. *Fowler's Zoo and Wild Animal Medicine.* 8th Ed. Elsevier Saunders; 2015:674-681.

3. Saito H. Storage iron turnover from a new perspective. *Acta Haematologica.* 2019;141:201-208.

4. Smith JA. Passeriformes (songbirds, perching birds). In: Fowler ME, Miller RE, eds. *Fowler's Zoo and Wild Animal Medicine.* 8th Ed. Elsevier Saunders; 2015;236-246.

5. Emanuele D, Tuason I, Edwards QT. HFE-associated hereditary hemochromatosis: Overview of genetics and clinical implications for nurse practitioners in primary care settings. *J Am Assoc Nurse Pract.* 2014;26: 113-122.

6. Ganz T, Nemeth E. Iron homeostasis and its disorders in mice and men: potential lessons for rhinos. *J Zoo Wildl Med.* 2012;43.

7. Conrad M, Kagan VE, Bayir H, et al. Regulation of lipid peroxidation and ferroptosis in diverse species. *Genes Dev.* 2018;32:602-619.

8. Koutsos E, Gelis S, Echols MS. Advancements in nutrition and nutritional therapy. In: Speer BL, ed. *Current Therapy in Avian Medicine and Surgery.* St. Louis, MO: Elsevier; 2016;142-176.

9. Klopfleisch R, Olias P. The pathology of comparative animal models of human haemochromatosis. *J Comp Pathol.* 2012;147:460-478.

10. Nemeth N, Gonzalez-Astudillo V, Oesterle P, et al. A 5-year retrospective review of avian diseases diagnosed at the Department of Pathology, University of Georgia. *J Comp Pathol.* 2016;155:105-120.

11. Klasing KC, Dierenfeld ES, Koutsos EA. Avian iron storage disease: variations on a common theme? *J Zoo Wildl Med.* 2012;43:S27-34.

12. Péron F, Grosset C. The diet of adult psittacids: veterinarian and ethological approaches. *J Anim Physiol Anim Nutr.* 2014;98:403-416.

13. Perpiñán D. Problems of excess nutrients in psittacine diets. *Companion Anim* 2015;20:532-537.

14. O'Connor MR, Garner MM. Iron storage disease in african grey parrots (*Psittacus erithacus*) exposed to a carnivorous diet. *J Zoo Wildl Med.* 2018;49:172-177.

15. Pavone S, Salamida S, Pecorelli I, et al. Deadly outbreak of iron storage disease (ISD) in Italian birds of the Family Turdidae. *J Vet Med Sci.* 2014;76(9):1209-1212.

16. Ewbank AC, Strefezzi RF, Sacristán C, et al. Comparative morphometric evaluation of hepatic hemosiderosis in wild Magellanic penguins (*Spheniscus magellanicus*) infected with different *Plasmodium* spp. subgenera. *Rev Bras Parasitol Vet.* 2019;28:68-79.

17. Marrow JC, Hallager S, Sander SJ, et al. Comparison of circulating iron, total iron binding capacity, and percent transferrin saturation in wild and captive kori bustards (*Ardeotis kori*). *J Zoo Wildl Med.* 2018;49: 450-453.

18. Kekkonen J. Pollutants in urbanized areas: direct and indirect effects on bird populations. *Ecol Conserv Birds Urban Environ.* 2017:227-250.

19. Ma XY, Liu SB, Lu L, et al. Relative bioavailability of iron proteinate for broilers fed a casein-dextrose diet. *Poult Sci.* 2014;93:556-563.

20. Kim Y-A, Lee B-R, Kim Y-B, et al. Hepatic hemochromatosis in a toco toucan (*Ramphastos toco*). *J Vet Clin.* 2013;30:329-331.

21. Leone AM, Crawshaw GJ, Garner MM, et al. A retrospective study of the lesions associated with iron storage disease in captive egyptian fruit bats (*Rousettus aegyptiacus*). *J Zoo Wildl Med.* 2016;47:45-55.

22. Anderson GJ, Frazer DM. Current understanding of iron homeostasis. *Am J Clin Nutr.* 2017;106:1559S-1566S.

23. Faccioni AdO, da Silva B, Bueno E, et al. Hemochromatosis in Ramphastidae: pathogenesis and treatment, a review. *PUBVET.* 2014;8.

24. Lennox A, Clubb S, Romagnano A, et al. Monoclonal hyperglobulinemia in lymphosarcoma in a cockatiel (*Nymphicus hollandicus*) and a blue and gold macaw (*Ara ararauna*). *Avian Dis.* 2014;58:326-329.

25. Phillips BE, Venn-Watson S, Archer LL, et al. Preliminary Investigation of bottlenose dolphins (*Tursiops truncatus*) for hfe gene-related hemochromatosis. *J Wildl Dis.* 2014;50:891-895.

26. Bhat IA, Bashir S, Rather W, et al. Acute phase proteins and their clinical significance in veterinary medicine: an overview. *Pharma Innov.* 2018;7:104-108.

27. Raidal S. Laboratory diagnostics for birds. In: Heatley JJ, Russell KE, eds. *Exotic Animal Laboratory Diagnosis.* Hoboken, NJ: Wiley-Blackwell; 2020:429-436.

28. Rosenwax A, Peacock L, Sydney GS. Lameness in a sulphur-crested cockatoo associated with hepatic iron accumulation. *Proc Annu Conf Assoc Avian Vet Austr Comm.* 2012;125-128.

29. Cudd SK, Garner MM, Cartoceti AN, et al. Hepatic lesions associated with iron accumulation in captive kori bustards (*Ardeotis kori*). *Vet Pathol.* 2022;59:164-168.

30. Mainez M, Cardona T, Such R, et al. Bilateral renal tubular neoplasm in a channel-billed toucan (*Ramphastos vitellinus*). *J Avian Med Surg.* 2015;29:46-50.

31. Sandmeier P, Clauss M, Donati OF, et al. Use of deferiprone for the treatment of hepatic iron storage disease in three hornbills. *J Am Vet Med Assoc.* 2012;240:75-81.

32. Cray C. Reference intervals in avian and exotic hematology. *Vet Clin North Am Exot Anim Pract.* 2015;18: 105-116.

33. Jones MP. Avian hematology. *Clin Lab Med.* 2015;35:649-659.

34. Kushwaha S. Heavy metal concentrations in feathers of critically endangered long-billed vultures (*Gyps indicus*) in Bundelkhand region, India. *Int J Life Sci Scienti Res.* 2016;2:365-375.

35. Jalili N, Zaeemi M, Razmyar J, et al. Comparison of iron chelators used alone or in combination with phlebotomy in common mynahs (*Acridotheres tristis*) with experimental iron storage disease. *Avian Pathol.* 2021;50:350-356.

36. Taschetto D, Vieira SL, Angel CR, et al. Iron requirements of broiler breeder hens. *Poult Sci.* 2017;96: 3920-3927.

37. Parni B, Verma Y. Biochemical properties in peel, pulp and seeds of Carica papaya. *Plant Arch.* 2014;14: 565-568.

38. Tanwar B, Andallu B, Modgil R. Influence of processing on physicochemical, nutritional and phytochemical composition of Ficus carica L.(fig) products. *Asian J Dairying Foods Res.* 2014;33:37-43.

39. Czech A, Zarycka E, Yanovych D, et al. Mineral content of the pulp and peel of various citrus fruit cultivars. *Biol Trace Elem Res.* 2020;193:555-563.

40. Adams S, Sello CT, Qin G-X, et al. Does dietary fiber affect the levels of nutritional components after feed formulation? *Fibers.* 2018;6:29.

41. Lavin SR. Plant phenolics and their potential role in mitigating iron overload disorder in wild animals. *J Zoo Wildl Med.* 2012;43:S74-82.

42. Rasidi EK, Cornejo J. Managing the health of captive flocks of birds. *Vet Clin North Am Exot Anim Pract.* 2021;24:521-530.

43. Lim ZJK, Xie S. Application of epidemiology and principles of herd/flock health for the exotic animal veterinarian. *Vet Clin North Am Exot Anim Pract.* 2021;24:495-507.

Diabetes Mellitus in Psittacine Birds

Brynn McCleery, DVM, DABVP (Avian Practice)

Session #2644

Affiliation: From Red Bank Veterinary Hospital, 197 Hance Avenue, Tinton Falls, NJ 07722, USA.

Abstract: This master class serves as an introduction and review of diabetes mellitus in psittacine birds. A brief overview of pancreatic anatomy and physiology is followed by discussion of signalment, clinical signs, diagnosis, and treatment of diabetes in psittacine birds. Cases of diabetes mellitus in parrots from the literature are summarized.

Introduction

Pancreatic anatomy and physiology

The pancreas is located in the duodenal loop in birds.[1-3] In most species, the pancreas consists of dorsal, ventral, and splenic lobes.[2-4] The ventral lobe has a longitudinal cleft.[4] As in mammals, the pancreas contains both exocrine and endocrine components; however, the endocrine tissue mass is larger in birds than mammals at 1-2%.[1-4] There are three types of islets: dark, light, and mixed. Light islets contain alpha and delta cells. Dark islets contain beta and delta cells, and mixed islets contain alpha, beta, and delta cells.[5] Islets contain a higher percentage of alpha cells than in mammals (e.g., 50% in birds compared to 20% in humans). Most alpha cells are located in the splenic lobe of the avian pancreas.[4] Overall, the distribution of A, B, D, and F (PP) cells is more random in birds compared to mammals.[1,3,5] In avian species undergoing subtotal pancreatectomy, the effects varied. For example, ducks became transiently diabetes while geese became permanently diabetic.[1] This may be due to the different distribution of islet cells in the remaining pancreas after the surgical procedure.[3]

The avian pancreas is under neural, exocrine, endocrine, paracrine, and humoral control.[1,2] The endocrine function of the pancreas involves the balance and negative feedback of 3 major hormones: insulin, glucagon, and somatostatin.[1,6]

Insulin

Although the main hormone of glucose regulation in mammals, insulin plays a lesser role in granivorous birds such as parrots.[1] Insulin is present in 1/10 the concentration in both the plasma and pancreatic tissue;[4] however, despite being present in lower amounts, avian insulin is considered a more potent stimulator for glycogenesis and hypoglycemia in birds.[1-3] Insulin is produced by beta cells and functions to decrease blood glucose via uptake by cells, storage of glycogen and fat, synthesis of proteins, and by decreasing gluconeogenesis.[1] Insulin has also been shown to regulate glucagon secretion or action in birds.[2] For this reason, diabetic birds with normal insulin levels may still benefit from insulin therapy. As opposed to granivorous birds, carnivorous birds are considered more insulin-dependent.[7]

Beta cells secrete insulin secondary to numerous inputs from the exocrine system, endocrine system, paracrine system, neural system, and hormones. Insulin secretion by these cells is not triggered by glucose directly. Beta cells are particularly responsive to glucagon, cholecystokinin, and absorbed amino acids.[1,2,6]

Glucagon

Unlike in mammals, glucagon is considered the dominant hormone of carbohydrate metabolism in granivorous birds.[1,2,7] The islets of granivorous birds contain approximately 50% alpha cells, whereas human islets contain 20% alpha cells.[8] This results in an 8-10 times higher pancreatic and plasma level of glucagon in granivorous birds compared to mammals (6-8x higher than humans).[1-4] The plasma glucagon to insulin ratio is reported to be 5-10 times higher in birds compared to mammals.[8] Birds also have a higher number of glucagon receptors.[2] Glucagon functions to increase blood glucose and free fatty acids through gluconeogenic, glycogenolytic, and lipolytic methods.[1] The catabolic function of glucagon helps birds meet the high metabolic demands of egg laying, sustained flight, and fasting. Alpha cells secrete glucagon secondary to free fatty acid and cholecystokinin stimulation, and secretion is inhibited by glucose.[1,2,6]

Somatostatin

Somatostatin is produced by delta cells and polices the concentration and ratio of insulin and glucagon.[1,2,4] Pancreatic levels of somatostatin in birds are up to 10 times higher than in mammals, and somatostatin suppresses the secretion of both insulin and glucagon up to 150 times more than in mammals.[1-3] Somatostatin secretion is controlled by the neural, hormonal, humoral, and paracrine systems.[1]

Other hormones

Other hormones play a lesser role in pancreatic endocrine function including pancreatic polypeptide and amylin. F/PP cells secrete avian pancreatic peptide (APP).[1,2,6] Pancreatic polypeptide occurs in higher levels in birds compared to mammals (20-40x) and functions to inhibit gastrointestinal motility and secretions.[1-3] Amylin has been shown to increase satiety in animals.

Diabetes Mellitus

Classification

Alteration in glucose homeostasis can lead to diabetes mellitus. Diabetes mellitus (DM) is classified into the following two types in small animal medicine:

Type I: Insulin dependent, secondary to destruction of beta cells

- Beta cell destruction may be due to genetic, environmental, or autoimmune factors[2]

Type II: Non-insulin dependent, due to a relative insulin deficiency or peripheral insulin resistance

- Associated with deposition of islet amyloid peptide (IAPP) in humans and felines[2,8]

In humans, this classification has been expanded further to better describe pathogenesis of the disease.

Birds may not fit the small animal classification well due to their differences in regulation. Diabetes mellitus could theoretically occur due to a lack of insulin or an excess of glucagon in birds. Furthermore, the ratio of insulin:glucagon and regulation to somatostatin may also be a critical factor in the development of disease.[3] Typing in avian patients is usually unclear except in cases when obvious pancreatic tissue destruction has been documented; however, in the literature, this often does not occur until post-mortem examination.

Causes

Theorized causes of diabetes mellitus in birds include the following: [2,3,5,6,8-11]

Type I (beta cell destruction): pancreatitis (primary vs. secondary to female reproductive pathology), coelomitis, iron storage disease, viral infection (Herpesvirus, Paramyxovirus, Adenovirus, Polyomavirus), toxins (zinc toxicity causing pancreatic acinar degranulation, fungal toxins, selenium), neoplasia, idiopathic

Type II (insulin resistance): obesity, high corticosteroids (including iatrogenic)/growth hormone/epinephrine/glucagon, high iron

Although islet amyloidosis has not yet been reported in birds, multiple cases of diabetes mellitus with concurrent hemosiderosis have been reported in parrots.[12,13] In humans, an inherited genetic condition called excessive iron storage (EIS) has been associated with diabetes mellitus. In affected individuals, two sets of mutated HFE gene occur leading to hemochromatosis. Excessive absorption of iron from the diet leads to storage of iron in the liver, pancreas, and skin and ultimately results in secondary diabetes mellitus (aka bronze diabetes).[14] In humans, as well as in mice and rat models of the disease, iron accumulates in the beta cells leading to oxidative stress and apoptosis. Beta cell destruction results in reduced insulin secretion. The iron also causes increased peripheral insulin resistance and increased glucagon levels.[12] In humans, treatment consists of phlebotomy, chelation therapy, and treatment of the DM.[12,14] Similar treatment regimens have been reported in psittacine birds with hemosiderosis and DM.[12]

Viral associated autoimmune destruction of islet cells is a reported cause of Type I diabetes mellitus in humans, with both Reovirus and mumps associated.[15] In birds, Herpesvirus and Adenovirus are reported causes of pancreatitis and could lead to the development of diabetes mellitus over time.[5,9,15] Herpesvirus has also been associated with the development of papillomatous masses of the biliary and pancreatic ducts in some avian species.

Pancreatic islet carcinoma has been reported in a budgerigar, and pancreatic adenocarcinoma was previously reported in a blue and gold macaw.[9,16] Other tumor types affecting the pancreas could lead to DM in avian species.

Signalment and Clinical Signs

Diabetes mellitus has been reported infrequently across multiple avian orders including Accipitiriformes, Anseriformes, Columbiformes, Galliformes, Passeriformes, Piciformes, Sphenisciformes, and Psittaciformes.[2,7,10,17,18] In psittacine birds, diabetes mellitus has been reported previously in macaws, African grey parrots, an Amazon parrot, cockatiels, budgerigars, conures (Nanday, sun conure), and an Eclectus parrot.[1,2,4,10,13,19] Overall, cockatiels, budgerigars, macaws, and toucans appear to be over-represented in the literature. There is no known age, sex, or hereditary predisposition.[2]

Symptoms in birds are similar to those reported in mammals and include polyuria and polydipsia, weight loss despite a good appetite, polyphagia, lethargy, regurgitation, and dehydration.[2,7,11,15]

Diagnosis

Suspicion of diabetes mellitus may occur based on clinical signs and an elevated blood glucose level. The blood glucose level in most psittacine birds is normally between 180-350 mg/dl (10-19.4 mmol/L).[2,3] A persistent blood glucose of > 800 mg/dl (> 44.4 mmol/L) is suggestive of diabetes mellitus in psittacine birds.[1-3]

As primary diabetes mellitus is uncommon in parrots, a full systemic work-up is recommended to look for other causes of hyperglycemia (e.g., stress, acute pancreatitis, egg related coelomitis, glucagon secreting pancreatic neoplasia, and hepatopathy)[3] or an underlying cause of DM. This may include complete blood cell count, a comprehensive biochemistry panel including bile acids, radiographs, ultrasound, CT scan, infectious disease testing (include Herpesvirus, *Chlamydia psittaci*, Polyomavirus, Paramyxovirus-3, and Adenovirus), heavy metal testing including zinc, and liver and pancreatic biopsy. Diagnostic testing recommendations should be based on patient history, clinical symptoms, physical examination, and infectious disease exposure risk. Although not identified as a direct cause of DM, testing for *Chlamydia psittaci* should be considered in at-risk birds since this bacteria can cause hepatomegaly and hepatic necrosis in birds.

If blood glucose is persistently elevated, additional testing can be used for diagnosis and monitoring of DM and is similar to that used in mammals. These tests may include the following.

Urine dipsticks

Mammalian dipsticks can be used avian species. Healthy birds often test negative for glucose on dipsticks, whereas diabetic birds may have a glucose reading of 3-4+.[3,15] Presence of any ketones is considered abnormal.[15] Due to avian anatomy and the presence of a cloaca, false positives can occur in birds due to fecal contamination.[2,3]

Insulin and glucagon concentration

References ranges exist for some avian species (Table 1).[3,4,11,15] In one study looking at measurement with three different assays, two of the assays failed to detect insulin in healthy birds.[4] Therefore, if an insulin level is reported as low in a bird with suspected DM, it could indicate that the disease is not insulin-dependent or it could be artifactual due to limitations of the test. These tests often require a larger sample of blood and may, therefore, not be possible in smaller species.[10]

Table 1. Insulin and glucagon concentrations reported in healthy birds.

Species	Insulin µU/ml (pmol/L)	Glucagon pg/ml
Macaws (insulin n = 11, glucagon n = 19)[4]	4.18-12.4 (30-89)	299-1190 (mean 715, median 643)
Amazon (n = 1)[11]	7.7 (46.2)	604
Cockatiel (n = 2)[11]	5.8-8.6 (34.8-51.6)	780-964
Toco toucan[11]	7-8.2 (42-49.2) (n = 2) 24 (144) (n = 1)	758-1312 (n = 2) 90.7-109.5 (n = 2) 577-1368 (n = 4)
Chicken	22-40 (132-240)	
Duck	16-20 (96-120)	

Fructosamine

This test evaluates proteins resulting from an irreversible, nonenzymatic, and insulin-independent binding of glucose to serum proteins.[10] Although fructosamine provides insight on glucose regulation over the previous 1-3 weeks in dogs and cats, there is a shorter half-life for serum proteins in birds (e.g., only 3 days for albumin in chickens), and fructosamine may therefore reflect a much shorter time period in birds.[3,12] In 19 psittacine birds of various species, fructosamine levels ranged from 113-238 µmol/L.[10]

B-hydroxybutyric acid (BHBA)

Reference intervals exist and have been used in the diagnosis and management of DM in macaws. Levels in 20 birds ranged from 450-1422 µmol/L.[4]

Amylase

Amylase may be elevated in cases of avian pancreatitis, but a statistically significant difference was not found in a diabetic macaw compared to euglycemic macaws in one study.[3,4,6]

Commercial assays for measuring insulin, glucagon, fructosamine, BHBA, and amylase have not been validated in avian patients and results should be interpreted with this in mind.[3] As is common in avian medicine, reference ranges for these values may also differ between avian species.

Blood glucose curves can be used for monitoring patients with DM.[2] In dogs and cats, the ideal range of blood glucose is between 100-300 mg/dl (5.55-16.65 mmol/L) with a nadir of approximately 80-150 mg/dl (4.44-8.32 mmol/L) and peak of < 300 mg/dl (16.65 mmol/L). If the nadir is too low (< 80 mg/dl, 4.44 mmol/L), insulin dosing is often decreased by 25%. If the nadir is too high (> 150 mg/dl, 8.32 mmol/L), insulin dosing is increased by 10-20%. If the nadir is appropriate but the average blood glucose is too high (> 250 mg/dl, 13.87 mmol/L), switching insulin types may be considered.[20] The same guidelines can be applied to avian patients. As in mammals, blood glucose measurements in the hospital may be affected by stress and be higher than values at home. Other complicating factors include body size and repeated venipuncture. Depending on the size of the patient and the frequency of sampling, use of a laboratory analyzer may not be possible. Instead, a handheld glucometer can be utilized; however, handheld glucometers have not been validated in psittacine birds. In Hispaniolan Amazon parrots, three analyzers underestimated blood glucose by 17-32% with inconsistent bias.[21] Therefore, values should be interpreted with caution. Furthermore, dose adjustments are more difficult to make in avian patients based on the glucose curve results without dilution or compounding insulin, each of which has risks. For smaller patients where glucose curves with a handheld glucometer may not be practical, monitoring using urine dipsticks may be more feasible. Monitoring of clinical signs (e.g., reduced urine output, reduced thirst) can also help guide therapy and should be performed in all patients.

Treatment

Treatment goals include: 1) reduction of clinical signs associated with DM, 2) achievement of glycemic control without inducting hypoglycemia, and 3) reduction of the development of long-term complications. In small animal patients, long term complications of patients with DM include cataract development, diabetic neuropathies, and secondary infections. Although the first two have not been reported in parrots with DM, fungal meningoencephalitis was noted at necropsy of a diabetic chestnut fronted macaw, a pulmonary abscess was noted in an African grey parrot, and overwhelming infection was noted in a Nanday conure.[8,10,15] DM can lead to impaired function of leukocytes and impaired humoral and cell mediated immunity.[10]

Treatment options include those listed below.

Diet

A high fiber, low sugar, low fat, and low iron diet is recommended.[2,3]

Glipizide

Glipizide is a sulfonylurea medication used in type II DM in humans and cats.[2] This medication stimulates insulin production and secretion by beta cells, and therefore requires functioning beta cells;[3,9] however, other reported mechanisms of action include inhibition of hepatic gluconeogenesis and increased peripheral insulin sensitivity.[2] Although reported to be effective in smaller psittacine birds such as cockatiels and budgerigars, limited success has been noted in other avian species, and the dosing rage is quite large.[2,3,18,19] In cats, side effects can include vomiting, anorexia, elevated liver enzymes, icterus, and hypoglycemia.[22] Safety studies have not been performed in birds. Commonly reported dosing for birds is 0.3-0.5 mg/kg PO q12h and up to 1 mg/kg PO q12-24h.[2,3,9,23]

Somatostatin

Synthetic somatostatin may be effective if the DM is due to elevated glucagon levels. This hormone will help control the insulin:glucagon ratio and inhibit secretion from islet cells. Use has been reported in two birds. In a sulfur breasted toucan, symptoms improved despite a persistently elevated glucose and glucagon level. In a blue and gold macaw, the therapy was ineffective.[10,24]

Metformin

Metformin is an oral medication that decreases hepatic gluconeogenesis, decreases glucose absorption in the gastrointestinal system, and increases insulin's ability to transport glucose across cell membranes. It does not work directly to affect insulin production or secretion. In cats and dogs, side effects include lethargy, GI upset, weight loss, vomiting, and anorexia. This medication is considered contraindicated in cases with hepatic, renal, or cardiovascular disease.[22] Although used anecdotally in birds, there are no reports of use in avian species in the literature. The author has had some success in improving clinical symptoms of DM in smaller psittacine birds using a combination of glipizide and metformin; however, patients remained unregulated on BG curves.

Thiazolididinediones

These medications increase glucose transporter expression and fatty acid uptake. In obese cats, thiazolididinediones have been shown to increase insulin sensitivity and lipid metabolism.[6,22] There are no reports of use in birds.

Alpha glucosidase inhibitors

These medications inhibit sucrase, maltase, and lactase in the small intestinal brush border. Side effects in dogs include diarrhea and weight loss.[22] No reports exist of use in birds.

Insulin (see below)

Insulin is considered the mainstay of DM treatment in humans, dogs, and cats. Insulin lowers blood glucose by stimulating peripheral glucose uptake, inhibiting hepatic gluconeogenesis, inhibiting lipolysis in adipocytes, inhibiting proteolysis, enhancing protein synthesis, inhibiting the progression of beta cell destruction by reducing glucose toxicity to pancreatic cells, and preventing formation of amyloid deposits derived from amyloid peptide (cats).[1,2,4] Furthermore, insulin has been shown to regulate glucagon secretion or action in birds, contributing to lowered glucose levels. It is through this mechanism that birds with normal insulin levels may still benefit from insulin therapy.[1,2]

Types of insulin include:

a. Short-acting: regular

b. Intermediate: NPH, porcine lente (Vetsulin, Caninsulin)

c. Long-acting: protazmine zinc iletin (ProZinc), insulin glargine (Lantus), insulin detemir (Levemir)

Regular, NPH, porcine lente, protamine zinc, and insulin glargine use has been reported in birds with mixed success rates. Overall, porcine lente and PZI have the best success rates reported in the literature;[2,12] however, resistance to porcine insulin has been reported over time.[12] Dosing ranges are very wide in birds (0.002-3.3 IU/kg q12-48h) with dosing in parrots normally reported between 0.1-1.3 IU/kg q12h.[2,3,10,23] A starting dose of 0.1-0.2 IU/kg is most common. Insulin can be administered intramuscularly or subcutaneously in avian patients.[2]

The goal of insulin therapy in birds should be to keep the patient mildly hyperglycemic to avoid life-threatening hypoglycemia. The author typically hospitalizes patients for 48 hours while starting insulin therapy and recommends a blood glucose curve 1 week after initiation of therapy, followed by every 2-3 months once regulated.

Compounded and diluted insulins both have inherent risks.[7] Insulin particles do not mix well in water which may lead to uneven suspensions and irregular dosing of patients. Some insulins, including insulin glargine and insulin detemir, should not be diluted. These insulins form microprecipitates in the neutral pH of blood or subcutaneous tissues after injection which allow for slow absorption leading to a rapid onset and long duration of action. Dilution prior to injection may alter the pH and cause microprecipitates to form prior to injection.[25] The shelf life of diluted insulin has not been well studied. In humans, short-acting insulin lispro was diluted and was considered stable for 32 days;[26] however, other insulins have not been evaluated.

Prognosis

As in cats and dogs, management of a diabetic bird requires a dedicated owner. For smaller birds, management with oral antiglycemic medications may require less investment on the part of the owner; however, these medications are often less effective compared to insulin therapy, and clinical symptoms may persist or worsen, often leading to death or euthanasia within months. Survival times for larger parrots receiving insulin therapy vary from days to years (Table 2) and may be affected by time of diagnosis and the underlying cause of the DM. Due to larger patient size allowing for easier insulin dosing adjustments and blood glucose monitoring, cases of DM in larger parrots are easier to manage in the author's opinion and overall prognosis is fair with a possible survival time of multiple years; however, the owner must be committed to q12-24h injectable insulin dosing, regular in-hospital monitoring, and continued at-home monitoring for successful management of avian DM. If the owner cannot commit to this level of care and oral antiglycemic medications are not controlling symptoms leading to a poor quality of life, euthanasia should be considered.

Table 2. Cases of diabetes mellitus reported in psittacine birds in the literature.

Species	Treatment	Outcome
Blue and gold macaw[27] *Low insulin, normal glucagon*		Died 4 days after starting insulin Necropsy: Decreased beta cells, alpha cell hypertrophy, hemosiderosis of liver
Blue and gold macaw[4] *Normal insulin, low glucagon*	Diet change Somatostatin NPH	No change with diet/somatostatin 7 months on insulin
Military macaw[12] *Ketonuria, ketonemia* *Elevated fructosamine*	Low iron diet, phlebotomy PZI	29 months+ Necropsy: hepatic & pancreatic hemosiderosis
Chestnut fronted macaw[12] *Ketonuria, ketonemia*	Low iron diet, deferoxamine, phlebotomy Caninsulin→PZI	Died 24 months after diagnosis Biopsy: hepatic hemosiderosis (*no endocrine pancreas biopsied)*
Chestnut fronted macaw[8] *Ketonuria*	NPH	Died 5 months later Necropsy: hepatic hemosiderosis, islet cell vacuolar degeneration, lymphocytic pancreatitis, fungal meningoencephalitis
Cockatiel[9]	Glipizide	Euthanized 9 days later Necropsy: chronic herpesvirus pancreatitis
African grey parrot[15] *Low insulin*		Euthanized Necropsy: lymphocytic pancreatitis, islet cell hypertrophy
Nanday conure[10] *Low insulin, ketonuria*	HumulinR, Caninsulin, HumulinU	Died 1 month later Necropsy: Type I (insulin dependent, depletion of endocrine pancreas), death from second infections

Cases in the Literature

Diabetes mellitus is uncommonly reported in avian species compared to cats and dogs. Most reports in the literature include single cases or a small number of cases (Table 2).

Additional reports from proceedings include the following:

- Cockatiel: DM was suspected to be secondary to obesity. The patient was treated with diet change and glipizide 0.05mg PO q12h. Glipizide was discontinued 1 year later, and the patient had improved.[18]
- Amazon: An adult, male, blue-fronted Amazon parrot was euthanized shortly after diagnosis of DM. Endoscopic pancreatic biopsies had shown severe pancreatic atrophy and fibrosis. These changes were confirmed on necropsy examination.[13]

- African grey: A 1-year-old male African grey was diagnosed with DM and tested positive for *Chlamydia psittaci*. Pancreatic biopsies showed mixed mononuclear infiltrate. Amylase and lipase were elevated on biochemical analysis. The bird was treated with bovine PZI.[13]

- Eclectus: The patient was initially managed with oral antiglycemics ineffectively and later managed with insulin glargine.[28] A cause was not initially apparent on systemic work-up. Over the course of therapy, bile acids became elevated. Liver biopsy showed severe chronic hepatopathy with marked hepatocellular atrophy, marked biliary hyperplasia, and mild multifocal lymphoplasmacytic and heterophilic hepatitis. Aerobic, anaerobic, and fungal culture of the liver was negative. Hepatic iron concentration was considered elevated compared to references in chickens and macaws. The patient was euthanized 47 months after diagnosis (20 months after starting insulin therapy). Pancreatic islet cell atrophy was noted on necropsy.

References

1. Hazelwood RL. Pancreas. In: Whittow GC, ed. *Sturkie's Avian Physiology.* 5th Ed. San Diego, CA: Elsevier; 2000:539-555.

2. Pilny AA. The avian pancreas in health and disease. *Vet Clin North Am Exot Anim Pract.* 2008;11:25-34.

3. Orosz SE, Monks D, de Matos R. Clinical endocrinology of the protein hormones. In: Speer BL, ed. *Current Therapy in Avian Medicine and Surgery.* St. Louis, MO: Elsevier; 2016:378-399.

4. Bonda M. Plasma glucagon, serum insulin, and serum amylase levels in normal and hyperglycemic macaw. *Proc Annu Conf Assoc Avian Vet.* 1996;77-88.

5. Schmidt RE, Reavill DR. Lesions of the avian pancreas. *Vet Clin North Am Exot Anim Pract.* 2014;17:1-11.

6. Hudelson KS, Hudelson PM. Endocrine considerations. In: Harrison GJ, Lightfoot TL, eds. *Clinical Avian Medicine.* Vol II. Palm Beach, FL: Spix Publishing; 2006:541-557.

7. Bartlett SL, Bailey R, Baitchman E. Diagnosis and management of diabetes mellitus in a Bali mynah (*Leucopsar rothschildi*). *J Avian Med Surg.* 2016;30:146-151.

8. Pilny AA, Luong R. Diabetes mellitus in a chestnut-fronted macaw (*Ara severa*). *J Avian Med Surg.* 2005;19:297-302.

9. Phalen DN, Falcon M, Tomaszewski EK. Endocrine pancreatic insufficiency secondary to chronic herpesvirus pancreatitis in a cockatiel (*Nymphicus hollandicus*). *J Avian Med Surg.* 2007;21:140-145.

10. Desmarchelier M, Langlois I. Diabetes mellitus in a Nanday conure (*Nandayus nenday*). *J Avian Med Surg.* 2008;22:246-254.

11. Lumeji JT. Endocrinology. In: Ritchie BW, Harrison GJ, Harrison LR, eds. *Avian Medicine: Principles and Application.* Lake Worth, FL: Wingers Publishing Inc; 1994:582-606.

12. Gancz AY, Wellehan JFX, Boutette J, Malka S, Lee SE, Smith DA, Taylor M. Diabetes mellitus concurrent with hepatic haemosiderosis in two macaws (*Ara severa, Ara militaris*). *Avian Pathol.* 2007;36:331-336.

13. Ganz AY, Wellehan JFX, Boutette J, et al. Diabetes mellitus in large psittacines: a possible relationship with excessive iron storage. *Proc Annu Conf Assoc Avian Vet.* 2005;267-269.

14. Mitchell TC, McClain DA. Diabetes and hemochromatosis. *Curr Diab Rep.* 2014;14:488.

15. Candeletta SC, Homer BL, Garner MM, Isaza R. Diabetes mellitus associated with chronic lymphocytic pancreatitis in an African grey parrot (*Psittacus erithacus erithacus*). *J Assoc Avian Vet.* 1993;7:39-43.

16. Ryan CP, Walder EJ, Howard EB. Diabetes mellitus and islet cell carcinoma in a parakeet. *J Am Anim Hosp Assoc.* 1982;18:139-142.

17. Wallner-Pendleton EA, Rogers D, Epple A. Diabetes mellitus in a red-tailed hawk (*Buteo jamaicensis*). *Avian Pathol.* 1993;22:631-635.

18. Pollock CG, Pledger T. Diabetes mellitus in avian species. *Proc Annu Conf Assoc Avian Vet.* 2001;151-154.

19. Wissman MA. Budgerigar and cockatiel clinics. *Proc Annu West Vet Conf.* 2003.

20. AAHA. Blood glucose curves. Available at: https://www.aaha.org/aaha-guidelines/diabetes-management/monitoring/blood-glucose-curves/ Accessed April 4, 2022.

21. Acierno MJ, Mitchell MA, Schuster PJ, et al. Evaluation of the agreement among three handheld blood glucose meters and a laboratory blood analyzer for measurement of blood glucose concentration in Hispaniolan Amazon parrots (*Amazona ventralis*). *Am J Vet Res.* 2009;70:172-175.

22. Plumb DC. *Plumb's Veterinary Drug Handbook.* 9th Ed. Stockholm, WI: John Wiley & Sons; 2018.

23. Hawkins MG, Guzman DS, Beaufrere H, et al. Birds. In: Carpenter JW, ed. *Exotic Animal Formulary.* 5th Ed. St. Louis, MO: Elsevier; 2018;287-316.

24. Kahler BA. Sandostatin (synthetic somatostatin) treatment for diabetes mellitus in a sulfur breasted toucan (*Ramphastus sulfuratus sulfuratus*). *Proc Annu Conf Assoc Avian Vet.* 1994;269-273.

25. Barone JV, Tillman EM, Ferry RJ. Treatment of transient neonatal diabetes mellitus with subcutaneous insulin glargine in an extremely low birth weight neonate. *J Pediatr Pharmacol Ther.* 2011;16(4):291-297.

26. Stickelmeyer MP, Graf CJ, Frank BH, et al. Stability of U-10 and U-50 dilutions of insulin lispro. *Diabetes Technol Ther.* 2000;2(1):61-66.

27. Shivaprasad HL, Bonda M. Diabetes mellitus in and adult blue and gold macaw (*Ara ararauna*). *Proc Annu Conf Assoc Avian Vet.* 2009;341-342.

28. McCleery V. Management of diabetes mellitus in an eclectus parrot (*Eclectus roratus*) using insulin glargine. *Proc ICARE.* 2019.

Avian Endocrinology

Yvonne van Zeeland, DVM, MVR, PhD, DipECZM (Avian), DipECZM (Small Mammal), CPBC, and
Nico Schoemaker, DVM, PhD, DipECZM (Small Mammal), DipECZM (Avian), DABVP (Avian Practice)

Session #2659

Affiliation: From the Division of Zoological Medicine, Department of Clinical Sciences, Faculty of Veterinary Medicine, Utrecht University, Utrecht, 3584 CM, The Netherlands.

Manuscript previously published in van Zeeland Y, Schoemaker N. Avian endocrinology. Proc Annu Conf Assoc Avian Vet. 2017;247-253 and van Zeeland Y, Schoemaker N. Avian endocrinology. Proc ExoticsCon. 2018;297-303.

Manuscript originally published in van Zeeland YRA. Endocrine diseases in parrots. Br Small Anim Vet Assoc Congr Times. 2016. Reprinted with permission.

Abstract: Compared to mammals, birds, including parrots, seem to seldom suffer from endocrine disorders. Hypo- and hyperadrenocorticism, as well as hyperthyroidism and thyroid neoplasia, have rarely been recognized in live birds. Hypothyroidism is suggested to occur more often, but in many cases lack of appropriate testing prevents a definite diagnosis from being made, thereby keeping the number of confirmed cases low. Secondary nutritional hyperparathyroidism, on the other hand, is among the most common endocrine diseases seen in parrots that are fed diets low in calcium and/or vitamin D. In addition, a poor calcium:phosphorus ratio may result in osteodystrophy or hypocalcaemia. Imbalanced diets may furthermore result in iodine deficiency and goiter, most commonly seen in budgerigars. Pituitary neoplasia and diabetes mellitus are also recognized in this species, but do also occur in other parrot species. In many cases, ante mortem diagnosis and treatment of endocrine disease remains challenging because of the limited pathophysiologic knowledge, unavailability of validated tests and lack of information on appropriate dosing regimens in birds; however, by remaining alert to the possibility of a potential underlying endocrine disorder and using the knowledge obtained from other companion animals, endocrine disease in parrots may be more readily recognized, diagnosed and treated.

Introduction

In birds, the endocrine system consists of the hypothalamic-pituitary complex and pineal gland, carotid bodies, the thyroid, parathyroid, ultimobranchial, and adrenal glands, the pancreas and endocrine cells of the gastrointestinal tract, and the gonads.[1,2] All of these organs secrete hormones into the bloodstream that help to maintain homeostasis and regulate internal processes such as growth and development, metabolism and reproductive activity. Disease usually occurs as a result of a hypofunction or hyperfunction of the gland. Although endocrine diseases do occur in psittacine birds, their incidence appears to be relatively low, with most of the available information derived from single case reports or small case series; however, the low incidence may (in part) be explained by the lack of knowledge and appropriate tests to diagnose endocrine abnormalities. As a result, endocrine disease may occur more often and should certainly not be ruled out when compiling a differential diagnosis.

Of the various endocrine diseases that have been reported in parrots, secondary nutritional hyperparathyroidism, resulting in hypocalcaemia and/or metabolic bone disease appears most common, especially in birds that are fed an all-seed diet.[3] Similarly, goiter may be seen, most frequently in budgerigars, as a result of iodine deficiency.[4] Other endocrine diseases appear less common in parrots, with reports of their incidence being limited to a single or small number of birds.

Diseases of the Pituitary Gland

The pituitary gland, also referred to as the hypophysis, is a small gland that is located at the base of the brain, just below the hypothalamus. It consists of an anterior part (adenohypophysis) and posterior part (neurohypophysis) that both produce and secrete hormones (i.e., luteinizing hormone [LH], follicle stimulating hormone [FSH], thyroid stimulating hormone [TSH], adrenocorticotropic hormone [ACTH], growth hormone [GH], prolactin, arginine vasotocin [AVT] and mesotocin, respectively[1,2]). Disease may occur as a result of various processes, most commonly as a result of a (benign) neoplastic growth. Such pituitary tumors have most commonly been described in budgerigars,[5-7] but may also occur in other species.[8-12] Generally, clinical signs will develop as a result of the mass pressing against the surrounding tissues, thereby causing neurologic signs, exophthalmos, and/or blindness.[13] Similarly, pressure against the normal pituitary cells may result in undersecretion of one or more of the pituitary hormones. Occasionally, oversecretion of a specific hormone may be noted, leading to clinical signs of Cushing's disease (as a result of overproduction of ACTH[12]) or diabetes insipidus (as a result of overproduction of AVT [14]). Although a presumptive diagnosis can be made based on the clinical signs and ruling out of other diseases, definite diagnosis requires visualization of the pituitary mass using advanced imaging techniques such as CT or MRI. In addition, hormonal hypersecretion may be detected using hormone-specific assays or function tests (eg, water deprivation test[14-16]); however, size of the patient and tumor may prevent ante-mortem diagnosis of the tumor, thereby rendering post-mortem examination as the only option to obtain a final diagnosis. Thus far, no treatment has been reported for these type of tumors, although one report has demonstrated long-term management of central diabetes insipidus in a grey parrot using desmopressin.[14]

Diseases of the Thyroid Gland

Disease of the thyroid gland may result from hyperplasia, neoplasia, infection, inflammation or atrophy, with clinical signs occurring as a result of enlargement, hypo- or hyperfunction of the gland.[13] The number of documented cases of thyroid tumors and hyperthyroidism in psittacines is scarce, with most cases involving non-functional thyroid adenomas or adenocarcinomas.[10,17,18] Clinical signs will usually result from the compression of the surrounding tissues by the tumor and may include dyspnea, regurgitation, and dysphagia.[10,17,18] These clinical signs may more commonly be observed, especially in budgerigars,[4] in cases of thyroid hyperplasia (goiter). This disease, which is most often caused by an iodine deficiency, is usually diagnosed based on the history, clinical signs and response to dietary iodine suppletion.[10]

Hypothyroidism has often been suspected as the cause of feather loss (due to a lack of feather growth) and obesity in parrots.[19,20] In many cases, diagnosis was based on the finding of a low plasma T4 and/or a positive response to thyroxine suppletion;[10,13,21] however, definite diagnosis requires the use of a TSH stimulation test, with hypothyroid birds showing less than two-fold increases in plasma T4 levels compared to baseline.[10,13,22-24] In addition, scintigraphy may be helpful to diagnose hypothyroidism, but is not widely available.[25] Following confirmation of the diagnosis, treatment may be initiated with oral thyroxine suppletion.[13,20]

Diseases of the Parathyroid Glands

Thus far, disease of the parathyroid glands in birds has primarily been linked to hyperplasia resulting from dietary calcium, phosphorus and/or vitamin D deficiencies or imbalances.[3] These dietary imbalances will subsequently result in a disturbance of the mineral homeostasis, which stimulates the parathyroid glands to produce more PTH to correct these imbalances (i.e., secondary nutritional hyperparathyroidism[26]). In particular, grey parrots appear prone to developing disease, potentially resulting from an increased dependency on ultraviolet light for their calcium metabolism.[26,27] Clinical signs often include weakness, ataxia, tetany and/or seizures. In young, growing birds, bone deformities and pathologic fractures of the long bones and spinal column are frequently seen, whereas female birds may present with egg binding or production of poorly calcified or deformed eggs. Diagnosis is usually based on low (ionized) calcium levels in the blood,[28] whereas radiographs may aid in the diagnosis of bone abnormalities.[13,26] Treatment consists of oral and parenteral calcium and/or vitamin D_3 supplementation, dietary corrections and exposure to UV-B radiation through natural sunlight or artificial lighting.[3,26]

Aside from hypocalcemia, hypercalcemia has also been reported to occur in parrots. Most often, this is part of a physiologic phenomenon in egg laying females, with increased levels of estrogen resulting in increased calcium absorption in the gastrointestinal tract, increased production of calcium-binding proteins (e.g., albumin), and mobilization of calcium from the medullary bone.[29] Pathologic hypercalcemia also occurs, most common in response to long-term provision of a diet containing excessive amounts of calcium and/or vitamin D_3.[29,30] The disease is commonly seen in young, growing birds (particularly macaws) that are fed excessive amounts of vitamin and mineral supplements.[29,30] Clinical signs include poor weight gain in growing animals, depression, anorexia, nausea, polyuria/polydipsia, ataxia, muscle weakness and joint pain.[29,30] Diagnosis is usually based on the clinical signs combined with high (ionized) blood calcium levels. In addition, uric acid and/or vitamin D_3 metabolites may be measured.[29,30] Radiography is often useful to assess bone density, renal size and detect tissue mineralization.[29,30] In case of advanced cases (i.e., presence of tissue calcifications and visceral or articular gout), prognosis is generally poor. In mild cases, prognosis may be more favorable following timely treatment with forced diuresis and dietary corrections.[29,30]

Diseases of the Adrenal Glands

The avian adrenal glands differ from those in mammals with regard to the lack of a distinct cortex and medulla and the production of corticosterone rather than cortisol as the predominant glucocorticoid hormone.[21,31] Similar to other animals, disease may result from hyper- or hypofunction of the adrenal glands, resulting in excess or lack of one of the adrenal hormones (glucocorticoids, mineralocorticoids, [nor]epinephrine). Thus far, the number of birds diagnosed with spontaneous hyperadrenocorticism has been scarce,[12,31,32] with only one case with an ante mortem diagnosis of Cushing's syndrome as a result of a primary adrenal gland tumor reported in a psittacine bird.[33] In this Senegal parrot, clinical signs included polyphagia, polydipsia, mild alopecia, obesity and hepatic lipidosis. Following diagnosis, which was achieved using a modified dexamethasone suppression test, the bird was successfully treated with trilostane for over one year.[33] Aside from primary adrenal hyperfunction, hyperadrenocorticism may also result from overstimulation by a pituitary tumor (i.e., Cushing's disease) or exogenous corticosteroid administration, resulting in similar signs as those described above.[12,18] In addition, exogenous corticosteroids exert profound effects on the immune system and may suppress both humoral and cell-mediated immunity, resulting in increased susceptibility to infections, in particular aspergillosis.[31,34,35]

Thus far, there have been no reports documenting the presence of spontaneous hypoadrenocorticism in pet birds. It is, however, likely that adrenal hypofunction occurs in these species. Similar to other animals, adrenal cell atrophy or necrosis may be induced by neoplasia, bacterial or viral infections, immune-mediated disease, amyloidosis, trauma, or drugs (e.g., mitotane[18]). Diagnosis may be achieved using an ACTH stimulation test;[31,36,37] however, since the condition of animals with hypoadrenocorticism may quickly deteriorate as a result of hyponatremia-associated dehydration and shock, and hyperkalemia-associated cardiac disturbances, it is highly likely that fatal complications develop too rapidly in birds, without the possibility of obtaining an ante mortem diagnosis or initiating appropriate treatment.

Diseases of the Pancreas

In birds, glucagon, secreted by the α cells of the pancreatic islands, is considered the major glucose-regulating hormone, with insulin playing a less well-defined role in the glucose metabolism.[38] As such, glucagon excess or inadequate insulin:glucagon ratios rather than hypoinsulinemia has been proposed as the cause of avian diabetes.[39] Histologically, destruction of the β cells (consistent with type 1 diabetes in humans) has been documented in birds, with underlying causes varying from paramyxovirus or herpesvirus infections to hemochromatosis (iron storage disease[40,41]). Similarly, causes for peripheral insulin resistance (consistent with type 2 diabetes), such as obesity, high levels of endogenous or exogenous corticosteroids, and increased levels of other hormones (i.e., glucagon, growth hormone, epinephrine), have also been identified.[40-42] Neoplasia associated with diabetes mellitus has been diagnosed in one bird.[43]

In parrots, diabetes mellitus is most commonly reported in budgerigars and cockatiels, although it may occur in larger species as well.[18,40-42,44,45] A presumptive diagnosis can usually be made based on the clinical signs (polyuria, polydipsia, weight loss despite good appetite) in combination with the finding of a persistent hyperglycemia and glycosuria.[13,41,44] The normal range for plasma glucose is significantly higher than that of mammals and may range from 180-503 mg/dl (10.0-27.9 mmol/L), dependent on the species, with values > 800 mg/dl (> 44.4 mmol/L) considered diagnostic for diabetes mellitus.[44] Elevated fructosamine levels and/or ketonuria may also be found, supporting the diagnosis.[13] Furthermore, determination of plasma glucagon and insulin has been recommended, but remains challenging due to the lack of appropriately validated tests.[13,46] Treatment of birds with diabetes mellitus can be challenging, especially given the small size of many patients and the lack of clear-cut therapeutic protocols.[13] Prior to initiating treatment with glucose-regulating drugs, attempts should be made to rule out and treat any underlying or concurrent conditions. Similarly, birds may require supportive care in the form of fluids or force feeding. In birds with persistent hyperglycemia, treatment may be attempted with insulin and/or oral hypoglycemic agents (e.g., glipizide), although the success of this treatment may vary from case to case.[13,40,44,46,47] Hospitalization of the bird is often recommended to enable gradual titration and regular monitoring of the patient (e.g., serial blood glucose monitoring or urinalysis to check for glycosuria) until the bird is stable, following which treatment can be continued at home.[13,44]

References

1. Orosz SE. Anatomy and physiology of the endocrine system-protein hormones. In: Speer BL, ed. *Current Therapy in Avian Medicine and Surgery.* St. Louis, MO: Elsevier Saunders; 2016:378-385.

2. Ritchie M, Pilny AA. The anatomy and physiology of the avian endocrine system. *Vet Clin North Am Exot Anim Pract.* 2008;11:1-14.

3. Stanford M, Keen J, Campbell-Ward ML. Hypocalcemia in other species. In: Rand J, ed. *Clinical Endocrinology of Companion Animals.* Ames, IA: Wiley-Blackwell; 2013:335-355.

4. Schmidt RE. Avian thyroid metabolism and diseases. *Semin Avian Exot Pet Med.* 2002;11:80-83.

5. Bauck L. Pituitary neoplastic disease in nine budgies. *Proc Annu Conf Assoc Avian Vet.* 1988;87-89.

6. Langohr IM, Garner MM, Kiupel M. Somatotroph pituitary tumors in budgerigars (*Melopsittacus undulatus*). *Vet Pathol Online.* 2012;49:503-507.

7. Schlumberger HG. Neoplasia in the parakeet I. Spontaneous chromophobe pituitary tumors. *Cancer Res.* 1954;14:237-245.

8. Curtis-Velasco M. Pituitary adenoma in a cockatiel (*Nymphicus hollandicus*). *J Assoc Avian Vet.* 1992;6: 21-22.

9. Latimer KS. Oncology. In: Ritchie BW, Harrison GJ, Harrison LR, eds. *Avian Medicine: Principles and Applications.* Lake Worth, FL: Wingers Publishing Inc; 1994:640-672.

10. Lumeij JT. Endocrinology. In: Ritchie BW, Harrison GJ, Harrison LR, eds. *Avian Medicine: Principles and Applications.* Lake Worth, FL: Wingers Publishing Inc; 1994:582-606.

11. Romagnano A, Mashima TY, Barnes HJ, et al. Pituitary adenoma in an Amazon parrot. *J Avian Med Surg.* 1995;9:263-270.

12. Starkey SR, Morrisey JK, Stewart JE, Buckles EL. Pituitary-dependent hyperadrenocorticism in a cockatoo. *J Am Vet Med Assoc.* 2008;232:394-398.

13. De Matos R, Monks D. Diseases of the endocrine system-protein hormones. In: Speer BL, ed. *Current Therapy in Avian Medicine and Surgery.* St. Louis, MO: Elsevier Saunders; 2016:385-399.

14. Starkey SR, Wood C, de Matos R, et al. Central diabetes insipidus in an African grey parrot. *J Am Vet Med Assoc.* 2010;237:415-419.

15. Alberts H, Halsema WB, De Bruijne JJ, Lumeij JT. A water deprivation test for the differentiation of polyuric disorders in birds. *Avian Pathol.* 1988;17:385-389.

16. Lumeij JT, Westerhof I. The use of the water deprivation test for the diagnosis of apparent psychogenic polydipsia in a socially deprived African grey parrot (*Psittacus erithacus erithacus*). *Avian Pathol.* 1988;17: 875-878.

17. Doukaki C, Kik MJL, Ducatelle R, et al. Thyroid adenocarcinoma in two parrots. *Vet Rec.* 2014;2:e000101.

18. Rae M. Endocrine disease in pet birds. *Semin Avian Exot Pet Med.* 1995;4:32-38.

19. Kritchevsky JS. Hypothyroidism in other species. In: Rand J, ed. *Clinical Endocrinology of Companion Animals.* Ames, IA: Wiley-Blackwell; 2013:278-290.

20. Oglesbee BL. Hypothyroidism in a scarlet macaw. *J Am Vet Med Assoc.* 1992;201:1599-1601.

21. Oglesbee B. Diseases of the endocrine system. In: Altman RB, Clubb SL, Dorrestein GM, Quesenberry K. *Avian Medicine and Surgery.* Philadelphia, PA: Saunders; 1997:482-488.

22. Greenacre CB. TSH testing in birds using canine, bovine, or human TSH. *Proc Annu Conf Assoc Avian Vet.* 2011;49.

23. Lothrop CD, Loomis MR, Olsen JH. Thyrotropin stimulation test for evaluation of thyroid function in psittacine birds. *J Am Vet Med Assoc.* 1985;186:47-48.

24. Zenoble RD, Kemppainen RJ, Young DW, Clubb SL. Endocrine response of healthy parrots to ACTH and thyroid stimulating hormone. *J Am Vet Med Assoc.* 1985;187:1116-1118.

25. Harms CA, Hoskinson JJ, Bruyette DS, et al. Technetium-99M and iodine-131 thyroid scintigraphy in normal and radiothyroidectomized cockatiels (*Nymphicus hollandicus*). *Vet Radiol Ultrasound.* 1994;35:473-478.

26. Stanford M. The effect of UV-B lighting supplementation in African grey parrots. *Exot DVM* 2004;6:29-32.

27. Stanford M. Calcium metabolism. In: Harrison GJ, Lightfoot TL, eds. *Clinical Avian Medicine.* Vol 1. Palm Beach, FL: Spix Publishing; 2006:141-151.

28. Stanford M. Clinical pathology of hypocalcaemia in adult grey parrots (*Psittacus erithacus*). *Vet Rec.* 2007;161:456.

29. Campbell-Ward ML. Hypercalcemia in other species. In: Rand J, ed. *Clinical Endocrinology of Companion Animals.* Ames, IA: Wiley-Blackwell; 2013:385-395.

30. Schoemaker NJ, Lumeij JT, Beynen AC. Polyuria and polydipsia due to vitamin and mineral oversupplementation of the diet of a salmon crested cockatoo (*Cacatua moluccensis*) and a blue and gold macaw (*Ara ararauna*). *Avian Pathol.* 1997;26:201-209.

31 De Matos R. Adrenal steroid metabolism in birds: anatomy, physiology, and clinical considerations. *Vet Clin North Am Exot Anim Pract.* 2008;11:35-57.

32. Cornelissen H, Verhofstad A. Adrenal neoplasia in a scarlet macaw (*Ara macao*) with clinical signs of hyperadrenocorticism. *J Avian Med Surg.* 1999;13:92-97.

33. Van Zeeland YRA, Bastiaansen P, Kooistra H, Schoemaker NJ. Diagnosis and treatment of Cushing's syndrome in a Senegal parrot. *ExoticsCon.* 2015;119.

34. Verstappen FALM, Dorrestein GM. Aspergillosis in Amazon parrots after corticosteroid therapy for smoke-inhalation injury. *J Avian Med Surg.* 2005;19:138-141.

35. Westerhof I. Pituitary-adrenocortical function and glucocorticoid administration in pigeons. *Vet Q.* 1997;19:60-61.

36. Lothrop CD, Olsen JH, Loomis MR, et al. Evaluation of adrenal function in psittacine birds, using the ACTH stimulation test. *J Am Vet Med Assoc.* 1985;187:1113-1115.

37. Walsh MT, Beldegreen RA, Clubb SL, Chen CL. Effect of exogenous ACTH on serum corticosterone and cortisol concentrations in the Moluccan cockatoo (*Cacatua moluccensis*). *Am Vet Res.* 1985;46:1584-1588.

38. Hazelwood RL. Pancreatic hormones, insulin/glucagon molar ratios, and somatostatin as determinants of avian carbohydrate metabolism. *J Exp Zool.* 1984;232:647-652.

39. Sitbon G, Laurent F, Mialhe A, et al. Diabetes in birds. *Horm Metab Res.* 1980;12:1-9.

40. Gancz AY, Wellehan JF, Boutette J, et al. Diabetes mellitus concurrent with hepatic haemosiderosis in two macaws (*Ara severa*, *Ara militaris*). *Avian Pathol.* 2007;36:331-336.

41. Pilny AA. The avian pancreas in health and disease. *Vet Clin North Am Exot Anim Pract.* 2008;11:25-34.

42. Candeletta SC, Homer BL, Garner MM, Isaza R. Diabetes mellitus associated with chronic lymphocytic pancreatitis in an African grey parrot (*Psittacus erithacus erithacus*). *J Assoc Avian Vet.* 1993;7:39-43.

43. Ryan CP, Walder EJ, Howard EB. Diabetes mellitus and islet cell carcinoma in a parakeet. *J Am Anim Hosp Assoc.* 1982;18:139-142.

44. Campbell-Ward ML, Rand J. Diabetes mellitus in other species. In: Rand J, ed. *Clinical Endocrinology of Companion Animals.* Ames, IA: Wiley-Blackwell; 2013:191-200.

45. Desmarchelier M, Langlois, I. Diabetes mellitus in a nanday conure (*Nandayus nenday*). *J Avian Med Surg.* 2008;22:246-254.

46. Bonda M. Plasma glucagon, serum insulin, and serum amylase levels in normal and hyperglycemic macaw. *Proc Annu Conf Assoc Avian Vet.* 1996;88.

47. Pilny AA, Luong R. Diabetes mellitus in a chestnut-fronted macaw (*Ara severa*). *J Avian Med Surg.* 2005;19:297-302.

Avian Parasitology Review

Cheryl B. Greenacre, DVM, DABVP (Avian Practice), DABVP (Exotic Companion Mammal Practice),
Kristina Vuong, DVM,
Michelle Liu, DVM,
Mee-Ja Sula, DVM, DACVP

Session #2650

Affiliation: From the Department of Small Animal Clinical Sciences (Greenacre, Vuong), Department of Biomedical and Diagnostic Sciences (Liu, Sula), College of Veterinary Medicine, University of Tennessee, 2407 River Drive, C-247, Knoxville, TN 37996, USA.

Abstract: The objective of this presentation is to describe the more common parasitic diseases of birds presenting to a veterinary hospital. The objective of these proceedings notes is to provide a thorough review of avian parasites for reference in the hospital. This review includes protozoa, helminths, nematodes, and mites, and it outlines their life cycles, clinical signs, testing and treatment. Species-specific doses of medications mentioned in the text can be found in *Exotic Animal Formulary*.[1]

Protozoa

Giardia

Giardia has been reported in many orders of birds including Psittaciformes (most commonly budgerigars, lovebirds, and cockatiels), Piciformes (toucan), Ciconiiformes (heron), Chadriformes (avocet), Passeriformes (shrike), and Falconiformes (kite). The cyst is shed in the feces of an infected host and is then ingested by a new host, thereby entering the GI tract. There are several methods of identifying the presence of *Giardia* including zinc sulfate flotation test of feces to detect the cyst form, direct saline smear (± addition of Lugol's iodine stain for enhanced contrast) of fresh feces to detect the motile trophozoites, ELISA test for *Giardia* antigen in feces (Cornell Diagnostic Laboratory), or PCR and DNA sequence assays (Animal Genetics, Inc., US). It is best to collect over multiple days due to intermittent shedding. Clinical signs may be none to diarrhea and weight loss. Treatment for giardiasis is metronidazole, carnidazole, or ronidazole. Use of nitroimidazoles in food producing animal species such as chickens or turkeys is prohibited. The *Giardia* genus contains many different species and genotypes, and vary in their ability to infect host species, including humans, therefore giardiasis from birds is potentially zoonotic. A 1:10 bleach solution is recommended to kill the parasites in the environment on non-porous surfaces.

Trichomonas

Trichomonas has been reported in many orders of birds including Columbiformes, Galliformes, Falconiformes, Psittaciformes, and Passeriformes. Transmission is through ingestion of trophozoites, or through the navel of neonates. Diagnosis is based on the presence of motile trophozoites on a direct saline smear of fresh feces. Clinical signs include whitish plaques in the oropharynx, esophogus, and crop, and anorexia, dysphagia, weight loss, and dyspnea. The plaques can sometimes spread, and *Trichomonas* can invade deeper structures. Treatment for Trichomoniasis is metronidazole.

Histomonas

Histomonas has been reported in many orders of birds including Galliformes and Rheiformes. The trophozoite is ingested by the host via contaminated food, water, earthworms or within *Heterakis gallinariuim* eggs. Diagnosis is based on a concomitant *Heterakis* infection and pathognomonic round green lesions on liver at necropsy. Clinical signs may be none to sulfur yellow feces, hepatomagaly, ascites, diarrhea and weight loss. Treatment for histomoniasis is metronidazole and treatment for the *Heterakis* infection. Use of nitroimidazoles in food producing animal species such as chickens or turkeys is prohibited. Also, eliminate access to earthworms, and do not house chickens with turkeys.

Coccidia

Eimeria and *Isospora*

Coccidia have been reported in probably all birds, but the genus and species of coccidia are very host specific. *Eimeria* is the most common coccidian of Galliformes and Columbiformes. *E. acervulina* is the most common and causes whitish patches in the upper half of the small intestine. *E. tenella* causes hemorrhagic cecal cores. Infection is by ingestion of oocysts. The presence of coccidia is best detected with the zinc sulfate flotation test of feces. Clinical signs may be none to diarrhea and weight loss. Treatment for coccidiosis is amprolium, sulfadimethoxine, diclazuril, or toltrazuril.

Cryptosporidium

Cryptosporidium has been reported in many orders of birds including Galliformes, Anseriformes, Psittaciformes and Passeriformes.[3] Transmission is through ingestion or inhalation of sporulated oocysts. Because of the small size and low shedding rate of Cryptosporidium it is difficult to detect but a Sheather's sugar flotation test is best. An acid-fast stain can also be used to detect the organism. There is an ELISA and a PCR available from Cornell Diagnostic Laboratory. Clinical signs may involve GI tract, respiratory tract or urinary tract, especially in immunosuppressed birds. Treatment for cryptosporidiosis in birds has not described. Some have tried paromomycin, azithromycin, or nitazoxinide.

Atoxoplasma

Atoxoplasma has been reported in many orders of birds including Passeriformes and Psittaciformes. Common examples include canaries and the Bali mynah. Transmission is through ingestion of oocysts. There are several methods of identifying the presence of *Atoxoplasma* including fecal float (best in adult), buffy coat smear (Romanowsky's stain to look for asexual stage in lymphocytes and monocytes), liver impression smear (Wright's stain) or PCR of feces. Clinical signs can be none to diarrhea, weight loss and signs of liver disease (dark spot of liver visible through the skin over the coelomic cavity in neonates). Clinical signs are more severe in young and include death of fledgling birds, whereas adults can remain asymptomatic. Treatment attempts include diclazuril, toltraurazil or sulfachorpyridazine.

Toxoplasma

Toxoplasma infection occurs in many species, but most commonly in lory, lorikeet and rosella. Infection is by ingestion of sporulated oocysts from felid feces. Also, cockroaches can act as a mechanical vector. Diagnosis is based on identifying the organism in tissues at necropsy. Most bids are asymptomatic (have antibodies) or it may be reactivated later in the lungs, heart or liver and accompanied by anorexia and weight loss. Treatment for toxoplasmosis in birds is not described.

Sarcocystis

Sarsocystis has been reported in many orders of birds, but the definitive hosts include the Falconiformes and Strigiformes, whereas the intermediate hosts include the Psittaciformes (Old World), Passeriformes and Columbiformes. The life cycle in the definitive host is by ingestion of bradyzoites in muscle tissue, whereas in the intermediate host it is by ingestion of sporocysts in opossum feces, which may have been moved mechanically by a cockroach. Usually the diagnosis is made at necropsy, grossly there are hemorrhagic lungs and on histopathology the parasite is identified in the lung tissue (or muscle tissue of definitive hosts). Clinical signs may be none to mild (New World psittacine birds) including diarrhea, neurological signs and weight loss, to severe (Old World psittacine birds) including dyspnea and death in the intermediate hosts, especially in the winter months. Antemortem diagnosis is based on serology (University of Miami Diagnostic Lab). Treatment can be attempted with diclazuril, toltrazuril or paramomycin. Elimination of cockroaches in the aviary is a helpful preventive.

Hemoparasites

Plasmodium

Malaria is highly pathogenic in the canary, penguin and gyrfalcon. *Plasmodium* is spread by mosquitos. Diagnosis is via Wright's or similar stain blood smear displaying RBC's with a displaced nucleus. Clinical signs may be none to weakness and death. Treatment for malaria, if needed, is with chloroquine or primaquine.

Haemoproteus

Haemoproteus has been reported in many orders of birds but is most common in raptors and pigeons. *Haemoproteus* is spread by the Hippobscid louse fly and the Culicoides mosquito. The number of circulating organisms increases with stress. Diagnosis is via Wright's or similar stained blood smear displaying RBC's with a nucleus that is not displaced. Clinical signs may be none to weakness. Treat underlying disease/stressors.

Leukocytozoon

Leukocytozoon is significant in Galliformes and Anseriformes. *Leukocytozoon* is spread by the black fly. Diagnosis is based on a blood smear stained with Wright's or similar stain displaying WBC's with a nucleus that is displaced by the parasite. Clinical signs may be none to weakness. There is no specific treatment for leukocytozooanosis.

Trematodes and Cestodes

Liver flukes (Trematodes)

Digenetic liver flukes are sometimes encountered in Old World Psittaciformes. The eggs pass in feces of a host, then a miracidium hatches out and bores into a snail intermediate host where the miracidium loses its ciliated coat and forms a sporocyst and produces cercaria (the final infective stage), which is either imbibed or ingested by a host. The cercaria enters the second intermediate host, an arthropod, which is then ingested by the definitive host. The term metacercaria is used after cercaria have encysted on vegetation or inside the second intermediate host. After metacercaria are ingested they encyst in the definitive host's intestinal tract and the immature stage migrates to a predilected site. Diagnosis is based on visualization of the operculated fluke eggs on a fecal float. Clinical signs of infection by liver flukes includes hepatomegaly, depression and anorexia. Treatment for liver flukes is praziquantel.

Tapeworms (Cestodes)

The most common pet birds that are hosts to tape worms include African grey parrots, finches and cockatoos. The egg hatches after being swallowed by the intermediate host and the embryo penetrates the intestinal wall to migrate to a suitable place to grow. Then it forms a cyst (i.e., bladderworm) and develops hooks and then the intermediate host is ingested by the final host. Proglottids develop after attachment. Diagnosis is based on visualization of the eggs on a fecal float. Clinical signs can be none to unthriftiness with diarrhea. Treatment of tapeworms is with praziquantel.

Nematodes

Ascarids

All species can be infected with ascarids that are usually species specific. Transmission is through ingestion of ova. Diagnosis is by identification of eggs on a fecal float. Clinical signs include none to anorexia, weight loss and death. Treatment is pyrantel pamoate, fenbendazole, or ivermectin, but to prevent repeat infection the eggs in the environment must be steamed or flamed.

Heterakis

The orders most commonly affected by *Heterakis* are the Galliformes and Anseriformes (poultry and ducks). Transmission is by ingestion of embryonated ova. Diagnosis is based on a fecal float. At necropsy the presence of cecal nodules is highly suggestive. This nematode is associated with the disease histomoniasis. Treatment of *Heterakis* is with pyrantel pamoate or fenbendazole.

Syngamus

The most common host orders for *Syngamus* are Galliformes and Anseriformes. The life cycle is direct with earthworms acting as a transport host. The eggs are coughed up and swallowed and passed in the feces. The infective larva develops in the egg starting at about 3 days and hatch in 9 days when they are subsequently ingested by the host directly or by the host eating an earthworm or snail that ate the infective larva. The larva can live months to years in the snail. After ingested, the larva are carried in the blood stream to the lungs where they undergo ecdysis. Then males and females migrate to larger bronchi to permanently copulate, then migrate to the trachea at about 7 days post-infection. The eggs are shed 7-20 days later. Males are smaller at 2-6 mm, whereas females are larger at 5-20 mm long. Diagnosis is based on visualization of the eggs on a fecal float, or direct visualization of the paired male and female worms shaped like a red, 1 cm "Y" in the trachea of the bird. Clinical signs include coughing, blood at ricti, head shaking and dyspnea. Treatment is with thiabendazole, fenbendazole, or ivermectin, or they can be retrieved endoscopically.

Capillaria

Capillaria is a common parasite and has been reported in Falconiformes, Galliformes, Columbiformes, Psittaciformes (budgie, macaw) and Passeriformes (canary). Transmission is by ingestion of embryonated ova. Diagnosis is by identifying the bipolar ova on a fecal float. Clinical signs include none to diarrhea, weight loss, dysphagia and melena. Treatment for capillariosis is usually with fenbendazole.

Chandlerella (filarid nematode)

Chandlerella has been reported in Passeriformes and cockatoos. Transmission is thought to be via a mosquito. Diagnosis is by identification of microfialria on a Wright's-stained blood smear. The organism was found in the heart (ventricles) of a Ducorp's cockatoo at necropsy. Clinical signs include none to possibly weakness. No treatment is reported for chandlerellosis or any microfilaremia in birds.

Pelicitus (filarid nematode)

Pelicitus is a filarid nematode and has been reported in Amazon parrots and conures. Transmission is not known. Diagnosis is by identification of eggs on aspiration of a subcutaneous mass on a distal pelvic limb or direct visualization after surgical removal. Clinical signs include a subcutaneous mass on a distal pelvic limb. Treatment is surgical removal.

Nematodes of the eye

Nematodes of the eye include *Thelazia*, *Oxyspira*, *Ceratospira*, and *Annulospira*.

Spirurida

Various species of spiruroid nematodes are reported in certain species of birds including *Tetrameres* in grackles, *Microtetrameres* in kakas and *Spiroptera* in cockatoos. Transmission is probably through an insect intermediate host. Diagnosis is based on a fecal float, or identification of masses within the proventriculus and ventriculus. Clinical signs include none to weight loss and melena. Treatment is with ivermectin.

Mites

Knemidokoptes pili and *laeris* mites

Knemidokoptes pili is most commonly reported in passerine birds such as canaries and finches, and Psittaciformes such as budgies and others. Transmission is via direct contact. Diagnosis of *K. pili* is based on typical pitted appearance to the legs and face of the bird and via visualization of the mite under the microscope after a tape prep or scrape of the affected area. *K. laeris* affects the feathers. Clinical signs include pitting of the surface tissue of the face and legs. Treatment is with *topical* ivermectin or moxidectin.

Sternostoma tracheocolum mites

Sternostoma is most commonly reported in finches, especially Lady Gouldian finches. Diagnosis can be made by visualizing dark specks in the trachea during transillumination of the trachea on a physical examination. Typical clinical signs include dyspnea and a characteristic clicking sound with each breath. Treatment is with topical ivermectin.

Poultry mites

Dermanyssus, (red fowl mite), and *Ornithonyssus* (white or Northern fowl mite) are commonly reported in poultry. Diagnosis can be made by identifying the mite under microscopy after a tape prep or skin scrape. Clinical signs may be none to pruritis and anemia. *Dermanyssus* feeds on the bird at night and is off the bird and in the environment and bedding during the day. *Ornithonyssus* spends its entire life on the bird. When treating the bird for *Dermanyssus*, the environment and enclosure must be treated as well. Treatment includes permethrin or carbaryl spray/powder, or fluralaner.

Recommended Reading

1. Carpenter JW, ed. *Exotic Animal Formulary.* 5th Ed. St. Louis, MO: Elsevier; 2018.

2. Bowman DD. *Georgi's Parasitology for Veterinarians.* 11th Ed. St. Louis, MO: Elsevier; 2021.

3. Swayne DE. *Diseases of Poultry.*,14th Ed. Hoboken, NJ: Wiley-Blackwell; 2020.

Avian Ophthalmology: Principles, Application and Practical Cases

Ruediger T. Korbel, Prof, Dr med vet habil, DipECZM (Avian)

Session #2643

Affiliation: From Center for Clinical Veterinary Medicine, Clinic for Birds, Small Mammals, Reptiles and Ornamental Fish, LMU University of Munich, Oberschleissheim, Germany.

Abstract: This master class covers five sections:
1) seeing the invisible: how birds see;
2) behind the scenes: anatomy and physiology;
3) nice gadgets to have: a practical hands-on workshop on using ophthalmologic instrumentation;
4) not a book with seven seals: the examination procedure; and
5) avian ophthalmology: the next step of diagnostics and therapy with an interactive review on clinical cases.

Participants submit clinical ophthalmologic cases including images and case histories for an interactive discussion on diagnostics and therapy within the lab.

Adapted from Korbel R. Avian ophthalmology: principles and application. Proc ExoticsCon. 2019;293-303, Korbel R. Avian ophthalmology: principles and application. Proc ExoticsCon. 2018;345-354, Korbel R. Avian ophthalmology: principles and application. Proc Annu Conf Assoc Avian Vet. 2017;201-210, Korbel R. Avian ophthalmology: principles and application. Proc ExoticsCon. 2016;173-180, Korbel R. Avian ophthalmology: principles and application. Vet Med Falconry 21st Century. 2014;33-38.

Introduction

Ocular disorders in birds are of great importance to vision and they commonly hint at numerous systemic disorders, which may even be transferable to man (e.g., salmonellosis, mycobacteriosis). This lab provides the practitioner with a practical review of suitable instrumentation following a step-by-step tutorial through the complete avian ophthalmologic examination procedure as a reliable basis for diagnoses and therapy of ocular disorders in birds. In the practical lab, diurnal and nocturnal raptors, both with healthy and diseased eyes, will be used to demonstrate both physiologic as well as pathologic situations.

Certain avian-specific capacities, like flying, are closely bound to fully functional vision underlining the importance of avian ophthalmology. In addition, ocular lesions in birds are an expression of systemic disorders more than in mammals and therefore represent an important diagnostic criterion. The ocular symptomatology frequently enables specific conclusions to be drawn on suspected disorders or it may even be pathognomonic for a certain disease. Thus, the avian eye may be seen, to a much larger extent than in mammals, diagnostic window. On average, more than 30% of all traumatized birds (incidence higher in raptors than in pet birds) suffer from ocular lesions, which are most often hidden within the inner structures of the eye with hemorrhages arising from the pectin oculi. Ophthalmoscopy (i.e., examination of the posterior eye segment) therefore is obligatory in traumatized birds. Avian ophthalmology is not a highly specific working field within avian medicine but should be an integral part of the general examination procedure. After a short review on anatomic and physiologic peculiarities, commonly and newly developed routine ophthalmologic examination procedures are described followed by a description of clinical pictures in the eye of various patient groups based on an etiologic schematization. Avian ophthalmology is becoming more and more important in terms of avian visual perception, which strikingly differs from that of man (including avian visual perception within the UV and the ability to resolve high frequency flicker rates). These aspects are important for animal welfare since commercially kept poultry are commonly kept under artificial light sources that meet human visual perception requirements but not those of avian vision.

In many birds the eye is the most important sensory organ. The capacities of the avian eye which exceed those of the mammalian eye in part are an adaptation to the specific way of life and habitats as well as physical activities that are closely bound to perfectly functioning vision (e.g., flying). Even partial impairment of vision that can be caused by any one of many known eye diseases, always has far-reaching consequences because compensation by other senses (including olfactory and acoustic sensory perceptions) is usually insufficient, if possible at all. As a result, ornitho-ophthalmology occupies an important position in avian medicine.

I. Anatomic and Physiologic Basics

Basic principles in avian ophthalmology are similar to mammalian ophthalmology. A broad variety of ophthalmologic, physiologic, and anatomic peculiarities should be considered within the class Aves representing 8,800 species and more than 28,000 subspecies. One of the most important aspects include the striated rather than smooth intraocular musculature, making it necessary to use bird specific techniques for inducing mydriasis. Additionally, the fundus oculi includes numerous peculiarities such as the pectin; an anangiotic retina; none, one or bifoveate retina; lack of the retinal pigment epithelium (RPE); and more. The presentation includes a section on avian visual capacities, which are of greatest importance of keeping pet birds, raptors as well as commercial kept poultry under artificial light sources and thus are of animal welfare importance.

Significance of visual capacities

Visual function in raptors is essential for flying, surviving in the wild and reproduction. The eye as the main sense organ in birds and its visual capacities have no general superiority compared to mammals but shows a highly specialization as an adaption to living conditions. Thus, visual acuity is two to eight times higher than mammals. Visual fields are up to 360°. Stereopsis ranges from 0-70°. Maximum spatial frequency (the ability to dissolve a certain movement into single frames) is up to 160 frames/sec (10-15 frames/sec in man) and minimum detection of movements is up to 15°/hour (very slow movements). Ultraviolet-perception (UV-perception between 320 nm and 680 nm) is an ability common in diurnal birds bound to special UV-sensitive rods within the retina and an aspect not investigated very well to date. UV-perception probably plays an important role in inter- and intraspecific communication based on plumage-UV-reflection, even in birds that appear monomorphic for the human eye, for the identification and assessment of fruit ripeness based on varying UV-reflection of fruit wax layers, for phenomena of camouflage, orientation, and others. On the other hand, even complete blindness is not a reason for euthanasia in pet birds. Canaries, suffering from blindness due to cataract formation—a condition that occurs often in this species—act normal, as the interior of the cage or aviary is not modified.

Anatomic peculiarities

Although there are numerous anatomic and physiologic differences, like striated rather than smooth intraocular musculature, the anangiotic fundus oculi, and the pecten oculi, the approach to avian ophthalmology is similar to that employed in mammalian ophthalmology. Some anatomic peculiarities with relevance for the ophthalmologist are:

Size and weight of the avian eye

- Axial length 8 mm (kiwi, *Apteryx* sp.) up to 50 mm (ostrich, *Struthio camelus*).
- Weight of oculus sinister (OS) and oculus dexter (OD), in man 1%, fowl 7% (adult) resp. 12% (juvenile) compared to overall head weight.

Shape

- Determination by anulus ossicularis sclerae (~10 single bony plates).
- Flat type (diurnal birds with narrow heads, for example Columbifomes).
- Conical type (diurnal birds with broad heads, for example Falconiformes).
- Tubular type (nocturnal birds, for example Strigiformes).

Adnexal structures

- Eyelids.
  - Palpebra inferior larger than palpebra superior (exclusively owls).
  - Lower eyelid with tarsus, no Meibomian glands.
  - *Membrana nictitans* highly mobile, regulation of precorneal tear film, protective function for cornea, snorkel mask effect (compensation for loss of refractive power during underwater vision).

- Lacrimal glands.
  - Small or absent glandulae lacrimaliae.
  - Large glandula lacrimalis membrana nictitantis (vel. Harderian gland).
  - Replacement by nasal salt gland (aquatic species).
  - No musculus retractor bulbi.
  - Complete decussation of chiasma opticus (no consensual pupillary reflex).

- Anterior eye segment.
  - Striated intraocular musculature (highly mobile iris corresponding to external stimuli, no sensitivity to parasympatholytics/sympathomimetics, no reliable pupillary reflex corresponding to light stimuli).
  - Fundus diameter much larger than pupil diameter.
  - Ligamentum pectinatum/ciliary cleft with species specific peculiarities.
  - Lens with anulus pulvinus (Ringwulst), accommodation range 2 (owls) to 80 (waterfowl) D.

- Posterior eye segment.
  - Avascular (anangiotic) retina.
  - Afoveate, uni- or bifoveate retina.
  - Rods and cones (inclusive special UV-cones) in functional units.
  - UV-sensitive cones in most diurnal birds.
  - Slight retinal pigmentation in nocturnal, heavily pigmentated retina in diurnal birds.

- No retinal tapetum lucidum.
- Pecten oculi.
- Chorioidal, heavily pigmentated structure.
- Protruding into the vitreous.
- Obscuring papilla nervi optici.
- Pleated, vaned and conical type.
- 32 functional theories: nutritive, thermo- and pressoregulative function most obvious.

II. Instrumentation

This section on avian ophthalmology provides the practitioner with detailed information on specific instrumentation needed going through a routine ophthalmologic examination procedure in exotic animals. Specified instrumentation includes a Finoff transilluminator allowing examination of the anterior and posterior eye segment (using monocular direct ophthalmoscopy), a panoptic monocular ophthalmoscope (which meets specific needs for avian ophthalmology), or a combination of a head band ophthalmoscope in combination with each 30, 40, 78 and 90 diopter lenses. A review of luxmeter and spectral analytic devices is provided to address measurement o suitable artificial light conditions.

General examination

Minimum requirements for basic and general purposes:

- Focused light source with magnification lens (Finoff transilluminator).
- Instrumentation for manipulation of the eye lids (Graefe hook).
- Lacrimal cannula (Anel).
- Topical anesthetics proxametacaine, oxybuprocaine (duration of action ~7-8 minutes) or lidocaine (duration of action ~17 minutes).

Advanced equipment:

- Slit lamp (magnification x5-x15, better x20).
- Monocular direct panoptic ophthalmoscope, better.
- Head band ophthalmoscope with 30 and 78 D lens (additional aspherical 40, 60, 90 D).

General examination procedures for adnexal structures and anterior eye segment:

- Without restraint.
    - Assessment of vision via food intake, reluctance to fly, orientation.

- With restraint.
    - Examination of the ear opening.
    - Pupillary reflex.
    - Examination of the anterior eye chamber with lateral illumination.
    - Examination of the anterior eye chamber with lateral transillumination.
    - Examination of the anterior eye chamber with retroillumination.

Equipment and procedures for specific examinations:

- Slit lamp biomicrography.

- Gonioscopy (Lovac lens). Examination of the angulus iridocornealis with the pectinate ligamentum; etiologic assessment of primary/secondary glaucoma status.

- Tonometry. Estimation of the intraocular pressure (IOP). Use electronic short-acting tonometer or Schioetz tonometer in raptors. Standard reference values measured with an electronic tonometer calibrated for avian eyes, range from 9-22 mg Hg and are available for 42 species from 7 Orders. Standard reference values for psittacine birds range from 12-15 mm Hg for nocturnal birds from 9-12 mm Hg, for various poultry species IOP values are as high as 22 mm Hg. Minimum corneal diameter for reliable value is 9 mm. Reference values for newly developed tonometers (Tonovet and Tonolab) will be available soon.

- Schirmer tear test (STT). Test for the estimation of the lacrimal function. Use standardized filter strips of 2-, 3- and 5-mm width. Standard reference values using filter strips of various width for 42 species from 7 Orders showing a wide range of interspecific variations are available. Strigiformes show conspicuously low values.

- Phenol red thread tear test (PRTT). As an alternative for the standard STT the PRTT test may be used for evaluation of tear production. In principle, a cotton thread is placed into the lower conjunctival sac while wetting of the thread is visualized by phenol red as an indicator. Advantages of the PRTT over the STT include shorter reading times (15 seconds instead of 1 minute) and thus minimized stress to the bird and the possibility, using this test in birds even with very small eyelid fissures and low tear production rates, such as small psittacines.[1] Standard reference values are available for various psittacine,[1,2] raptor,[3] and other bird species. Values found in the literature are differing and strongly dependent on bird species and specific examination protocols used that have not been standardized.

- Electroretinography. Measurement of retinal function by recording electrical potentials after light stimulation. This technique gives no information about vision, only retinal function. Basic principles of electroretinography for routine examination have been established. Indications are retinal disorders and dioptric apparatus opacities.

- Optical coherence tomography (OCT). 2D and 3D examination of anterior (cornea, anterior eye chamber, iris, lens) and posterior (fundus) eye segment structures at almost histologic resolution in the anesthetized or non-anesthetized bird patient allow diagnosis even of several stages of diseases.[4]

- Laboratory examinations include bacteriologic examination of the conjunctival flora. Physiologic bacterial flora contains gram-positive bacteria, while gram-negative bacteria are an indicator for pathologic conditions. Standard reference values have been established for 42 bird species from eight orders.

III. Examination

Basic examination principles are similar to those in mammalian ophthalmology; however, the small size of avian eyes plus many anatomic and physiologic peculiarities required specific adaptations. These are demonstrated on a comparative basis to techniques in mammals including basics of patient restraint, examining adnexal structures and anterior eye segment, induction of mydriasis using curariforme drugs, mono- and binocular ophthalmoscopic techniques and visualizing the fundus (Figs 1 and 2). This lecture includes a review on the most recent and sophisticated diagnostic imaging techniques including ultrasonography, optical coherence tomography, digital scanning ophthalmoscopy.

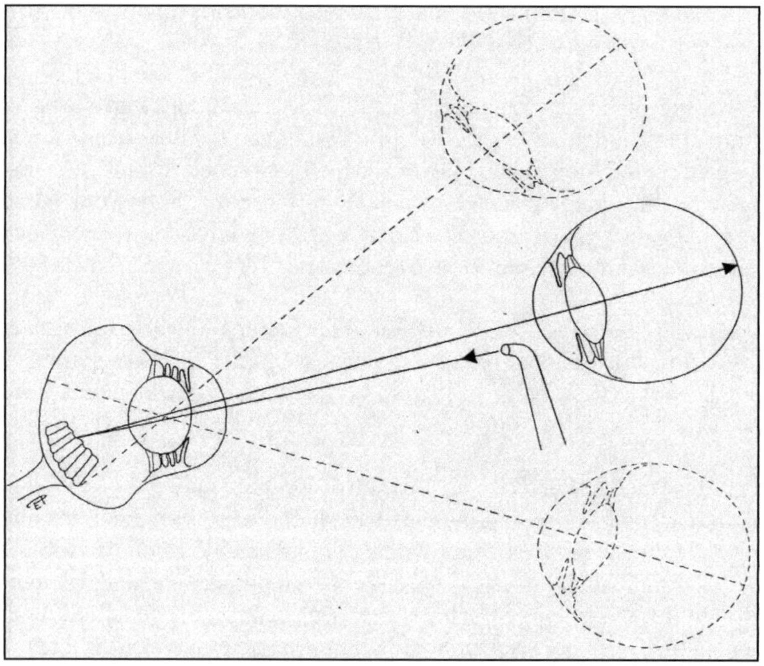

Figure 1. Optical principles of the (monocular direct) ophthalmoscopic examination of avian eyes. Per definition ophthalmoscopy is the examination of ocular structures situated behind the lens, using a focused light beam directed through the pupil and performing the examination using light, reflected from the fundus oculi. As a rule, a nearly coaxial direction of the light beam directed through the (dilated) pupil (mydriasis) of the patient's eye (left) on the one hand and the reflected light beam from the patient's fundus to the examiners eye (right; viewing axis) on the other hand is indispensable in order to project an image on the examiner's retina.

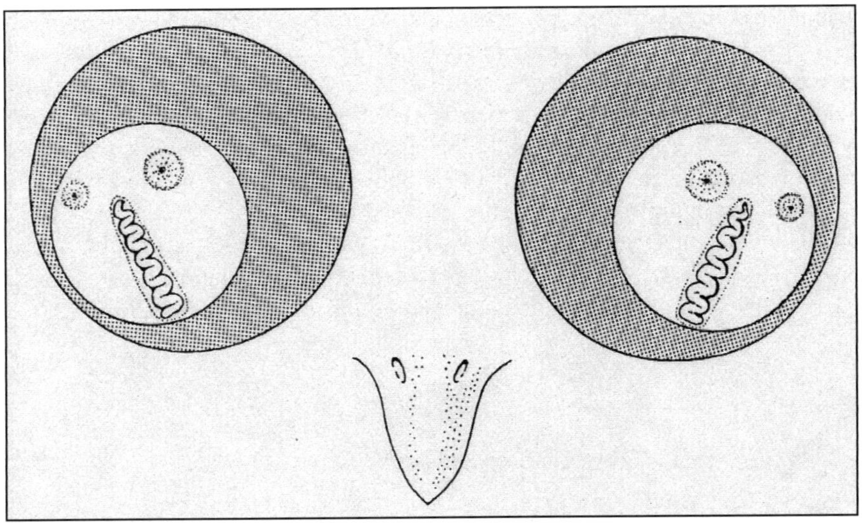

Figure 2. Topographic landmarks within the avian fundus with the pecten oculi and the central and temporal fovea. Accessible viewing field (bright area) using monocular direct ophthalmoscopy (taken from 4).

Mydriasis and air sac perfusion technique

Induction of mydriasis is indispensable for the examination of the posterior eye segment (ophthalmoscopy). A major difference between the mammalian and the avian eye is that the commonly used mydriatics of atropine and tropicamide have little effect in the avian patient due to a striated rather than smooth intraocular musculature. The iris is partly under voluntary control. It is essential to have a dilated pupil (mydriasis) to perform an ophthalmoscopy (i.e., examination of the posterior eye segment including the vitreous, the fundus and the pecten oculi).

Today the method of choice is conjunctival application of rocuronium (12-35 µl/eye)[5-7] resulting in a mydriasis with species-specific differences regarding lag of action after administration (up to 90 minutes in common buzzards and African grey parrots) and completeness of mydriasis and duration of action (20-360 minutes) have to be considered and require further scientific studies. Neuromuscular blocking agents such as d-tubocurarine (3%; 0.01-0.03 ml) may be used for long-term induction of a mydriasis and have been used in the past but necessitates intracameral application via paracentesis, as the drug penetrates the cornea insufficiently and must be administered directly into the anterior chamber by paracentesis using a 27- to 30-gauge needle. This technique includes substantial risk for injury of intraocular structures causing hyphaema, increasing intraocular pressure (IOP), transmission of conjunctival flora with consecutive uveitis and systemic side effects if larger doses than recommended are used. Therefore, use this technique just for therapeutic reasons (e.g., prevention of posterior or anterior synechia resulting from uveitis, secondary miosis and adhesions), where long-lasting mydriasis (12-18 hours) is necessary. More updated techniques to introduce mydriasis include the use of rocuronium.

An alternative for routine induction of a mydriasis as well as for intraocular surgery and surgery in the head area is air sac perfusion anaesthesia.[8] APA consists of a retrograde perfusion of the lung-airsac-system through a perfusion catheter via the left caudal thoracic airsac. As a carrier gas 0-3 L/min/kg BW of O_2 is used. Effects of nitrous oxide application include low potentiation of isoflurane of approximately 11% and thus improvement of the circulatory situation and release of the surgeon from isoflurane waste gases. Higher perfusion rates than recommended result in respiratory alkalosis due to a CO_2-wash-out-effect causing severe cardiac arrhythmias.

Isoflurane maintenance concentrations vary - dependent on different bird species - between 1.0 Vol. % to 2.4 Vol. % (*Columba livia* Gmel., 1789). Pulse oximetry is indispensable, as APA causes a reversible apnea due to reduced CO_2 partial pressure causing a missing stimulation of the respiratory center. Advantages of APA include an ability to provide long period anaesthesia, which is used for routine ophthalmoscopy, electroretinography and head surgery in birds, to free surgical access to the head for intraocular surgery, stable or decreased intra-ocular pressure and reversible apnea with an absolute immobilization of the patient. Achievement of mydriasis for ophthalmoscopy may be optimized by systemic administration of 0.2 mg/kg BW of the muscle relaxant vecuronium which allows a complete mydriasis and areflexia with a lag period of approx. 26 sec. and a duration of 256 min. in pigeons (*Columba livia* Gmel., 1789) and a reduction of isoflurane consumption of approx. 25% at the same time. This technique allows examination even of the very lens periphery with the annular pad and the extreme fundus periphery.

APA principles

- Anesthesia induced mydriasis and free surgical access to the head area.
- Perfusion of lung-air sac-system with oxygen-nitrous oxide-isoflurane or sevoflurane mixture.
- Retrograde perfusion via left caudal thoracic air sac using a specific air sac catheter.
- Induction of a reversible apnea due to lowering of the CO_2-partial pressure with subphysiologic values (below 48 mm Hg; no stimulation of respiratory center).
- Modified anesthetic machine with low flow flowmeter (0.01 L/min to 0.5 L/min).

APA performance

- Carrier gas O_2.
- Induction via head chamber.
- Maintenance via air sac catheter.
- Isoflurane concentration of 1.0 vol% to 2.7 vol%.
- Monitoring via pulse oximetry (no respiratory movements), reflex score, blood gases, body temperature.

APA advantages

- Routine method for ophthalmology and head surgery.
- Benefits of inhalation anesthesia and long period anesthesia.
- Free surgical access to the head.
- Stable or decreasing intraocular pressure.
- Reversible apnea, absolute immobilization of the patient.
- Achievement of mydriasis for ophthalmology.

Ophthalmoscopy, a technique to examine the fundus oculi (a clinical term, describing ocular structures, which are situated behind the lens) using a focused light beam reflected from the fundus, can be carried out by both monocular and binocular techniques and direct or indirect ophthalmoscopy in combination with double aspherical ophthalmoscopic lenses (at 30, 40, 60, 78 and 90 diopters (D, Volk Bio II) refractive power (Figs 3 and 4). In all cases, indirect binocular ophthalmoscopy using a head ophthalmoscope is advisable. A 30 D lens is used in birds with larger pupil diameters (nocturnal and diurnal raptors), ophthalmoscopy of pigeons and larger psittacine species requires a 78 D lens, those in small birds (e.g., canaries, budgerigars) the use of a 90 D lens. Alternatively, monocular indirect ophthalmoscopy may easily be performed in all birds, especially in smaller species. A long-term survey including more than 32% of traumatized birds showed haemorrhage in the vitreous body originating from lesions of the pecten oculi. Thus, ophthalmoscopy is obligatory in traumatized birds.

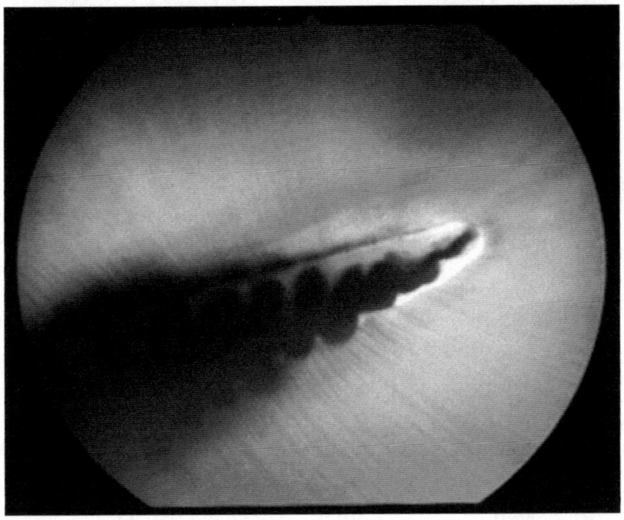

Figure 3. Fundus oculi (common buzzard [*Buteo buteo*], oculus sinister [OS]) as seen with the scanning digital ophthalmoscope (SDO), demonstrating the diurnal bird's fundus with typically anangiotic appearance and absent tapetum, the pecten oculi (pleated type), optic nerve head (white), and central fovea (1 o'clock position). This technique adapted to exotic animal ophthalmology allows taking still and video images through pupillary diameters as small as 4 mm.

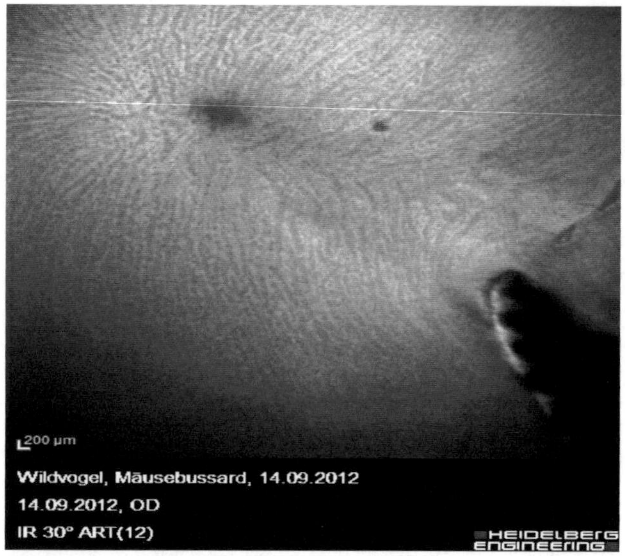

Figure 4. Fundus oculi (common buzzard [*Buteo buteo*], oculus dexter [OD]) as seen with the optical coherence tomograph (OCT) with the upper and lower vessel cascade, central fovea, pecten oculi (5 o'clock position), and retinal scar formation (central and 3 o'clock position).

IV. Ocular Disorders

Fully functional vision is indispensable in (falconry kept) raptors and wild birds, however of great importance even in captive pet birds. Most often avian ocular diseases reflect ocular manifestations of systemic disorders, frequently representing zoonoses such as salmonellosis, mycobacteriosis, and avian influenza, thus being of importance for both the patient and the patient owner. In addition, most emerging diseases, such as borna virus infections, are of high relevance. The presentation gives a broad overview of avian ocular disorders as well as diagnostic and therapeutic approaches on an interactive basis.

The ophthalmologist considers it logical to classify eye disorders per the various ocular structures involved. A complete review on avian ophthalmology discords based on a morphologic basis is given in this presentation and includes a literature review.[9-11] In a long-term investigation on ocular disorders in birds, an overall incidence of 7.6% of all the birds examined has been found.[10] The highest incidence of eye disorders compared with other patient groups was generally found among wild birds (11.7%), among which most eye disorders were seen in Accipitriformes (26.1%), Strigiformes (20.0%), and Falconiformes (19.7%). In contrast, a lower incidence was found in falconry raptors with 7.3 %, but again Accipitriformes (6.3%) were most frequently affected by eye disorders, followed by falcons (5.5%) and owls (5.0%). Among disorders of the posterior eye segment (fundus oculi) most often trauma related hemorrhages (most frequently arising from the pectin) may be diagnosed. In general, chronic lesions caused by disorders located within the upper part of the fundus result in a poor prognosis because birds, and especially raptors, are orientating themselves primarily using the upper part of the fundus, with the central and temporal fovea included.

References

1. Stadtbaeumer K, Korbel R. Untersuchungen zum phuenolrot fadentest (PRT) bei vöge *Proc Conf Avian Dis.* 2002;17-25.

2. Rajaei SM, Ansari Mood M, Khorram H, et al. Measurement of tear production using the phenol red thread test in the common mynah. *J Avian Med Surg.* 2015;29(2):146-148.

3. Smith SP, Barbon AR, Forbes. Evaluation of the phenol red thread tear test in falconiformes. *J Avian Med Surg.* 2015;29(1):25-29.

4. Korbel R, Schulze C, Sigg R, Rinder M. Comparative investigations in the healthy and diseased avian fovea region using ophthalmoscopy and optical coherence tomography (OCT). Proceed 4th iCare Conference on Avian, Herpetologial and Exotics (iCare). London, April 27 – May 2, 2019.

5. Barsotti G, Briganti A, Spratte J, et al. Safety and efficacy of bilateral topical application of rocuronium bromide for mydriasis in European kestrels (*Falco tinnunculus*). *J Avian Med Surg.* 2012;26:1-5.

6. Baine K, Hendrix DVH, Kuhn SE, Souza MJ, Jones MP. The efficacy and safety of topical rocuronium bromide to induce bilateral mydriasis in Hispaniolan Amazon parrots (*Amazon ventralis*). *J Avian Med Surg.* 2016;30:8-13.

7. Dongo PSJ, Pinto DG, Guimarães MB, et al. Effects of bilateral topical administration of repeated doses of rocuronium bromide as a mydriatic agent in Amazon parrots *(Amazona aestiva* and *Amazona amazonica). J Exotic Pet Med. 2021;39:11-16.*

8. Koenig HE, Korbel R, Liebich HG. *Avian Anatomy: Textbook and Colour Atlas.* 2nd Ed. Translated and revised by C. Klupiec. Sheffield, UK: 5M Publishing Ltd;2016:159-187.

9. Kern TJ. Exotic animal ophthalmology. In: Gelatt KN, Gilger BC, Kern TJ, eds. ed. *Veterinary Ophthalmology.* Vol 2. Ames, IA: Wiley-Blackwell; 2007:1370-1405.

10. Korbel RT. Disorders of the posterior eye segment in raptors: examination procedures and findings. *Raptor Biomedicine III.* Lake Worth, FL: Zool Educ Network; 2000:179-194.

11. Korbel RT, Bohnet N. Ophthalmological examination of the avian eye-an interactive CD-ROM. Munich, Germany; 2007.

Introduction to the Neurologic Examination and Neurologic Diseases of Birds

Seth C. Oster, DVM

Session #2568

Affiliation: From the Department of Clinical Sciences, Auburn University, College of Veterinary Medicine, 1220 Wire Road, Auburn, AL 36849, USA and the Southeastern Raptor Center, Auburn University, College of Veterinary Medicine, 1350 Pratt-Carden Dr, Auburn AL, 36849, USA

Abstract: This master class is designed for new practitioners to avian medicine. The course focuses on fundamentals of neurologic examination of birds and use of these findings to localize lesions. From there the master class discusses common neurologic diseases, their clinical signs, lesion localization, methods of diagnosis and treatment. Discussion includes use of various diagnostic modalities including radiography, advanced imaging, electromyography, and pathogen testing.

Introduction

Avian neurologic examination and diseases can be an intimating area of medicine for new practitioners. Avian patients have several unique anatomical and physiologic differences from mammal patients that present several challenges to a complete neurologic examination. The goal of a neurologic exam is to both localize the lesion within the nervous system and to monitor patient's progression once treatment is instituted. Lesion localization is vital in determining appropriate differentials and guiding a diagnostic plan.

Physical Examination

The neurologic examination of the avian patient should start with a distance observational examination. This is the practitioner's opportunity to evaluate the patient's mentation and posture without stimulating the patient.

Mentation can be defined as both the level and content of a patient's consciousness.[1] These can be affected by the patient's surroundings, so practitioners should take this into account during the assessment. Most birds in an unfamiliar environment, such an exam room, should be alert and aware of their surroundings. A bird presenting that is fluffed, hiding its head, or not responding to the presence of the practitioner would be considered to have a decreased level of consciousness. Changes in behavior such as lack of recognition of people, objects, or other animals familiar with the patient would be considered an altered content of consciousness. Level of consciousness is mediated by the brainstem and content is mediated by the cerebral cortex.

Level of consciousness can be categorized as alert, lethargic, depressed, obtunded, stuporous, or comatose.[1] Alert is a patient's normal level of awareness and response to surroundings. Lethargic is an aware patient with sluggish responses. Depressed patients have a decreased response to environmental stimuli such as sounds or interactions with the practitioner. These patients will often begin sleeping when left undisturbed in familiar surroundings. Obtunded patients are similar to depressed patients, but these patients will often sleep in unfamiliar surroundings such as the exam room. Obtunded patients can be easily roused from sleep. Stuporous patients require a noxious stimulus to response to their surroundings, such as a toe pinch. A comatose patient is unaware of its surroundings, does not respond to noxious stimuli, but reflexes remain present. Depression and lethargy can be commonly caused by systemic disease outside of the CNS, while obtunded, stuporous, and comatose more often indicate disease within the cerebrum or brainstem, although systemic disease is possible as well.

Content of consciousness can be harder to assess. Abnormal content can be defined as dementia or delirium.[1] Dementia is defined as little awareness of surroundings or little concern for their surroundings. This could manifest as patient flying into obstacles that it would normally be able to avoid. Delirium is defined as an overaction to familiar stimuli, such as the owner or other objects in the home environment. Either of these abnormalities could indicate structural or metabolic damage to the cerebrum. Identifying changes to a patient's content of consciousness requires a thorough history.

Cranial nerve assessment is very useful in localizing a lesion within the central nervous system (CNS). Several exam techniques can be used to evaluate the cranial nerves of an avian patient.

Evaluation of CN I (olfactory) can be difficult in birds. A lack of reaction to a noxious smell may indicate an abnormality of this nerve.[1,2]

The menace response can be used to assess CN II (optic) and CN V (trigeminal).[1,2] Remember that the menace response is a learned response and may be absent or inconsistent in young patients or stoic patients that have normal function of CN II and V.

Direct pupillary light responses can be used to test CN II and CN III (oculomotor). Birds have complete crossover of the fibers of the optic nerves at the level of the optic chiasm.[1,2] Due to this, they do not have indirect PLRs. In birds with large eyes, such as owls, this author has observed light passing from the examined glob into the other eye, illuminating it and causing a PLR. Birds also have striated muscle in the iris, so in some cases patients may be able to override this response. Performing a PLR early in the examination in a very dark room with a bright light source provides the most consistent results.

Vision can also be used to test CN II.[1,2] Evaluating vision can be done by watching the patient's ability to track moving objects such as a thrown cotton ball. Patients can also be offered food rewards in the exam room and the practitioner can observe the patient's ability to recognize and maneuver to these items. Obstacle avoidance can also be assessed. Remember that a patient in a decreased mentation may have abnormal responses to all of these examination techniques, but this may not indicate abnormalities with CN II.

Strabismus, or abnormal position of the globe, can indicate dysfunction with several nerves. CN III dysfunction can lead to ventrolateral deviation of a globe.[1] CN IV (trochlear) dysfunction will result in dorsolateral deviation. CN VI (abducens) dysfunction will lead to medial deviation of the globe.

Nystagmus can be complex to evaluate and localize a lesion correctly. A nystagmus can be caused by either a peripheral or central lesion. Vertical or rotary nystagmus can only be caused by a central lesion.[1,2] Horizontal can be caused by either peripheral or central lesions. Peripheral nystagmus lack any changes to the patient's proprioception or mentation. CN III, IV, and VI can all contribute to vertical or rotary nystagmus. A spontaneous horizontal nystagmus indicates dysfunction of CN VIII (vestibulocochlear) with the fast phase going away from the side of the lesion. Moving the head horizontally will induce a normal physiologic nystagmus, with a fast phase in the direction of the movement. This is a functional test of CN III, IV, VI, and VIII.

Palpebral reflex and facial sensation are methods to evaluate CN V.[1,2] Lack of palpebral response indicates dysfunction of the ophthalmic, maxillary, and mandibular branches. Sensation is mediated by all three branches, with the ophthalmic providing sensation to the upper eye lid, skin of the forehead, and rostral nasal cavity. The maxillary provides sensation to the lower eye lid and the lateral margin of the upper beak. The mandibular provides sensation to the skin around the lower beak along with the rictus and mucosa.

Facial expression is a function of CN VII (facial) and CN V.[1,2] The mandibular branch of CN V provides motor to the lower beak. Dysfunction can lead to a drooping jaw. CN VII provides sensation and motor to the face. Dysfunction can lead to a facial droop, asymmetry of facial features, or lack of response to touching of the patient's face.

Balance is a function of CN VIII.[1,2] Loss of balance or incoordination can indicate dysfunction of this nerve. Head tilts may also be observed. Lesions can be either peripheral or central. Evaluation of balance in conjunction with mentation and the presence of a nystagmus can help distinguish if the lesion is central or peripheral.

Taste is a function of CN IX (glossopharyngeal). Use of a bitter food item can help to evaluate the function of this nerve.[1,2] The patient may also have changes in diet preference recently.

The absence of a gag reflex indicates dysfunction of CN IX, and X (vagus).[1,2] Note that CN XII (hypoglossal) provides motor to the tongue, trachea, and syrinx, so it may contribute to this reflex as well.

Tongue position and motor function is a function of CN XII.[2] Observing the position of the tongue in the oral cavity and noting deviation from midline may indicate dysfunction. Manipulating the tongue gently with the practitioner's fingers can also be used to evaluate muscle tone.

A bird's posture is another important area to evaluate. Abnormal posture of the neck or limbs could indicate lesions in peripheral nerves or the CNS itself. A head tilt can be caused by injury to the cervical spine, lesions within CN VIII, or the cerebellum. Abnormal limb position could be caused by injury to the peripheral nerves, the corresponding spinal segments, or the brainstem. Evaluations of spinal reflexes and cranial nerves are needed to localize the lesion to the level of the spinal cord or peripheral nerve.

Postural reflexes can be used to evaluate coordination, proprioception, and sensation of the limbs. If the patient is observed to knuckle the feet, this could indicate lack of proprioception to the pelvic limbs.[2] The practitioner can hold the patient off the table and place a foot in a knuckled position. Note that this is inconsistent in some birds that are very resistant to handling. Another method to evaluate proprioception of the pelvic limbs is to have the patient stand with one foot on a towel or piece of paper and slowly pull the towel or paper laterally and observe how the patient returns the foot to a standing position.

Hopping or hemi walking can also be used to evaluate coordination and proprioception of the legs.[2] This is performed by restraining one leg near the body and moving the patient forward and side to side. Again, many avian patients may not tolerate this examination. Alternatively, the patient can be placed on a perch and gently pushed from one side to evaluate how the opposite leg compensates for the increase weight bearing.

Perching can be used as well to evaluate function of the pelvic limbs and can be done in two different manners. The patient can be presented a perch that they are able to visualize.[2] This will evaluate proprioception and coordination. The patient can also be blinded by the use of a hood in a raptor or by presenting the perch from an angle that the bird cannot see.[1] In this case, allow the perch to touch the foot gently and observe if the patient is able to grasp the perch without looking. This will evaluate sensation to the leg in addition to proprioception and coordination. Perching on a practitioners hand or arm also will allow the practitioner to judge symmetry of grip strength.

To evaluate proprioception of the wings, the practitioner can pull the wing into an extended position away from the body and the release the wing.[1,2] Observing the wing not returning to a normal flexed position can indicate a proprioceptive deficit. Wing droops are also commonly observed with orthopedic injuries of the shoulder.

Spinal reflexes are useful in conjunction with the postural exam to determine location of lesions in the CNS vs the PNS. A deep pain response can be used to test reflexes of both the extended leg or wing.[1,2] In the leg, a toe and be used for the stimuli while the metacarpus can be used in the wing. In both cases, the patient should retract the limb as part of a normal reflex. Hyperreflexia can indicate an upper motor neuron lesion while absent or hyporeflexia can indicate a lower motor neuron lesion. While preforming these examinations, it is important to observe for the presence of a cross extensor reflex in the opposite limb. This is a UMN lesion and indicates serve injury of the spinal cord proximal to the limbs being tested.

A cloacal reflex can be performed on a bird similar to how a perineal reflex is performed in a mammal.[2] Tactile stimuli to the peri-cloacal area should elicit contraction of the external sphincter muscle and ventral movement of the tail.

Patellar and gastrocnemius reflexes can also be evaluated in the pelvic limb of birds.[1,2] These is done in the same manner as the test is performed in a mammal. Not all patients will easily tolerate this type of examination.

Note that in birds, spinal nerves exit lateral of the spinal cord segment they originate from, rather than caudally as they do in mammals.[1,2] They also lack a caudate equina. This simplifies lesion localization in the spine of birds.

Sensation is valuable in determining location and severity of lesion. Loss of deep pain sensation, like in mammals, indicates severe spinal cord disease.[1,2] The loss of proprioception, motor, and then superficial pain precede the loss of deep pain. Deep pain response can be evaluated while performing withdrawal reflexes of the wings and legs. It is important to note that the patient must make a conscious response to the stimuli such as looking at the limb, vocalization, or intentional movement of the limb other than withdrawal. Remember, a withdrawal reflex maybe intact with an absent deep pain response if the lesion is cranial to the spinal cord segment being evaluated.

Diagnostics

Since a number of systemic or metabolic diseases can manifest with neurologic clinical signs, a minimum database of a complete blood count and serum chemistry is recommended. As the practitioner works through a differential list, specific infectious disease tests may be needed as well. These will be gone over with their respective diseases.

Radiographs are a very useful tool, especially in the cases of trauma. Orthopedic injuries can damage peripheral nerves. Pelvic and limb radiographs are useful in determine the extent of orthopedic damage and its location relative to nerve tracks.[2] Spinal fractures can be detected as well but are not always easily visible in this author's experience. If a spinal lesion is suspected, but well positioned spinal radiographs do not reveal any abnormalities, practitioners should consider the use of more advanced imaging such as CT or MRI.

Cerebrospinal fluid analysis is a common diagnostic used in small animal medicine to evaluate for degenerative, neoplastic, or inflammatory disorders of the brain or spinal cord. CSF collection is difficult in birds and carries significant risk of hemorrhage.[2] As such, it should be only performed on a patient that is completely anesthetized.

The collection site is through the foramen magnum between the cerebellum and the dorsal surface of the medulla oblongata.[2] The patient can be placed in lateral recumbency and the atlanto-occipital joint flexed in a 30-degree angle. Using a 25- to 27-gauge needle, approach the patient on midline just caudal to the occipital protuberance. The needle should be directed rostrally at a 45-degree angle to the horizontal axis of the head. Advance the needle slowly until clear fluid is noted in the hub. In most avian patients, 0.1-0.5 ml can be collected. There is significant risk of hemorrhage due to a large venous plexus near the sample site as well as the risk of cisternal puncture. Due to these risks, this diagnostic is not as commonly performed in avian neurology as it is in small mammal neurology.

CT and MRI imaging are very useful diagnostic tools and often offer a higher degree of accuracy than traditional radiographs.[2] CT is valuable in assessing boney structures as well as soft tissue structures when IV contrast agents are used. A dose of 2.22 mg/kg of iodinated IV contrast material is recommended.[1] Contrast can help identify soft tissue masses that have increased blood flow. MRI is also an excellent tool for the evaluation of the soft tissue of the spinal cord and brain. MRI offers higher resolution of these structures than CT offers, but requires significantly longer time to acquire the image than CT does. Due to the need for patients to remain still during the process of image acquisition, both CT and MRI require the patients to be under anesthesia. When suspecting a lesion of the brainstem or cerebrum, MRI is the most appropriate diagnostics for evaluating structural changes.

Electromyography (EMG) is used to evaluate the health of muscle cells, the motor unit, and the lower motor neuron.[2] This diagnostic can aid in the identification of peripheral nerve diseases as well as muscular diseases. This author has commonly used this diagnostic to evaluate for brachial plexus injuries in raptors after shoulder trauma. This diagnostic is performed by measuring the electoral activity of the affected muscles. Normal resting muscle does not show any observable electoral activity other than end-plate noise and a sharp wave from insertion of the needle. Prolonged insertion potentials can indicate injury to the motor unit. Spontaneous randomly occurring potentials, called fibrillation potentials, commonly are seen with denervation injuries. Time of onset of these fibrillations potentials is unknown for avian patients, but is reported to be seen 5-7 days after injury in dogs. This author has been able to positively identify fibrillation potentials with confirmed denervation at necropsy in raptor patients 2 weeks after presentation.

Nerve conduction studies can also be used to test peripheral nerve function.[2] This diagnostic evaluates conduction of an electrical potential by stimulating a nerve with an electrical impulse. M wave size is decreased in patients with axonal damage or dysfunction. Prolonged conduction times or velocity indicates demyelination of the nerve. This diagnostic can be more technically difficult to perform than an EMG but provides more direct evaluation of the peripheral nerve.

Common Neurologic Diseases

Seizures can be caused by a number of etiologies that are discussed in their specific diseases. For emergency control of seizures, treatment with either IV diazepam at 0.5-1 mg/kg or IM midazolam at 0.5-1 mg/kg[1] is recommended. These doses can be repeated every 2 minutes up to three times to control seizures. Long-term control can be achieved with phenobarbital at 2-10 mg/kg PO q12h. It is important to monitor plasma concentrations with phenobarbital. Therapeutic ranges have not been reported in birds, but the trend of plasma concentration over time can guide dosing decisions. Other medications have recently been reported for the long-term control of seizures in birds. Levetiracetam has been reported with a dose range of 10-100 mg/kg PO q8h. Zonisamide has also been reported with a dose of 20 mg/kg PO q8h.

Trauma is a common presenting issue for both wild birds and captive birds. Clinical signs may vary based on the portion of the nervous system affected (peripheral versus central) and the severity of the injuries. Clinical signs may include limb weakness, paralysis, loss of reflexes, decreased mentation, cranial nerve deficits, and seizures.[3] If the owner did not observe the trauma and the lesion is localized to the CNS, the practitioner should rule out other causes of intracranial disease. Radiographs, CT, and MRI can be utilized to evaluate these cases. Treatment focuses on reduction of swelling and pain management. This can be achieved through the use of a combined analgesic plan of NSAIDS and opioids. If cerebral edema is suspected, IV mannitol at 0.5-1 mg/kg over 20 minutes is recommended.[1]

Chlamydophila psittaci can present with a large list of clinical signs including lethargy, anorexia, conjunctivitis, sinusitis, dyspnea, diarrhea, opisthotonos, ataxia, seizures, and tremors.[4] A subclinical carrier state is also possible. PCR of a cloacal, choanal, and conjunctival swab combined with detection of circulating antibodies is the primary means of diagnosis. Due to intermittent shedding, PCR false negatives are possible. Histopathology of granulomas may also detect this organism. Treatment with doxycycline for 6 weeks is recommended. Although azithromycin and enrofloxacin have also been reported,[5] although their efficacy is questionable and are not recommend as a treatment currently.

Bayliascaris procyonis infections can occur when patients have contact with racoon feces.[5] Clinical signs of this infection include ataxia, weakness, torticollis, opisthotonus, and death. No antemortem diagnostics have been definitive, but MRI could be used to evaluate the brainstem and cerebrum. No treatment has been successful in reported cases, and prognosis is poor once neurologic signs develop. Prevention by avoiding contact with mammal feces is recommended.

Sarcocystis faculata has been reported to infect avian patients exposed to opossum feces or cockroaches.[5] Clinical signs include weakness, dyspnea, head tilt, ataxia, exophthalmos, nystagmus, and anorexia. Skeletal muscle biopsies can be used to identify the organisms as well as serum antigen testing. The rapid progression of this disease may often result in death prior to the confirmation of the diagnosis. Treatment with diclazuril, pyrimethamine, and trimethoprim-sulfamethoxazole has been reported to be successful.

Toxoplasmosis gondi infections can occur with patients exposed to cat feces.[5] Clinicals signs include dyspnea, ruffled feathers, diarrhea, emaciation, blindness, and torticollis. Serology can be used to evaluate antibody levels, but seropositive birds can be very common. Treatment with trimethoprim-sulfadiazine is recommended.

Avian bornavirus primary affects gangli and nerve fibers of the gastrointestinal system of companion birds. It can less commonly affect the central nervous system or other peripheral nerves.[6] Infection in some cases results in an immune-mediated inflammatory response to the affected nervous tissue. Clinical signs include whole seeds in the droppings, weight loss, regurgitation, enlarged proventriculus, muscle tremors, incoordination, ataxia, head shaking, opisthotonus, and seizures. Diagnosis can be made by histopathology of affected tissue. Detection of the virus via PCR along with antibody levels may aid in the diagnosis of ABV, but this does not always correlate with the diagnosis of proventricular dilatation disease. Treatment includes COX-2 inhibitors such as robenacoxib or celecoxib.

West Nile virus is a common infectious disease spread by mosquitoes. Raptors and passerines appear more susceptible than psittacine birds.[5] Clinical signs include chorioretinitis, feather loss, ataxia, muscle tremors, torticollis, circling, and seizures.[4] Definitive diagnosis can be difficult, as the viremia often precedes the development of clinical signs.

Whole blood PCR and serology are available. Treatment includes meloxicam and supportive care. Neurologic deficits may be permeant in some cases.

Organophosphate toxicity results in inhibition of acetylcholinesterase and is a common disease of wild birds.

Clinical signs include diarrhea, vomiting, salivation, lacrimation, dyspnea, bradycardia, depression, tremors, and seizures.[1,2,4] Diagnosis can be made by blood levels of cholinesterase. Treatment with atropine is recommend. Response to treatment is often a strong indicator of organophosphate toxicity when blood levels are not available or practical.[4]

Heavy metal toxicity is caused by the ingestion of either lead or zinc. This is a common disease of both wild and captive birds. In suspected cases, a thorough history should be taken to investigate possible sources. Clinical signs include leg weakness, plantigrade stance, toe knuckling, blindness, and seizures.[5,7] Lead has also been reported to cause hematuria in some psittacine. Diagnosis can be made with whole blood levels for lead and zinc. Radiographs may also aid in the identification of metallic foreign material in the gastrointestinal system. Treatment includes decontamination of the GI tract either by feeding of a bulky food item (e.g., peanut butter or psyllium) or via endoscopic removal. Gastric lavage can be attempted, but possesses a significant risk of aspiration, so the patient should be anesthetized and intubated during this process.[1] Chelation therapy with either CaEDTA or DMSA should also be instituted along with concurrent fluid therapy.

Atherosclerosis is a common disease of psittacine birds, with risk factors including obesity and a high-fat diet (i.e., usually an all seed diet).[1] Clinical signs may wax and wane. Signs include seizures, dull mentation, congestive heart failure, and exercise intolerance.[7] Diagnosis of atherosclerosis can be difficult in the antemortem patient. Severely affected vessels will calcify, making them visible on radiographs. Echocardiography can also find obstruction of the outflow tract. Treatment should be tailored based on the echocardiogram and electrocardiogram findings as well as the use of anticonvulsants if seizures are present.

Hypocalcemia has been reported in African grey parrots, egg-laying hens of any species, and patients kept on a low calcium diet.[1] Clinical signs include seizures, muscle tremors, ataxia, posterior paresis, and pathologic fractures. Radiographs can be used to identify these types of fractures and serum ionized calcium levels can be used to confirm the diagnosis. Treatment includes injectable and oral calcium supplementation and vitamin D.

References

1. Orosz S, Antinoff N. Clinical avian neurology and neuroanatomy In: Speer B, ed. *Current Therapy in Avian Medicine and Surgery*. 1st Ed. St. Louis, MO: Elsevier; 2016:363-377.

2. Clippinger T, Bennett R, Platt S. The avian neurologic examination and ancillary neurodiagnostic techniques: a review update. *Vet Clin North Am Exot Anim Pract.* 2007;10:803-836.

3. Scott D. *Raptor Medicine, Surgery, and Rehabilitation.* 2nd Ed, Boston, MA: CABI; 2016.

4. Wernery U. Infectious diseases In: Samour J, ed. *Avian Medicine*. 3rd Ed. St. Louis, MO: Elsevier; 2016:434-521.

5. Hedley J, Kubiak M. Neurologic diseases of birds and reptiles. *J Exot Pet Med.* 2015;24:6-20.

6. Wellehan J, Lierz M, Phalen D, et al. Infectious disease. In: Speer B, ed. *Current Therapy in Avian Medicine and Surgery*. 1st Ed. St. Louis, MO: Elsevier; 2016:22-106.

7. Guzman D. Disorders of the nervous system In: Samour J, ed. *Avian Medicine*. 3rd Ed. St. Louis, MO: Elsevier; 2016:421-533.

Clinical Importance of Radiographs in Raptor Rehabilitation

Dana Franzen-Klein, DVM, MS

Session #2606

Affiliation: From The Raptor Center, College of Veterinary Medicine, University of Minnesota, 1920 Fitch Avenue, St. Paul, MN 55108, USA.

Abstract: Radiographs have high value as a diagnostic tool in a raptor rehabilitation setting. They are commonly available and relatively cost-effective. From the intake exam to the pre-release exam, radiographs can reveal subtle or hidden problems that significantly change prognosis and treatment. This one hour masterclass will focus on clinical examples that highlight the importance of radiographs in a raptor rehabilitation setting. We will discuss general radiograph schedules for raptors as they move through the rehabilitation process. The case examples will focus on clinical presentations that are common in raptor rehabilitation, clinical conditions where the radiographic presentation may be difficult to interpret or identify, and situations where radiographs significantly change prognosis for release to the wild. This masterclass is focused on fundamentals and geared towards clinicians, nurses, technicians or students who are relatively new to raptor radiography.

Raptor Radiography

Human safety

Any person exposed to radiation should be wearing proper personal protective equipment (PPE). This includes a lead apron, lead thyroid guard and lead gloves. Ensure that no human body part is exposed to the x-ray beam during radiograph exposure. Dosimeters should be worn by each person exposed to radiation to measure radiation exposure levels.

Restraint for radiographs

In the majority of cases, sedation or general anesthesia is required to obtain adequately positioned radiographs and to decrease patient stress and discomfort. This can be accomplished with general anesthesia using isoflurane gas or with injectable sedatives. In birds with potentially painful injuries such as acute fractures or internal trauma, analgesics should be administered prior to handling the bird for radiographs.

However, in some cases the animal is not systemically stable for sedation due to concurrent medical conditions such as cardiovascular shock or severe head trauma. In those cases, radiographs should be postponed until the animal is more stable, or a single awake VD radiograph with minimal manipulation of the limbs can be taken.

Radiograph positioning

Straight, well-positioned radiographs are vital for accurate radiographic interpretation and appropriate medical decision making. Symmetry within a radiograph and consistent positioning between different individuals of the same species are two important characteristics that allow for proper radiographic evaluation. If radiographs are poorly positioned it is easy to over-interpret the image or to misdiagnose or not see an abnormality that would have been more readily apparent if the image was well positioned. A minimum of two orthogonal views are required to fully evaluate the patient.

Ventral-dorsal view (VD)

The bird is positioned in dorsal recumbency in the middle of the plate. The wings are partially and symmetrically extended laterally and taped in position (Figure 1). For larger birds that don't fit on the plate, separate VD views of both wings and the pelvic limbs should be taken. Masking tape works well for radiograph positioning because it causes little to no damage to the feathers but is secure enough to keep the limb in place. Ensure that the tape is going over the major metacarpal bone to secure the wing, not only over the primary feathers, that way the wings are more secure if the bird begins to struggle (Fig 1). The legs are fully extended caudally towards the tail and taped separately (Fig 1). The keel is positioned directly over the spine. The center of the radiograph beam should be level with the caudal point of the sternum. A straight, symmetrical, well positioned VD radiograph is vital for accurate assessment of orthopedic conditions, such as rotational alignment of a fracture after healing or shoulder girdle fractures or luxations, and for accurate assessment of internal abnormalities such as increased soft tissue opacity in the lung fields (Fig 2).

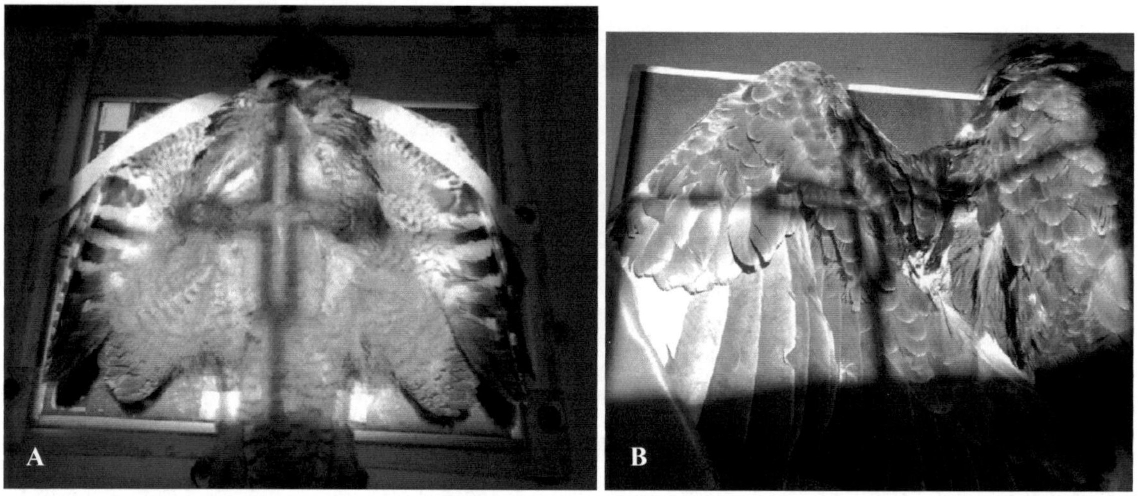

Figure 1. Positioning examples for a VD whole body radiograph in a great horned owl (A), and a VD wing radiograph in a bald eagle (B). Note that the pelvic limbs are extended caudally so that the stifle is pulled away from the side of the coelomic cavity, and that the tape is securing the major metacarpal bones and not just feathers.

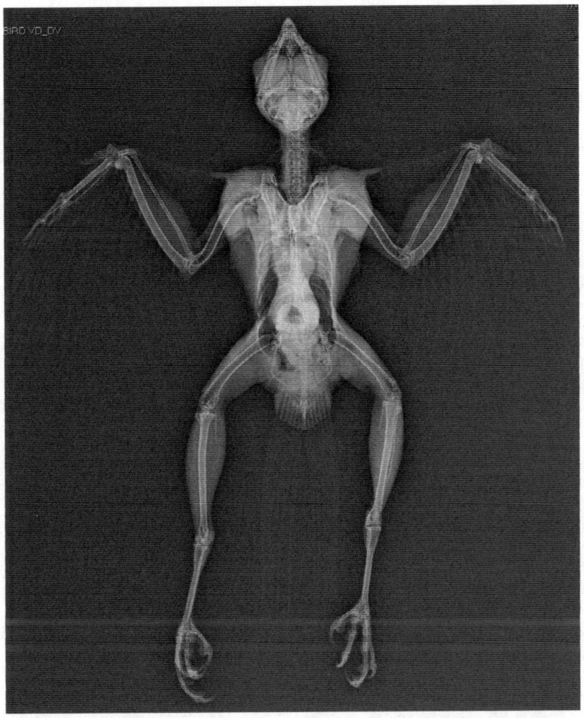

Figure 2. A correctly positioned VD whole body radiograph of a merlin. Not that the wings and legs are symmetrical, with equal flexion/extension in all joints. This is particularly important when assessing symmetry in the wings for complete evaluation of orthopedic conditions. In a straight VD view the keel is directly superimposed over the spine. In this image the keel is very slightly rotated to the right of midline relative to the spine.

Lateral view (Lat)

The bird is positioned in right lateral recumbency. The wings are extended dorsally above the back and the legs are extended caudally towards the tail (Fig 3). Ensure that the tape is securing the bones and not just soft tissue or feathers. The down leg (right) and down wing (right) are positioned slightly cranial to the upwards facing limbs (left) so the limbs are separated from each other and can be evaluated in the image. Having a convention of placing the down limb forward allows you to know which limb is the right versus left in the image. In a straight lateral radiograph the central ridge of the keel is parallel to the table. A common mistake is placing too much tension on the L wing when taping it to the plate, which causes rotation of the coelomic cavity. The beam of the radiograph should be centered just cranial to the caudal tip of the sternum.

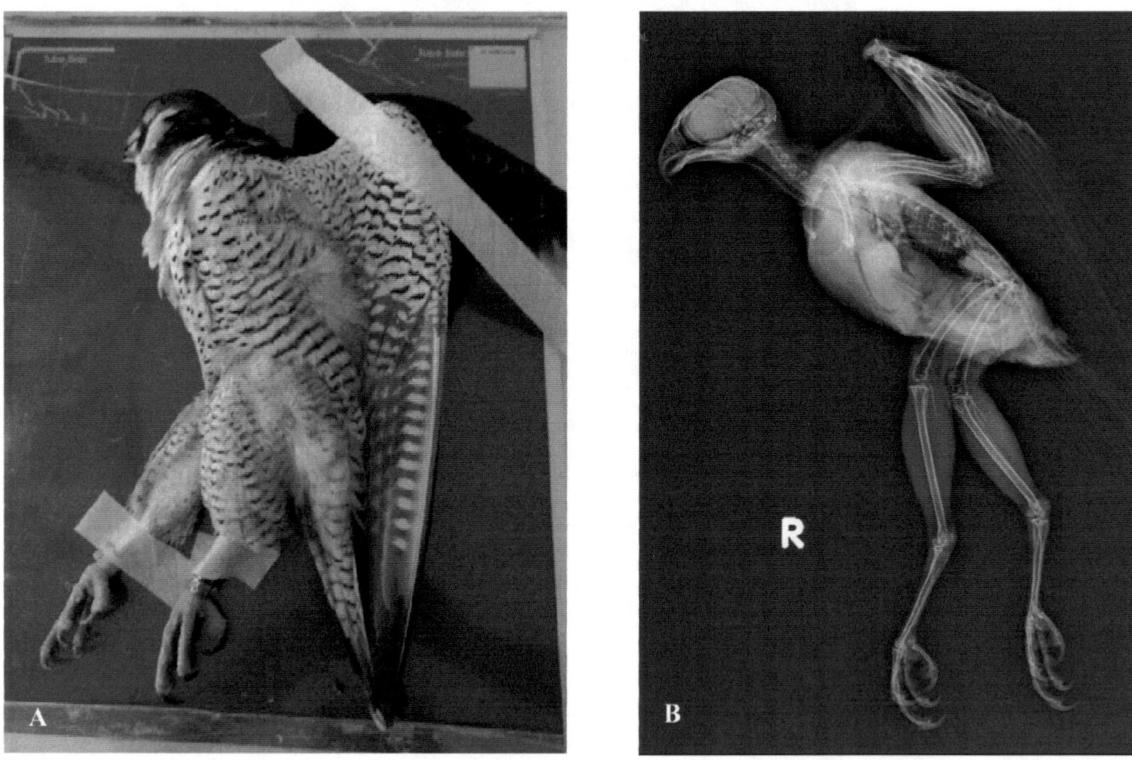

Figure 3. A) Positioning example of a lateral whole body radiograph in a peregrine falcon. Note that the tape is going across the major metacarpal bone of the left wing, the right wing is secured by resting against the plate. The pelvic limbs are extended caudally so the stifle is not overlying the coelomic cavity. B) Correctly positioned lateral radiograph of a merlin. Note that the coracoids and coxofemoral joints are superimposed in a straight lateral image.

Caudal-cranial view (Cd-Cr)

This is the orthogonal view used to evaluate the humerus, radius, ulna and major metacarpal bone along with the VD view. This view is also helpful for evaluating the shoulder girdle, especially the scapula and distal coracoid. To obtain this view, hold the bird by the pelvic limbs and lift the body towards the ceiling while holding the pelvic limbs away from the x-ray beam; hold the wing by the tip of the major metacarpal bone or the primary feathers, and position the wing so that the primary and secondary flight feathers are slightly curved away from you, the covert feathers will be pointing straight towards the ceiling, and the humerus, radius, ulna and major metacarpal bone are rotated 90 degrees to the plate compared to the VD view (Fig 4). A straight caudal-cranial view allows for accurate assessment of the articular surfaces and fracture alignment (Fig 5). An oblique caudal cranial view aids in the assessment of radius or ulna fractures by removing the superimposition (Fig 5). Slight rotation may also aid in complete assessment of the scapula (Fig 5).

Figure 4. Photographs demonstrating positioning for a caudal-cranial view of the right wing in a bald eagle. Note that the pelvic limbs are securely restrained and moved away from the x-ray beam.

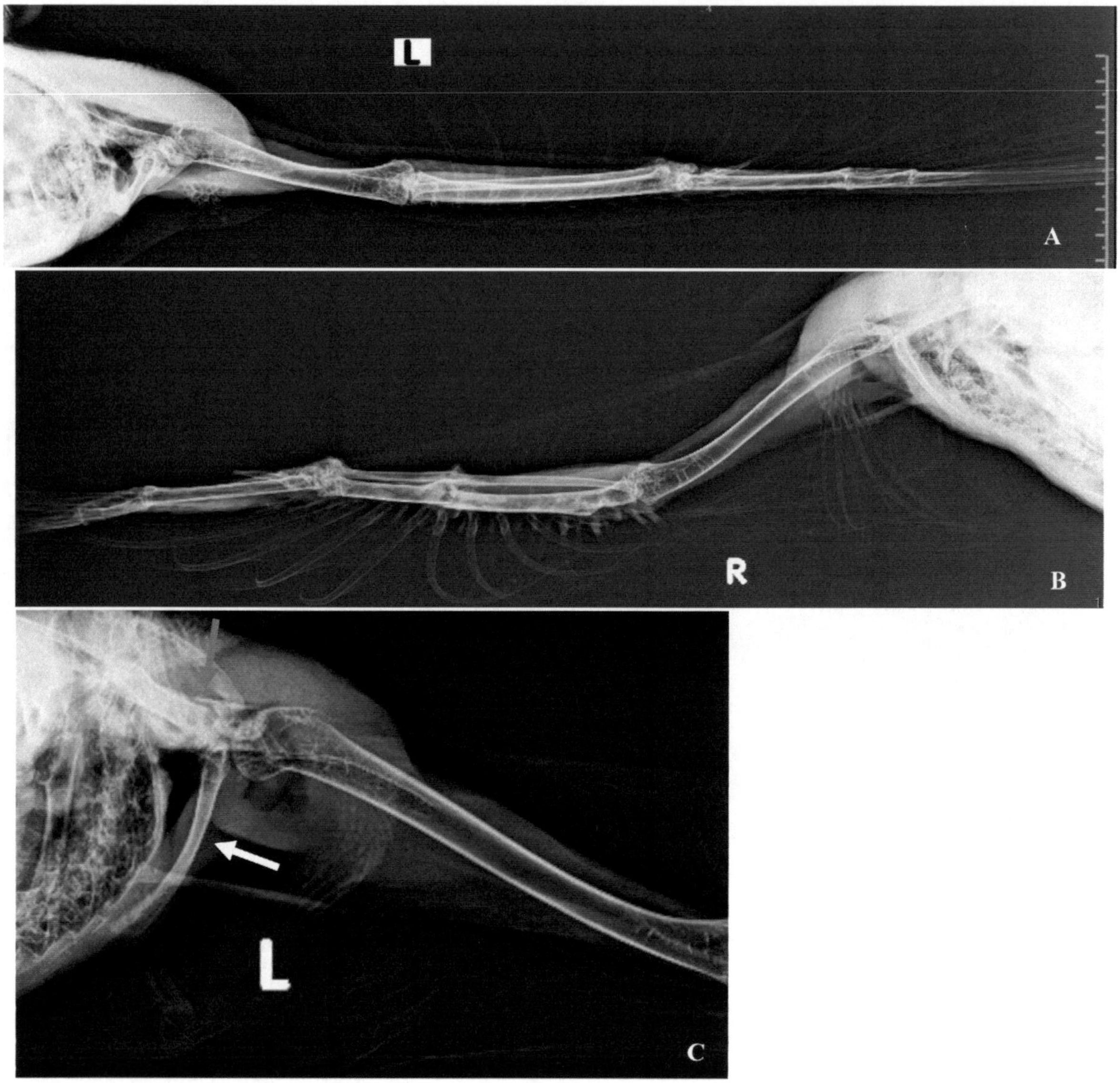

Figure 5. A) A straight caudal-cranial wing radiograph in a peregrine falcon. Note that the radius and ulna and superimposed. B) An oblique caudal-cranial radiograph in a red-tailed hawk used to separate the radius and ulna to allow for assessment of fracture healing. C) A caudal-cranial view in a barred owl that allows for full evaluation of the scapula (white arrow) and highlights the distal left coracoid fracture (red arrow).

Caudoventral-craniodorsal oblique view (H-view)

This view is helpful to evaluate the shoulder girdle, particularly the coracoid. The bird is positioned as for a VD view. Instead of taking the image with the x-ray beam at a perpendicular angle to the bird's body, the beam is adjusted so it is entering the bird's body at a 45-degree angle (Figs 6 and 7).[1]

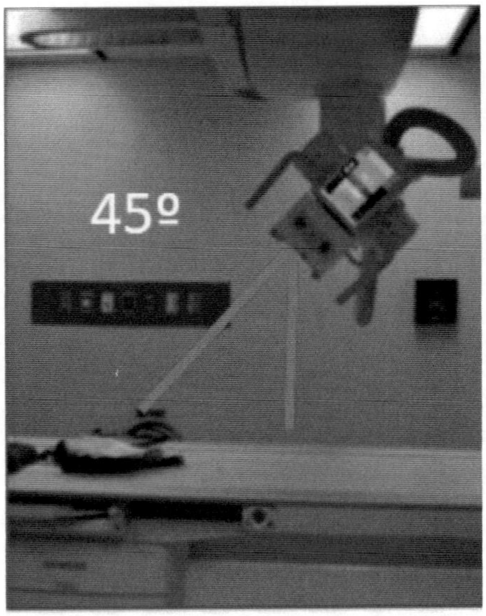

Figure 6. Image demonstrating the angled x-ray beam used to obtain an H-view; image modified from the original publication outlining this radiograph technique (Visser et. al., 2015).

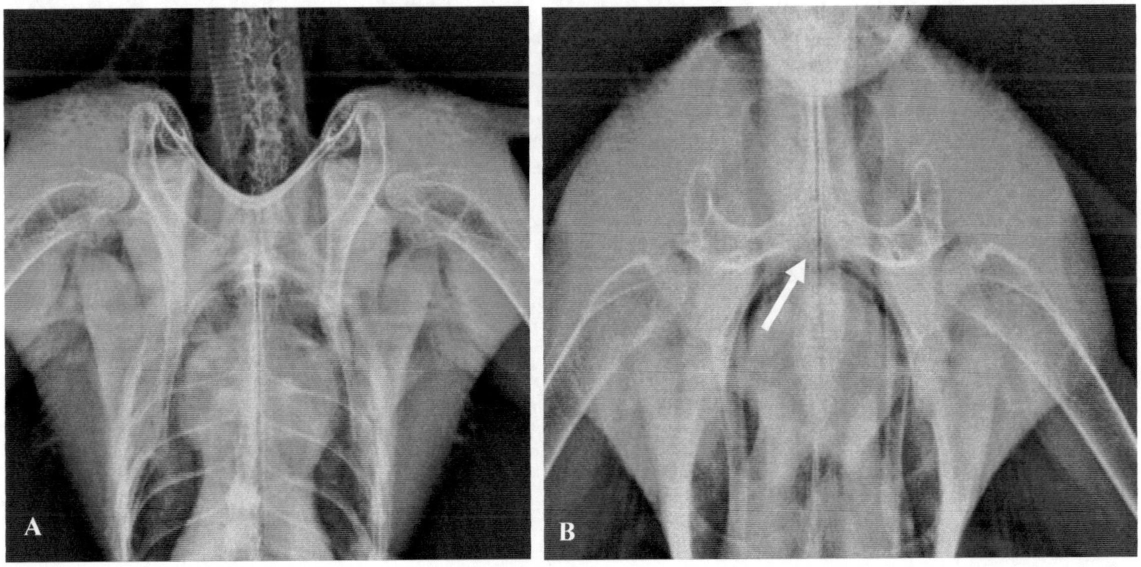

Figure 7. A paired VD (A) and H-view (B) of the shoulder girdle in a normal peregrine falcon. Note how the H-view particularly highlights the coracoids and their proximal insertion at the level of the keel (white arrow).

Radiographs of the feet

Dorsal-plantar and lateral radiographs of the feet can be obtained to fully assess the digits (Figs 8 and 9). A caudal-plantar view can provide additional assessment of the foot; this view is especially helpful for the first digit. The bird is positioned in ventral recumbency, and the foot is taped in plantar position against the plate with the digits taped separately.[2]

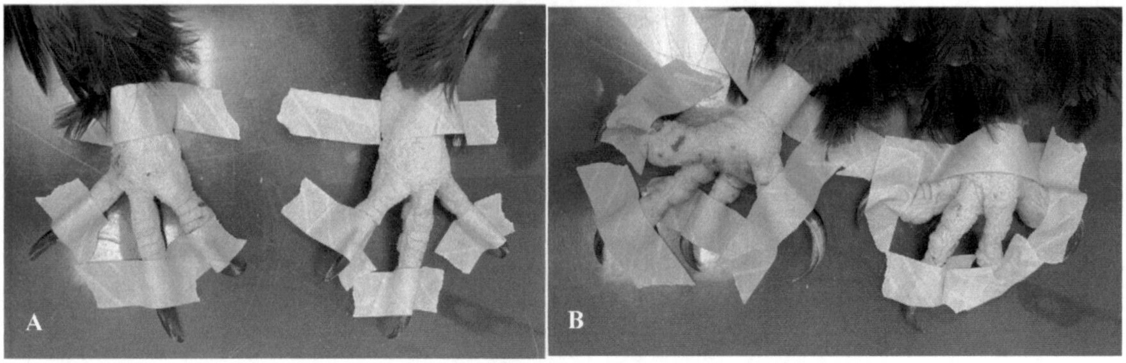

Figure 8. Image demonstrating the use of tape to separate the digits for a ventral-dorsal (A) and lateral (B) radiograph in a bald eagle. Note that in the VD view, the hallux is extended behind the tarsometatarsus.

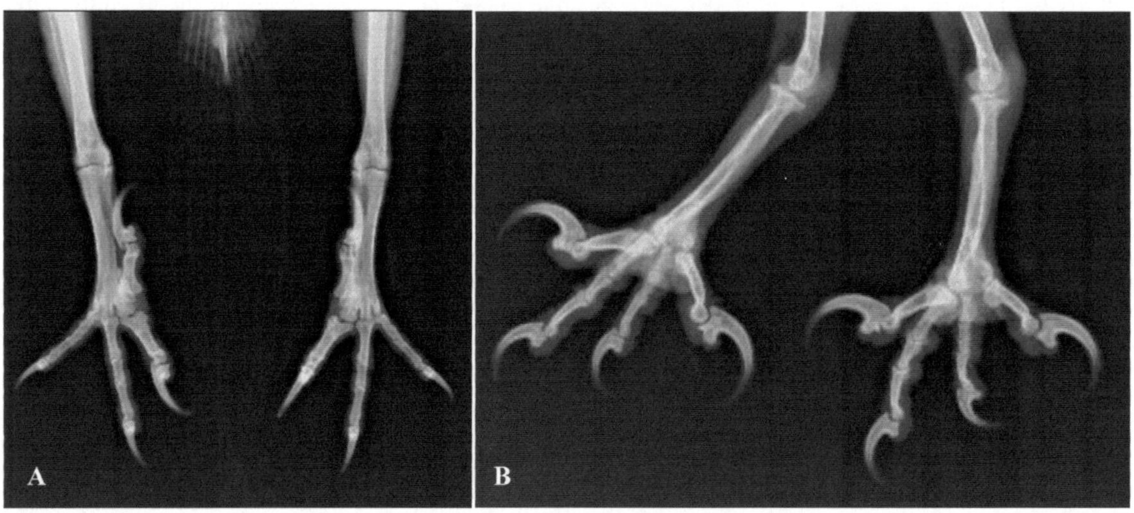

Figure 9. Correctly positioned VD (A) and lateral (B) foot radiographs in a bald eagle.

Stressed view of the shoulder joint

The bird is positioned as it would be for a ventral-dorsal radiograph, and the wings are fully extended cranially towards the head.[2] This allows for a more thorough assessment of the humeral head and the articular surfaces of the shoulder.[2]

Radiographs in a Rehabilitation Setting

Radiographs are commonly available, relatively cost-effective, and a valuable diagnostic tool in a raptor rehabilitation setting. Radiographs should be considered a baseline, required diagnostic at both the intake exam and the pre-release exam. Radiographs also have clinical value at several points during the medical treatment and rehabilitation process depending on the bird's primary injury.

Radiograph schedule

In a raptor rehabilitation setting, radiographs should be a core diagnostic that is part of your routine intake exam. If the bird is not systemically stable for sedation or anesthesia at the time of admission, radiographs may be postponed for a few days until the bird has improved. Alternatively, a single VD radiograph with minimal restraint can be obtained quickly with minimal stress on the bird and can provide helpful information that may change treatment or prognosis.

Recheck radiographs are indicated at various points throughout the rehabilitation process depending on the patient's medical conditions and response to treatment. Recheck radiographs two weeks post-admission occasionally reveal abnormalities that were missed or not yet apparent at the time of admission, such as a subtle incomplete fracture or neurogenic muscle atrophy. Recheck radiographs throughout medical treatment may reveal other unexpected complications, such as subtle changes associated with an aspergillosis infection in susceptible species. Recheck radiographs are also vital in the treatment of orthopedic injuries. Radiographs help dictate surgical approach, when surgical fixation may be removed, or when coaptation may be discontinued. Radiographs can also reveal orthopedic complications such as osteomyelitis and sequestrum formation that are not always apparent on physical exam.

Radiographs are also a vital tool and a core component of the pre-release exam. In our experience at The Raptor Center, pre-release radiographs have revealed abnormalities such as early degenerative joint disease that were not readily apparent on flight assessment or physical exam. Additionally, pre-release radiographs give you a collection of normal, healthy representatives of their species. This bank of normal images that is specific to your machine and radiograph techniques can be very helpful when trying to determine if something you are seeing in a different patient is normal or abnormal.

Radiograph value

Radiographs can reveal subtle or hidden problems that are not readily apparent on physical exam, can offer more detail or clarity to exam findings, and can provide information that may significantly change the prognosis or treatment for the individual. For example, some fractures are challenging to palpate on exam, especially in the very small owls or falcons that weigh 200g or less. Shoulder girdle fractures, especially scapula fractures and luxations, are very difficult to appreciate on physical exam and are best diagnosed with radiographs. Radiographs can reveal changes that are supportive of internal trauma that may not be apparent just on physical exam, such as cardio-hepatic separation, loss of serosal detail, or increased soft tissue opacity in the lung fields. Radiographic assessment of internal structures can also be a useful tool for the quiet "ADR" raptor patient without a clear cause for its lethargy. In this scenario, hepatomegaly or air sac lines might send you down the correct diagnostic and treatment path.

Clinical examples

This 1-hour master class focuses on case examples that highlight common conditions or presentations in a raptor rehabilitation setting, clinical conditions where the radiographic presentation may be difficult to interpret or identify, and situations where radiographs significantly changed the course of treatment or prognosis for release to the wild.

References

1. Visser M, Hespel A-M, De Swarte M, Bellah JR. Use of a caudoventral-craniodorsal oblique radiographic view made at 45° to the frontal plane to evaluate the pectoral girdle in raptors. *J Am Vet Med Assoc.* 2015;247(9):1037-041.

2. Samour J. Anatomical radiology. Association of Avian Veterinarians Online Course. Available at: https://www.aav.org/store/viewproduct.aspx?id=16100499. Accessed May 27, 2022.

Gross Lesion Recognition for the Avian Clinician

Michael M. Garner, DVM, DACVP

Session #2565

Affiliation: From Northwest ZooPath, 654 West Main Street, Monroe, WA 98272, USA.

Note: Abstract format selected by author(s). See Front Matter for description of conference proceedings manuscripts submitted as scientific abstracts.

Abstract: Gross lesion recognition, like other imaging modalities, is a bit of an art form. Vast zones of grey may confront the investigator when considering what the lesion is, what it could be, and what to do with it. And nobody is good at all of it. A seasoned livestock pathologist could possibly clear an entire day's necropsies without the help of a single histology slide. A bee keeper would recognize immediately the maggot stages that parasitize his colony. A dog-cat pathologist in a busy private laboratory or a companion animal clinician may not have a clue about gross lesions in livestock or bees. Historically, our attempts at accurately diagnosing disease by gross lesion recognition have been a humbling experience, and that has been the impetus for all further diagnostic specialties. So now that we have put gross pathology in its proper place, why even bother with such an imperfect science? Obviously, gross lesions, or lack thereof, are the first visual indication of what may be wrong with the patient. These are the lumps, the effusions, the asymmetrical oddities, the discolorations, the odiferous clues to disease that (hopefully) stimulate a scientific thought process culminating in a list of differential diagnoses. The purpose of this master class is to present images of common and not-so-common avian gross lesions in live and necropsy specimens, and in a participatory manner, establish a differential diagnosis and means for establishing a definitive diagnosis.

Recommended Reading

1. Campbell TW, Ellis C. *Avian and Exotic Animal Hematology and Cytology*. Ames IA: Iowa State University Press; 2007.

2. Garner MM. An overview of biopsy, cytology and necropsy techniques for clinical veterinarians. In: Aguilar RF, Hernandez SM, Divers SJ, Perpinan D, eds. *Atlas de Medicina de Animales Exoticos*. 2 da. Buenos Aires: Inter-Medica S.A. I.C.I.; 2010:437-443.

3. Latimer KS, Rakich PM. Necropsy examination. In: Ritchie BW, Harrison GJ, Harrison LR. *Avian Medicine: Principals and Application*. Lake Worth, FL: Wingers Publishing, Inc; 1994:356-376.

Section 7

Diagnostics 1

Sue Chen, DVM, DABVP (Avian Practice);
Sarah Sokolik, DVM, MPH
Session Chairs

Effects of a 0.3% Cholesterol Diet and a 20% Fat Diet on Plasma Lipids and Lipoproteins in Quaker Parrots (*Myiopsitta monachus*)

Hugues Beaufrère, DVM, PhD, Dipl ACZM, DABVP (Avian Practice), DipECZM (Avian),
Ken D. Stark, PhD,
R. Darren Wood, DVM, DVSc, DACVP

Session #2543

Affiliation: From the Department of Medicine and Epidemiology, School of Veterinary Medicine, University of California-Davis, Davis, CA 95616, USA (Beaufrère), Department of Pathobiology, Ontario Veterinary College, University of Guelph, Ontario, Canada (Wood), Department of Kinesiology, University of Waterloo, Ontario, Canada (Stark).

Note: Abstract format selected by author(s). See Front Matter for description of conference proceedings manuscripts submitted as scientific abstracts.

Abstract: Lipid disorders are common in captive psittacine birds, but associated changes in blood lipids and lipoproteins have not been well characterized. The Quaker parrot is prone to dyslipidemia and has been extensively used as an experimental model. The objectives of this study were to study the effects of a 0.3% cholesterol diet and a 20% fat diet on plasma lipids and lipoproteins in Quaker parrots. Two cross-over studies were performed with each diet. During each study, 12 parrots were divided into two groups fed the treatment or control diet for 2 weeks. After a 2-month wash-out period, the groups were reversed. At the end of each period, plasma lipidomics and lipoprotein profiling were performed. Data were analyzed by univariate tests adjusted for false discovery rate, volcano plots, and enrichment analysis. The cholesterol diet induced changes in many plasma lipids and lipoproteins. Total cholesterol and cholesteryl esters were significantly and markedly elevated. Ceramides were the second subclass of lipids that was elevated. Several glycerophosphocholines, sphingomyelins, and one diacylglycerol were also significantly elevated, albeit to a lesser magnitude. All lipoproteins were elevated with the greatest increase seen in non-HDL. The fat diet mainly resulted in a decrease in plasma glycerolipids and an increase in acyl-carnitines. Lipoprotein plasma levels remained unchanged. In conclusion, Quaker parrots fed a 0.3% cholesterol diet showed profound and complex dyslipidemic changes that may be used to further study lipid disorders in psittacine birds. A 20% fat diet higher in n-6 polyunsaturated fatty acids did not lead to dyslipidemia.

Comparison of Lipoprotein Analysis using Gel-Permeation HPLC and a Biochemistry Analyzer in Normolipidemic and Dyslipidemic Quaker Parrots (*Myiopsitta monachus*)

Hugues Beaufrère, DVM, PhD, Dipl ACZM, DABVP (Avian Practice), DipECZM (Avian), and Darren Wood, DVM, DVSc, DACVP

Session #2542

Affiliation: From the Department of Medicine and Epidemiology, School of Veterinary Medicine, University of California-Davis, One Shields Avenue, Davis, CA 95616, USA (Beaufrère), Department of Pathobiology, Ontario Veterinary College, University of Guelph, 50 Stone Road, Guelph, ON N1G2W1, Canada (Wood).

Note: Abstract format selected by author(s). See Front Matter for description of conference proceedings manuscripts submitted as scientific abstracts.

Abstract: Lipid accumulation disorders are common in psittacine birds and can be associated with changes in plasma lipoproteins, notably LDL and HDL; however, lipoprotein analysis using either standard laboratory analyzers or an indirect method of LDL estimation such as the Friedewald formula have not been validated in parrots. A research colony of 12 Quaker parrots were used to compare plasma values obtained from the Roche Cobas c501 for total cholesterol, total triglycerides, LDL cholesterol, and HDL cholesterol to Gel-Permeation HPLC (Liposearch®). To increase sample size and broaden analytical range to include dyslipidemic samples, two cross-over studies were performed using a 0.3% cholesterol diet and a 20% fat diet. Agreement between methods was assessed using linear mixed models on the bias and Bland and Altman plots. Calculated LDL values using the Friedewald formula and alternative formulas and the effects of triglycerides on the biases were also evaluated. A total of 45 plasma samples were used. The cholesterol diet induced a marked increase in cholesterol and all lipoproteins while the fat diet did not lead to dyslipidemia. Direct and indirect LDL measurements obtained with the clinical analyzer were not found to be in clinical agreement with GP-HPLC while HDL had acceptable agreement for normotriglyceridemic samples. Hypertriglyceridemic plasma samples were found to interfere with lipoprotein measurements. In conclusion, LDL measured by the Cobas c501 and indirect LDL estimations cannot be recommended in the Quaker parrot and non-HDL cholesterol should be used instead. Lipoprotein panels obtained from hypertriglyceridemic samples should be interpreted carefully.

The Effects of Time and Anticoagulant Type on Hematologic Values in Cockatiels (*Nymphicus hollandicus*)

Gretel Tovar-Lopez, MVZ,
Krista A. Keller, DVM, Dipl ACZM,
A. Russell Moore, DVM, MS, DACVP,
Sangeeta Rao, BVSc, MVSc, PhD,
Miranda J. Sadar, DVM, Dipl ACZM

Session #2702

Affiliation: From the Department of Clinical Sciences, College of Veterinary Medicine and Biomedical Sciences, Colorado State University, Fort Collins, CO 80523, USA (Tovar-Lopez, Rao, Sadar), Department of Veterinary Clinical Medicine, College of Veterinary Medicine, University of Illinois, Urbana, IL 61802, USA (Keller), Department of Microbiology, Immunology, and Pathology, College of Veterinary Medicine and Biomedical Sciences, Colorado State University, Fort Collins, CO 80523, USA (Moore).

Note: Abstract format selected by author(s). See Front Matter for description of conference proceedings manuscripts submitted as scientific abstracts.

Abstract: Avian hematologic characteristics complicate evaluation and, therefore, complete blood cell counts require manual analysis for best accuracy. Furthermore, quality of samples may be affected by prolonged shipment. The present study evaluated the effects of time and two anticoagulant types, dipotassium ethylenediaminetetraacetic acid (EDTA) and lithium heparin, on hematologic values in adult cockatiels (n = 21). Packed cell volume (PCV), total solids (TS), estimated white blood cell count (WBC), WBC differential (heterophils, lymphocytes, monocytes, eosinophils, basophils), hemolysis index (HI), and preparation quality were assessed and compared to baseline data at 24, 48, and 72 hours after venipuncture. Effects of time and anticoagulant type were evaluated with a 2-way ANOVA after log transformation. Statistical significance was defined at p-value ≤ 0.05 for all analyses. Heparinized samples had differences in WBC, heterophils, and lymphocytes at 48 hours, and PCV and TS at 72 hours. Results showed a positive proportional bias in PCV and TS from EDTA anticoagulant samples at 24, 48, and 72 hours, and from heparinized samples at 48 and 72 hours when compared to baseline. The EDTA samples had increased PCV and TS compared to heparin at all time points. The EDTA samples also had higher WBC and heterophil counts at 48 hours. These results support that the anticoagulant used and time spent in anticoagulant alters hematologic values in cockatiels, which may have significant clinical implications. These results emphasize the importance of sample processing within 24 hours, and that anticoagulant-specific reference intervals for hematologic variables should be considered.[1-5]

References

1. Moritomo T, Minami A, Inoue Y, et al. A new method for counting of quail leukocytes by flow cytometry. *J Vet Med Sci.* 2002;64:1149-1151.

2. Cray C, Zaias J. Laboratory procedures. *Vet Clin North Am Exot Anim Pract.* 2004;7:487-518.

3. Beaufrère H, Ammersbach M, Tully Jr TN. Complete blood cell count in Psittaciformes by using high throughput image cytometry: a pilot study. *J Avian Med Surg.* 2013;27:211-217.

4. Sheldon JD, Stacy NI, Blake S, et al. Comparison of total leukocyte quantification methods in free-living Galapagos tortoises (*Chelonoidis* spp.). *J Zoo Wildl Med.* 2016;47:196-205.

5. Campbell T. Blood sample collection and preparation in birds. In: Campbell T, ed. *Exotic Animal Hematology and Cytology.* 4th Ed. Ames, Iowa: Wiley Blackwell; 2015:165-172.

Section 8

Surgery

Sue Chen, DVM, DABVP (Avian Practice);
Sarah Sokolik, DVM, MPH
Session Chairs

Surgical Repair of Bilateral Mandibular Ramus Fractures in a Mallard Duck (*Anas platyrhynchos*)

Holly Galusha, DVM, and
Brian Speer, DVM, DABVP (Avian Practice), DipECZM (Avian)

Session #2662

Affiliation: From the Medical Center for Birds, 3807 Main Street, Oakley CA, 94561, USA.

Note: Abstract format selected by author(s). See Front Matter for description of conference proceedings manuscripts submitted as scientific abstracts.

Abstract: An approximately 5-year-old male mallard duck was presented for evaluation and surgical repair of bilateral fractures of the mandibular beak. The bird was unable to prehend food in this condition. At examination there were complete, open, and ventrally displaced rostral fractures at the rostral third of both mandibles with fragmented rhamphothecal keratin. Interramal soft tissue structures and all other skeletal structures were palpably intact. Intraoperatively, small rongeurs were used to debride bone and debris. On both the left and right sides, a 0.045 mm non-threaded Steinman pin was threaded retrograde into and out the rostral mandibular fragment, and then normograde into the caudal fragment. This pin was advanced to just rostral to the caudal angle of the mandible. The exposed rostral ends of the pin were cut and bent ventrally at a 90-degree angle. Pins (0.035 mm threaded IM) were placed ventrally, directed upwards into the mandibular ramus at the angle of the mandible, and driven until palpably engaging both ventral and dorsal mandibular cortices. A Type I extra skeletal fixator was then fashioned using intravenous tubing which was then filled with methylmethacrylate. An additional cross bar of IV tubing filled with methylmethacrylate was applied distally over the two most rostral pins, providing inter-fixator stability. The patient was housed in a protected environment with an immediate return to normal feeding behavior. Thirty-eight days following surgery, the bird was found to have removed the fixator. At repeat examination, there was excellent palpable stability of the fracture site and functional healing. This bird remains fully functional 9 months post surgery.

References

1. Lejnieks D. Treatment of a mandibular fracture using a steel plate in a lesser sulfur-crested cockatoo. *Exotic DVM.* 2004;6(4):15-17.

2. Carrasco DC, Dutton TAG, Shimizu N, et al. Distraction osteogenesis correction of mandibular ramis fracture malunion in a juvenile mute swan (*Cygnus olor*). *J Avian Med Surg.* 2016;30(1):30-38.

3. Anderson K, Arango J, Dugat D, et al. Distraction osteogenesis for correction of scissor beak in a chicken (*Gallus gallus domesticus*). *Proc ExoticsCon.* 2018;451.

4. Rocken Jun FE, Gobe T. Treatment of a mandibular fracture in a common buzzard (*Buteo buteo*). *Tierarztl Prax.* 1996;24(3):256-260.

5. Visgalli G, Cappelletti A, Nuvoli S. Beak osteosynthesis in a magpie. *Exotic DVM.* 2009;11(1):3.

6. Speer, BL, Fecchio R, Holliday C. The beak: form, function, and problems. In: *Current Veterinary Therapy in Avian Medicine and Surgery.* 2nd Ed. In press.

Use of a Harmonic Scalpel for Excisional Biopsy on a Moluccan Cockatoo (*Cacatua moluccensis*)

Patrick Sullivan, DVM, DABVP (Avian Practice)

Session #2666

Affiliation: From Angell Animal Medical Center, 350 S. Huntington Ave, Boston, MA 02130, USA.

Note: Abstract format selected by author(s). See Front Matter for description of conference proceedings manuscripts submitted as scientific abstracts.

Abstract: A 30-year-old Moluccan cockatoo (*Cacatua moluccensis*) presented for evaluation of a chronically ulcerated cutaneous mass approximately 1-2 centimeters to the right of the cloaca. Fine needle aspirate of the mass was inconclusive due to the absence of tissue cells. Blood work changes, which included a decrease in AST, uric acid, phosphorus, heterophils, lymphocytes, and PCV, were consistent with chronic inflammation. Ultrasound and radiography identified the mass as cutaneous in origin. Minor cardiac abnormalities were found on cardiac evaluation. The mass was refractory to medical therapy and was removed surgically using the Harmonic scalpel (Ethicon Endo-Surgery, Inc., Cincinnati, OH). The patient recovered well, without complication and complete healing was noted four weeks post-operatively. The tumor was most consistent with a fibrosarcoma based on histopathology. Use of the Harmonic scalpel in avian and exotic animal medicine has previously been described on a limited basis. Although other instruments have become more commonly used for mass removal, the Harmonic scalpel should be considered for use in a variety of avian surgeries as it has been shown to reduce adhesion occurrence rates while maintaining effective hemostatic control and wound healing time. To the author's knowledge, this is the first reported use of the Harmonic scalpel in this species.

Interdigital Web Sparing Partial Digit Amputation in a Gander (*Anser anser*) for Management of Chronic Pododermatitis

Holly Galusha, DVM, and
Brian Speer, DVM, DABVP (Avian Practice), DipECZM (Avian)

Session #2664

Affiliation: From the Medical Center for Birds, 3807 Main Street, Oakley CA, 94561, USA.

Note: Abstract format selected by author(s). See Front Matter for description of conference proceedings manuscripts submitted as scientific abstracts.

Abstract: A 4-year-old Emden gander (*Anser anser*) presented for evaluation of swelling associated with pododermatitis and mild lameness. Physical examination revealed a 4 cm × 4.5 cm fibrous callus at the left plantar proximal digital pad of digit III. Radiographs showed chronic medial luxation of the proximal interphalangeal joint of left digit III with lysis and osteophytosis of articular surfaces. Hematology demonstrated mild leukocytosis. Web sparing digit amputation was elected to palliate and in effort to preserve natural palmate form and swimming function. A plantar incision was made over the fibrous callus, which was reflected to expose the proximal interphalangeal joint. Flexor tendons were retracted. The first phalanx of digit III was rongeured from its distal epiphysis proximally to the mid-diaphysis. The second phalanx was dissected and removed in toto. The callus was surgically resected saving enough for primary closure with monofilament. Cultures obtained from the surgical site intraoperatively grew a non-hemolytic *Staphylococcus* spp. The patient recovered uneventfully from the procedure and continues to do well 8 months post-operatively with no observed lameness or alteration in swimming abilities. Pododermatitis is a common disease of waterfowl, with surgical treatment frequently complicated by large body size and biomechanics of these species. Euthanasia is a common recommendation for many cases, as amputation of pelvic limbs or digits may preclude the ability to restore and maintain acceptable welfare. Web sparing digit amputation offers the potential for survival of some of these patients with minimal impact to quality of life.

References

1. Olsen G. Web sparing amputation of the third digit in a duck. *Proc Annu Conf Assoc Avian Vet.* 2015;169.

2. Blair J. Bumblefoot: a comparison of clinical presentation and treatment of pododermatitis in rabbits, rodents, and birds. *Vet Clin North Am Exot Anim Pract.* 2013;16(3):715-735.

3. Backues K. Anseriformes. In: Miller RE, Fowler ME, eds. *Zoo and Wild Animal Medicine.* Vol 8. St. Louis, MO: Elsevier Saunders;2015:116-126.

Diagnosis and Treatment of Hypothyroidism in a Pet Pigeon

Nicola Di Girolamo, DMV, MSc (EBHC), PhD, DipECZM (Herpetology), Dipl ACZM,
Kelsea Studer, DVM,
João Brandão, LMV, MS, DipECZM (Avian),
Ian Kanda, RVT, VTS (Exotics),
Robert A. Wagner, VMD, DABVP (Exotic Companion Mammal),
Federico Fracassi, DVM, PhD, DipECVIM (Companion Animal),
Kellie Fecteau, PhD

Session #2725

Affiliation: From the Department of Clinical Sciences, College of Veterinary Medicine, Oklahoma State University, Stillwater, OK 74075, USA (Di Girolamo, Studer, Brandão, Kanda), Division of Laboratory Animal Resources, University of Pittsburgh, Pittsburgh, PA 15261, USA (Wagner), Department of Veterinary Medical Sciences, University of Bologna, Bologna 40064, Italy (Fracassi), Department of Biomedical and Diagnostic Sciences, College of Veterinary Medicine, University of Tennessee, Knoxville, TN 37996, USA (Fecteau).

Note: Abstract format selected by author(s). See Front Matter for description of conference proceedings manuscripts submitted as scientific abstracts.

Abstract: A 6-month-old, 330-gram, pet domestic pigeon (*Columba livia domestica*) was presented for an inability to properly molt and lethargy. The pigeon was obese, had anatomical characteristics of a chick, including cere and plumage, and a small caudal coelomic hernia. A CBC and plasma biochemistry panel were unremarkable. A CT scan confirmed excessive fat deposition in the coelom and a mass adherent to the liver. A fine needle aspirate of the mass indicated fat accumulation. A thyroid panel showed low tT4, and a thyroid stimulating hormone (TSH) stimulation test was planned for this pigeon and three presumably euthyroid pigeons. The four pigeons were administered 0.09 ml (80 µg) of recombinant human TSH (rhTSH) (Thyrogen, Sanofi). Blood was drawn at time 0, 3 and 6 hours after administration of rhTSH. Samples were analyzed for tT4 both by submission to the University of Tennessee Endocrinology Laboratory and by use of a specific rotor of a point-of-care analyzer (VetScan, Abaxis). Following rhTSH administration, the control pigeons showed a 4 to 21-fold increase in tT4, while all values of this pigeon were < 0.12 µg/dl. The pigeon was started on 20 µg of compounded L-thyroxine q12h. Following the administration of L-thyroxine, the pigeon molted and developed adult features. The caudal coelomic hernia resolved without treatment. There was a decreased amount of body fat in repeated CT images. The L-thyroxine dose was decreased based on additional tT4 testing to 2.5 µg q24h. The pigeon has been under treatment with L-thyroxine for over 2 years without any apparent adverse effect.

Section 9

Case Reports

Sue Chen, DVM, DABVP (Avian Practice);
Sarah Sokolik, DVM, MPH
Session Chairs

Diagnosis and Surgical Treatment of a Large Fibromatous Nasal Polyp in a Blue and Gold Macaw (*Ara ararauna*)

Holly Galusha, DVM, and
Brian Speer, DVM, DABVP (Avian Practice), DipECZM (Avian)

Session #2665

Affiliation: From The Medical Center for Birds, 3807 Main Street, Oakley, CA 94561, USA.

Note: Abstract format selected by author(s). See Front Matter for description of conference proceedings manuscripts submitted as scientific abstracts.

Abstract: A 26-year-old male blue and gold macaw (*Ara ararauna*) was referred for diagnosis and potential management of a chronic upper respiratory condition characterized by wheezing and nasal breathing sounds, accompanied by bilateral nasal discharge for approximately one year. Computed tomography (CT) imaging reviewed a large, well circumscribed mass-effect in the middle caudal nasal passageways bilaterally at their communication with the rostral choanal region. Endoscopic evaluation of this area confirmed the presence of this mass-effect; however, diagnostic full thickness endo biopsies were unable to be obtained. Eighteen months later, the patient returned for further assessment. There had been a mild progression of the same previously noted clinical signs over this timeframe. A repeat CT evaluation affirmed the persistence of the previously noted caudal nasal passageway mass-effect. A right infraorbital sinusotomy was performed using a transection of the jugal to facilitate exposure. Following an incision through the medial wall of the infraorbital sinus, the mass-effect was exposed in the rostral-most aspect of the choanal opening. This mass was relatively firm and was excised successfully in toto. The jugal was opposed using absorbable monofilament, and closure was routine. Histopathology confirmed the mass to be a fibromatous polyp of the nasal cavity. The patient continues to remain free of any further clinical signs 9 months later. This case report describes an effective use of the infraorbital sinusotomy to access and excise deeper structures in a large parrot,[1-2] in this case, a large and otherwise non-accessible fibromatous polyp in the caudal nasal passageway.

References

1. Speer BL, Echols MS. Surgical procedures of the psittacine skull. *Proc Annu Conf Assoc Avian Vet.* 2013; 99-109.

2. Speer, BL, Fecchio R, Holliday C. The beak: form, function, and problems. In: Speer BL, VanZeeland YRA, eds. *Current Veterinary Therapy in Avian Medicine and Surgery.* Vol 2. In press.

Section 10

Infectious Disease

Stephanie Lamb, DVM, DABVP (Avian Practice);
Hannah Attarian, DVM
Session Chairs

PaBV Infection of Embryonated Eggs Produces Asymptomatic, Viral Shedding Ducks

Caitlin Mencio, PhD,
Serene Yu, BS, DVM,
Ty Tortella, BS,
Kelsey Williams, BS,
J. Jill Heatley, MS, DVM, DABVP (Avian Practice), Dipl ACZM,
Jeffrey Musser, DVM, PhD

Session #2674

Affiliation: From the College of Veterinary Medicine & Biomedical Sciences (Mencio, Yu, Tortella, Williams, Heatley, Musser), Schubot Center for Avian Health (Mencio), Texas A&M University, College Station, TX 77843, USA.

Note: Abstract format selected by author(s). See Front Matter for description of conference proceedings manuscripts submitted as scientific abstracts.

Abstract: Parrot bornavirus (PaBV) is an etiologic agent for proventricular dilatation disease (PDD), a widespread avian disease affecting over 50 species. PaBV infection can impair neurologic function and gastrointestinal functions, which may lead to death. Vertical transmission of PaBV occurs; however, the effect of PaBV infection on embryos is unknown. This study models vertical transmission of PaBV in ovo and documents effects of infection on developing embryos and hatchlings. Embryonated mallard (*Anas platyrhynchos*) eggs were infected at day 5 via virus injection into the yolk. Infection had little effect on embryo development, and 15 of 16 hatchlings from the infected group (93%) showed no symptoms after hatching; however, 7 of 16 infected ducks (44%) were shedding virus by 2 weeks post hatch and all infected ducks had viral RNA in brain tissue. Viral RNA was also detected in 6 of 8 (75%) unhatched or dead-in-shell embryos from infected eggs. These results suggest vertical transmission of PaBV may produce primarily asymptomatic carriers of PaBV which could, in turn, increase viral exposure of captive and wild birds. In ovo infection was repeated in fertile chicken (single comb white leghorn, *Gallus domesticus*) eggs harvested prior to hatching. Again, PaBV injection failed to obviously affect development; however, unlike ducks, viral RNA was not prevalent in nervous tissue. These results imply a possible species susceptibility to PaBV vertical transmission and infection.

Chlamydia Management in a Large Mixed Aviary

Gabrina Goh, BVSc (Hon)

Session #2506

Affiliation: From Mandai Wildlife Group, Conservation, Research and Veterinary Department, 80 Mandai Lake Road, Singapore 729826, Singapore.

Abstract: In a large aviary of approximately 250 mixed free-flying birds there were seven mortalities over four weeks, with no specific post-mortem findings. Polymerase chain reaction tests for *Chlamydia psittaci* were positive on various organs such as heart, liver, lung, and spleen for four of these birds on post-mortem. Combined conjunctiva, choanal, and cloacal swabs from other birds in the aviary were collected for *C. psittaci* PCR testing. These results were negative; however, birds continued to be found dead or sick in the aviary and three birds thereafter tested positive on *C. psittaci* PCR. These birds only included Columbiformes and other orders of birds seemed unaffected. As such, the zookeepers were instructed to bring any Columbiforme bird they caught to hospital where they were treated with doxycycline orally or in water for at least 42 days. As five Columbiforme birds died despite the treatment plan, it was decided to treat all birds in the 0.2-hectare aviary. Birds were treated with doxycycline diluted in all available drinking water for two months in the aviary. At the beginning of treatment, two laughing doves (*Spilopelia senegalensis*) presented for the first time with non-specific signs of illness but were *C. psittaci* PCR negative. When treatment ended, birds from various orders were opportunistically caught for collection of conjunctival, choanal, and cloacal swabs and repeat *C. psittaci* PCR was performed. The repeated PCR tests have all been negative for *C. psittaci*, indicating resolution of the outbreak 142 days post the first death from *C. psittaci* infection in the aviary.

Introduction

C. psittaci, the cause of chlamydiosis or ornithosis is a common disease found all over the world and has been found in over 465 avian species. Outbreaks amongst psittacine birds and poultry have caused significant economic damage as it can lead to severe disease and death.[1] Clinical signs are nonspecific and depend on the bird's age or species. Affected birds may present with lethargy, dyspnea, nasal, or ocular discharge, green urates and fecal matter, diarrhea, vomiting, anorexia, pyrexia, and death.[2] Transmission may occur when birds are in contact with infected asymptomatic or symptomatic birds, from any discharge such as from the beaks, eyes, feces, or urine, via inhalation or ingestion.[2]

Case Report

At Jurong Bird Park, in Singapore, there are over 3000 birds and several large aviaries with mixed species housed together, therefore contagious diseases such as chlamydiosis can cause large numbers of deaths and spread rapidly in an aviary or to various parts of the park. One large walk-in aviary at Jurong Bird Park, Jungle Jewels, houses approximately 250 birds belonging to various orders including Anseriformes, Bucerotiformes, Columbiformes, Passeriformes, Piciformes, Galliformes, and Psittaciformes. The majority of the aviary consists of Passeriformes (approximately 40%), Columbiformes (approximately 30%), and Psittaciformes (approximately 20%). In December 2018, seven birds of various species died in Jungle Jewels. These birds included two tambourine doves (*Turtur tympanistria*), one white-fronted amazon (*Amazona albifrons*), three blue-spotted wood doves (*Turtur afer*), and a crestless fireback pheasant (*Lophura erythrophthalma*). Each post-mortem showed variious non-specific signs such as air sacculitis, hematoma from trauma, or endoparasitism.

The following month, 31 days after the first death in Jungle Jewels, a Malagasy turtle dove (*Nesoenas picturata*) chick carcass was found dead and had lung congestion on post-mortem. Another tambourine dove carcass then presented on day 33 with splenomegaly and organisms suspicious of *Chlamydia* sp. were observed on histopathology, which were also found in the lungs and liver (Fig 1). This was confirmed to have a *C. psittaci* infection on polymerase chain reaction (PCR) tests. This led to the re-testing of all previous carcasses, of which four out of eight carcasses were positive for *C. psittaci*. All birds that tested positive were Columbiformes. Birds that previously had been moved from Jungle Jewels to various aviaries in the park in the prior month also were swabbed and tested for *Chlamydia* PCR. All results from these ten birds were negative.

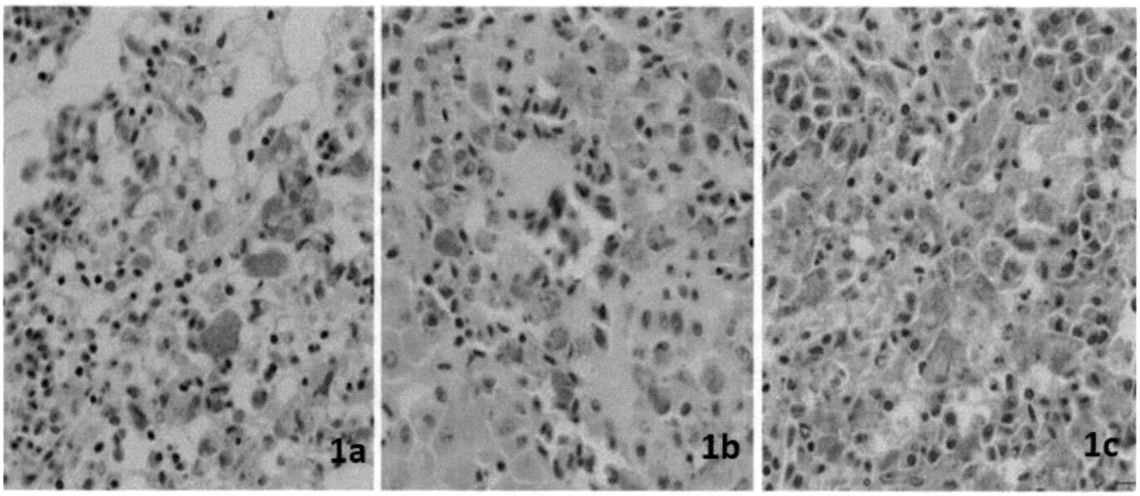

Figure 1. Lung (1a), liver (1b) and spleen (1c). Chlamydial organisms seen on histopathology ×100 magnification on the lung, liver and splenic samples.

On day 50, keepers began trapping birds in the aviary for further testing. Conjunctival, choanal, and cloacal swabs were taken from each bird for *Chlamydia* PCR. All ten birds caught were sampled on day 54 and tested negative on *Chlamydia* PCR, so no further trapping and testing were performed. Another two laughing doves (*Spilopelia senegalensis*) were brought into hospital and tested positive for *C. psittaci* on PCR and it was decided to proceed to the next treatment plan, which was to catch and hospitalize any Columbiformes caught for treatment.

The keepers started opportunistically trapping birds again in the aviary from day 57 and the Columbiformes were brought to the avian hospital for isolation, testing and treatment while all other species were swabbed for *Chlamydia* PCR, ringed for identification and released back into the aviary. Nine Columbiformes were trapped over 2 weeks, which included three laughing doves, two blue ground doves (*Claravis pretiosa*), two blue-spotted wood doves, a Malagasy turtle dove, and a Peruvian pigeon (*Patagioenas oenops*). These birds were brought to the hospital and treated with doxycycline (200 mg/L in water, Doxyvet liquid, Vetafarm, New South Wales, Australia) for 42 days. Clinical birds that were brought to the hospital, including a laughing dove and an African olive pigeon (*Columba arquatrix*) were treated with doxycycline (50 mg/kg PO BID, Doxyvet liquid, Vetafarm) doxycycline and subcutaneous Hartmann's solution (Compound sodium lactate, B.Braun, Penang, Malaysia) for 42 days. During this time over 2 weeks, another five laughing doves and an African olive pigeon were found dead in the aviary.

On day 71, the entire aviary was treated with doxycycline. This treatment proved challenging due to cost and environmental conditions. It was decided to treat the birds with doxycycline in water. All water reservoirs in the aviary were emptied as much as possible with continued draining daily and large tubs and metal trays of water were placed at the holding area in the aviary as a substitute for a water source (Fig 2). In the first two weeks, doxycycline liquid (250 mg/L of water, Doxyvet liquid, Vetafarm) was added into the water. Approximately 72 liters of water were used per day and divided into multiple trays and tubs in the aviary. After two weeks, the doxycycline liquid was changed to doxycycline powder (5 g/L of water, Psittavet, Vetafarm, New South Wales, Australia) due to product availability for the remaining 6 weeks. A total of 21 1-liter bottles of liquid doxycycline and 32 bottles of 450 g doxycycline powder were used in the 2 months of treatment in the aviary.

Figure 2. Medicated water placed in small water bowls for various species of birds at feeding stations in the aviary (2a) and medicated water placed in large tubs for Anseriformes and other birds (2b).

Since commencing doxycycline treatment in the aviary, one laughing dove was found dead in the aviary on day 82 and another laughing dove was found lethargic and was then euthanized. Both carcasses tested negative for *C. psittaci* on PCR and no further deaths were reported during the 2 months. Once treatment ended on day 135, the zookeepers were told to start trapping birds again opportunistically with trap cages and these birds were then swabbed for *Chlamydia* PCR tests. Approximately 15% of the population in Jungle Jewels was trapped over one week, of which 20% of the Columbiformes population, 5% of the psittacine population, 40% of the Anseriformes population and 10% of the other passerines in the aviary were tested. This included seven laughing doves, one Peruvian pigeon, four blue ground doves, five black-bellied whistling ducks (*Dendrocygna autumnalis*), three African olive pigeons, one spotted wood dove (*Turtur chalcospilos*), one crested quail dove (*Geotrygon versicolor*), five chestnut weavers (*Ploceus rubiginosus*), three dusky parrots (*Pionus fuscus*), one red-fronted cardinal (*Paroaria coronate*), one hoopoe (*Upupa epops*), one white-fronted amazon (*Amazona albifrons*), three saffron finches (*Sicalis flaveola*) and one Malagasy turtle dove. All test results came back negative and the aviary was declared negative a total of 142 days from the first death due to *Chlamydia* recorded in Jungle Jewels. There have been no new mortalities associated with *C. psittaci* infections in this aviary since the cessation of doxycycline treatment in the aviary.

Discussion

The source of infection of the aviary birds remains unidentified but there are several possibilities. The aviary is covered by a large mesh and wild birds are prevalent and often land on the mesh. These wild birds could potentially carry the organism and spread it as such. Interestingly, on day 80, one dead wild spotted neck dove (*Spilopelia chinensis*) was found less than ten metres from the aviary and was positive on PCR for *C. psittaci* and another emerald dove (*Chalcophaps indica*) that was not part of the collection was incidentally found within the aviary on day 64 and was also tested PCR positive for *C. psittaci*. This emerald dove was suspected to have found its way into the aviary from the wild. Another possible source would be guests in the park or zookeepers that may have picked up the infection from outside the aviary and spread it while walking through the aviary. Prior to the outbreak of chlamydiosis, a new trap cage was being built within the aviary which involved digging of the soil and loud noises, which would have significantly stressed the birds in the aviary and contributed to increased shedding and spread of the disease.

Chlamydiosis is a zoonotic disease and necessary precautions were taken during the outbreak. Zookeepers and cleaners were required to wear personal protective equipment such as N95 respirators and gloves. Zookeepers from other aviaries were not to enter the aviary, the aviary's walkway and handrails were regularly disinfected with F10 veterinary disinfectant and all were briefed about clinical symptoms of psittacosis in humans and were informed to report if they experienced any clinical signs. Signs to remind visitors to wash their hands and not to consume food or drinks were placed at the entrance of the aviary, along with a foot bath with F10 veterinary disinfectant for visitors to step in upon entering and exiting. There were no reported cases of zoonotic spread during the outbreak. The genotype of the *C. psittaci* in this case was not determined but as only Columbiforme birds were affected by the disease, it was thought to be more likely of genotype B which is less associated to zoonosis as compared to the other genotype, genotype A.

There were many difficulties encountered with treating the entire aviary with doxycycline. These difficulties included the following:

- There were multiple sources of water in the aviary, including five large ponds and a 3-metre tall waterfall where all the birds in the aviary mainly drank and bathed in. Draining the ponds proved difficult due to poor construction of the ponds, the ground was pebbled and contained sludge, it was also level instead of being sloped downwards which made drainage more difficult (Fig 3). In addition, it was monsoon season and rained almost daily, filling the ponds each time along with other puddles of water from which the birds would choose to drink. A day was set aside for staff to help with cleaning of the ponds, large water pumps had to be used and the ground had to be cemented eventually to create a smooth sloping surface.

- The walk-in exhibit was kept open and guests were continuously walking through which meant the aviary had to remain aesthetically pleasant. As such, the ponds could not be covered and water troughs had to be placed away from guests view.

- Thorough cleaning and disinfecting of the aviary were also difficult as there were ponds, graveled walkways, large areas of soil and tall plants throughout the aviary.

- Considerations for the various animals also had to be met, such as a large tub of water for all ducks in the aviary had to be provided which meant more water and more medications per day.

- Large amounts of doxycycline had to be ordered in a short amount of time which was costly. A total of more than 50 large bottles of doxycycline were required to treat the entire aviary population.

- It was a concern that birds that were already sick might not eat or drink and thus not be taking in the antibiotics sufficiently. This would mean sick birds might continue to shed the organism into the environment, furthering spread and more deaths may occur despite treatment.

- Blanket treatment of the entire aviary of approximately 250 birds that may or may not carry the disease and may not be taking the antibiotic course regularly could also lead to increased antibiotic resistance amongst the birds, especially with a 2-month course of doxycycline.

Figure 3. Water source in the aviary, multiple tiered waterfall with deep crevices (3a) and water flow through the aviary, large ponds with pebbled floors (3b).

Post treatment, the only concern raised was calcium deficiency in several doves. A lemon dove (*Columba larvata*) from Jungle Jewels laid an egg in the aviary during treatment which was infertile and had a soft translucent shell and a white-crowned pigeon (*Patagioenas leucocephala*) laid another egg with an abnormal shell integrity 2 months post-treatment. Three months post-treatment, a second lemon dove presented to hospital and was diagnosed with salpingitis and laid an egg similar to the one before, with a soft shell and inspissated content (Fig 4). This was attributed to the long-term doxycycline treatment since tetracyclines are known to bind calcium. In humans and young dogs tetracycline induced yellowish staining of dentin, and enamel hypoplasia are widely recognised due to the strong binding affinity of tetracyclines for calcium[3] and had probably affected the egg quality in these birds. In another study on mice, it was shown that long term administration of doxycycline for 35 days had caused a significant decrease in the serum calcium levels.[4] Besides soft-shelled eggs in these three birds, no other effects of long-term doxycycline treatment have been noted in the aviary.

Figure 4. An egg with an abnormal appearance passed by a lemon dove diagnosed with salpingitis; histopathology was not done.

Conclusions

Treatment of chlamydiosis in a large open-air aviary in a zoological institution presents many challenges, not only from a medical point of view but also of an operational point of view. There have been no further cases post cessation of doxycycline treatment of the aviary birds but we remain vigilant and continue to test any birds in the aviary or wild birds exhibiting clinical signs of infection by *C. psittaci*.

Acknowledgments: I thank my colleagues at the Jurong Bird Park avian hospital and animal care team, and the Mandai Wildlife Reserve pathologists who have all contributed to the management of this outbreak.

References

1. OIE World Organisation for Animal Health. Available at: https://www.oie.int/fileadmin/Home/eng/Health_standards/tahm/3.03.01_AVIAN_CHLAMYD.pdf Accessed March 18, 2022.

2. Welleham JFX, Lierz M, Phalen D, et al. Infectious diseases. In: Speer BL. *Current Therapy in Avian Medicine and Surgery.* St. Louis, MO: Elsevier; 2016:22-106.

3. Riond, JL, Riviere JE. Pharmacology and toxicology of doxycycline. *Vet Human Toxicol.* 1998;5:431-443.

4. Basak BC, Awal MA, Hasan Q, et al. Effect of calcium supplementation on clinical and hemato-biochemical parameters following long term administration of doxycycline in mice. *J Anim Vet Adv*. 2004;6:348-352.

An Outbreak of Respiratory and Cloacal Cryptosporidiosis in Attwater's Prairie Chickens (*Tympanuchus cupido attwateri*)

Michael M. Garner, DVM, DACVP,
Julie Swenson, DVM, Dipl ACZM,
Holly J. Haefele, DVM,
Lauren K. Schmidt, DVM,
April L. Childress,
James F. X. Wellehan, Jr, DVM, PhD, Dipl ACZM, Dipl ACVM (Virology),
Dipl ACVM (Bacteriology and Mycology)

Session #2561

Affiliation: Northwest ZooPath, 654 West Main Street, Monroe, WA 98272, USA (Garner), Fossil Rim Wildlife Center, Glen Rose, TX 76043, USA (Haefele, Swenson, Schmidt), Department of Comparative Diagnostic and Population Medicine, College of Veterinary Medicine, University of Florida, 2015 SW 16th Ave, Gainesville, FL 32608, USA (Childress, Wellehan).

Note: Abstract format selected by author(s). See Front Matter for description of conference proceedings manuscripts submitted as scientific abstracts.

Abstract: The Attwater's prairie chicken is an endangered subspecies of greater prairie chicken native to Coastal Texas and Louisiana. During the 2019 breeding season, large losses were experienced at a captive facility due to respiratory and enteric cryptosporidiosis, which has not been previously described in this species, nor diagnosed in the Northwest ZooPath (NWZP) archives. Various treatments, including antiprotozoal drugs, were ineffective. A total of 70 birds died (27.5% mortality) with clinical signs and gross lesions consistent with cryptosporidiosis. Eighteen were submitted to NWZP for histologic examination. Clinical signs included swollen cloaca, poor weight gain, oculonasal discharge, respiratory stridor, and ataxia or opisthotonos. Gross lesions included thin body condition, swollen cloaca, enlarged bursa, reddened ceca, swollen conjunctiva and nasal sinuses with mucoid or purulent exudate, congested or consolidated lungs, and purulent material within the facial sinuses and air cells of the calvarium. Histologically, cryptosporidia lined the apical aspects of nasal, sinusoidal, tracheal, bronchial, conjunctival, enteric and bursal mucosal epithelium with mucosal hyperplasia, erosions, ulcers and necrosis. Moderate to severe lymphoplasmacytic inflammation accompanied the infection, and heterophilic to pyogranulomatous inflammation was present in lesions that had secondary bacterial, mycotic or flagellate infections. Cloacal and sinus swab samples submitted for PCR and sequencing identified *Cryptosporidium avium* (formerly *Cryptosporidium* avian genotype V). This organism belongs to the non-gastric *Cryptosporidium* clade, and has been associated with disease in birds and squamates, often primarily centered on the cloaca and urinary tract. This report extends the known host range of *C. avium*, and demonstrates broader tissue tropism like the related *C. baileyi* and *Cryptosporidium* avian genotype II.

Section 11

Wildlife 2

Stephanie Lamb, DVM, DABVP (Avian Practice);
Hannah Attarian, DVM
Session Chairs

Stabilization and Long-Term Outcome of a Tibiotarsal Fracture in a Turkey Vulture (*Cathartes aura*) Using a Supracutaneous Plating Technique

Anthony Barillo, DVM,
Marina Liles, DVM,
Rodrigo Roca, DVM,
Nicola Di Girolamo, DMV, MSc (EBHC), PhD, DipECZM (Herpetology), Dipl ACZM,
João Brandão, LMV, MS, DipECZM (Avian)

Session #2730

Affiliation: From Department of Veterinary Clinical Sciences, College of Veterinary Medicine, Oklahoma State University, Stillwater, OK 74078, USA.

Note: Abstract format selected by author(s). See Front Matter for description of conference proceedings manuscripts submitted as scientific abstracts.

Abstract: A juvenile suspected male turkey vulture (*Cathartes aura*) presented after being found unable to walk. Physical exam revealed crepitus and instability in the left tibiotarsus and grade 4 lameness in the limb. Radiographs demonstrated a highly comminuted mid-diaphyseal fracture with early callus formation despite overt clinical instability. Based on the radiographs, the fracture was suspected to be less than one week old. The patient was anesthetized and underwent fracture reduction under fluoroscopic guidance. The fracture was stabilized with a 12-hole 2.4-mm locking compression plate (LCP) placed in a supracutaneous fashion with three bicortical locking screws placed both proximally and distally. The patient had a grade 3 lameness immediately following surgery, and lameness had resolved to a grade 2 by day 3 following surgery. Four weeks after surgery, radiographs demonstrated loss of cortical bone density and the construct was dynamized by removing the screw immediately proximal and immediately distal to the fracture site. At 6 weeks after surgery, radiographs revealed a bridging callus and all implants were removed. Following an additional 2 weeks of cage rest, recheck radiographs confirmed continued fracture healing and increased cortical density. At week 9, the patient was transitioned to a pre-release flight cage where he was noted to be able to set flight, land, grasp, walk, and perch normally. The patient was released at week 12 post surgery. This case report documents the successful functional outcome of a closed, highly comminuted mid-diaphyseal fracture of the tibiotarsus in a minimally invasive manner utilizing a supracutaneous plating technique.

Bobwhite Quail Vasectomies

Stephanie Lamb, DVM, DABVP (Avian Practice)

Session #2539

Affiliation: From Arizona Exotic Animal Hospital, 744 N Center St #101, Mesa, AZ 85201, USA.

Introduction

Vasectomies are a procedure where a surgical transection is made through the vas deferens to break down this anatomical segment and prevent the passage of sperm from the body. This is often done with the intention of preventing offspring while avoiding the more involved procedure of removing the testicles to render a male sterile. This procedure can be done for a variety of reasons, one of which is to help in conservation efforts for endangered species.

Bobwhite Quail

The masked bobwhite quail (*Colinus virginianus ridgwayi*) is an endangered species in the southernmost part of the state of Arizona and extending into the northern border of Sonora, Mexico. Due to habitat alterations their numbers became greatly reduced and efforts began in the 1930's to reintroduce them to their native habitats; however, the populations were not recovering and in the 1970s the management program changed in an attempt to improve outcomes. It was determined that young birds being released needed parents to learn how to survive in nature. Although adult masked bobwhite quail could be released with the young, they were needed in captivity for further breeding and population genetics. Therefore, a subspecies was selected to act as a surrogate parent to the captive bred individuals. The Texas Northern bobwhite (*Colinus virginianus texanus*) was chosen and wild adults were captured and collected; however, it was important that this subspecies not breed with the masked bobwhite quail once they were sexually mature as that would alter the genetic stock and render the reintroduction program pointless. Therefore, vasectomies were performed and a protocol was created to sterilize the adult Texas Northern bobwhite males before release with captive raised masked bobwhite young.[1]

The populations of masked bobwhite are currently being managed by the U.S. Fish and Wildlife Service with a captive breeding program occurring at the Buenos Aires National Wildlife Refuge in southern Arizona. The program was halted for a period of time but efforts to release birds were started again in 2017.

Vasectomy Procedure

The original program used a lateral flank approach to identify the vas deferens and perform the vasectomy. The patient was fasted for 12 hours, hooded, and then had a rubber band secured around the legs and wings. The bird was attached by hooks 25 cm apart to an operating board and feathers were plucked on the lateral side of the body. Alcohol was used to prepare the surgery site and a vertical incision was made cranial to the thigh with a scalpel. While the thigh was held posteriorly blunt scissors were inserted through the intercostal muscles between the last two ribs. A retractor was then inserted into this location and a high intensity light was directed into the cavity. The testes were identified as well as the vas deferens. A blunt probe was used to lift the posterior base of the testis and a sharp probe was then utilized to carefully tease the vas deferens off the common iliac vein. Next, the vas deferens was clamped with a hemostat at the base of the testis and the vas was torn from the testis with a blunt probe. It was suggested that about 5-20 mm of vas deferens be removed by tearing it away, but more typically 3-15 mm was removed. Nitrofurazone was then applied to the surgical site. The procedure was repeated on the opposite side and surgery was reported to be done in approximately 7 minutes.[1]

When the program was restarted the biologists running it initially experimented with 10 birds to see if an external vasectomy approach could be utilized. This procedure involves removal of the caudal most segment of the vas deferens where it inserts into the cloaca laterally. Although this approach can be used for budgerigars, zebra finches, and society finches, it is not possible to use this approach for all species. Species in which this can be done have the caudal vas deferens become enlarged and convoluted during breeding season, resulting in a structure called the seminal glomera that appears as a rounded swelling on either side of the cloaca. An incision can be made at this location and a segment of the vas deferens can be grasped and severed.[2] This was attempted in the masked bobwhite quail and it was determined to be unsuccessful and no birds were vasectomized by this approach.

The following year in 2018, 17 birds were used for the program and the original protocol created in the 1970s was attempted but anesthesia (isoflurane) and pain control (robenacoxib 10 mg/kg IM once) were utilized. In addition, an operating board was not used and instead patients were taped to a surgical table. On initial attempts at the procedure the tearing of the vas deferens from the testicle resulted in a tear of the common iliac vein and fatal hemorrhaging. This was a repeatable problem, so a caudal approach was instead used in the 15 remaining birds. An incision was made through the skin into the coelomic cavity at a spot halfway between the pubis bone and the tail base. The vas deferens was visualized here and transected. Although this approach seemed promising, within 24 hours, 10 of the 15 (66.6%) birds passed away and on necropsy it was found that one or both of the ureters had been cut in these birds when the vas deferens was cut as well.

Another set of 15 birds were scheduled to have the procedure performed but due to the poor success rate a new approach was needed. Six Northern bobwhite quail (*Colinus virginianus*) that were captive raised were used and a modification of the original procedure was performed. Birds were anesthetized with isoflurane via mask induction and were maintained on anesthesia. A bird was then placed in lateral recumbency, and the wings were pulled up and the legs were pulled back and taped to the table. The feathers were plucked from the lateral body wall ventral to the spine and cranial to the thigh in an area about 2 × 2 cm. The site was prepared with chlorhexidine solution. A vertical incision was made with a scalpel blade cranial to the thigh and hemostats were used to enter the coelomic cavity. A nasal speculum that was attached to a light producing handle (Welch Allyn) was inserted into the coelomic cavity to act as a retractor and a light source. The testicle was identified along with the vas deferens. Occasionally a cotton tipped applicator would be needed to push intestines ventrally and allow for visualization and isolation of the testicle and vas deferens. Next, a monopolar electrocautery unit was used to cauterize and cut the vas deferens. In some cases, both vas deferens could be cut from a single approach; however, the majority required a bilateral approach. With this new approach, five out of six (83.3%) birds survived the procedure and were vasectomized. One did suffer fatal hemorrhage.

After this new approach was used, in 2018 the remaining 15 birds had the procedure performed. Ten of the 15 survived the procedure (66.6%), three out of 15 (20%) suffered fatal hemorrhage and two died post operatively within a few days. One died from internal parasites but the other died of an unknown cause, though hemorrhage was ruled out based on post-mortem evaluation.

In 2019, 41 quail were used for the program. The first three birds had the original procedure from the 1970s protocol performed and all died from fatal hemorrhage. The remaining 38 had the technique using electrocautery and three out of 38 (7.9%) died from fatal hemorrhage but 35 (92%) had survived.

This same procedure was used in 2020 with 57 birds and two of 57 (3.5%) died from fatal hemorrhaging. One survived the surgery but had spinal trauma occur on recovery and was euthanized. The success rate for that season was 94.7%. Again this procedure was use in 2021 with 31 birds and two birds died from fatal hemorrhage and one died from unknown causes. Twenty-eight of 31 survived and the success rate for that year was 90.3%.

The original publication for vasectomies in 195 Texas Northern bobwhites reported a 92% success rate of birds surviving the procedure;[1] however, in conversation with an author of this paper it was reported that these statistics were determined after the technique was perfected. Had all birds been included in the study the success rate would have been lower. In our experience with this program, of the 177 total birds from 2017-2021, 30 (17%) passed away and 147 (83%) survived. Of those that died, 16 (53.3%) died from fatal hemorrhagic directly related to the procedure, 10 (33.3%) from cut ureters, and four (13.3%) died of caused not directly related to the procedure. If the statistical analysis is isolated to once the procedure using electrocautery was used, of 147 birds there were 16 (10.9%) deaths and 110 (89.1%) survived. Of the deaths 11/16 (68.8%) were due to fatal hemorrhaging.

Statistics for 2022 are included during the conference presentation. The practical application of this procedure is discussed and includes its use for other conservation projects or for population control for other means.

References

1. Ellis DH. A technique for vasectomizing birds. *J Field Ornithol*. 1981;52(1):69-71.

2. Lierz M, Petritz O, Samour J. Reproduction. In: Speer BL ed. *Current Therapy in Avian Medicine and Surgery*. St. Louis, MO: Elsevier; 2016:433-460.

Assessment of Raptor Blood Lead Concentrations Relative to Full Implementation of the California Lead Ban

Krystal M.T. Woo, DVM,
Michelle G. Hawkins, VMD, DABVP (Avian Practice),
Katie M. Beckmann, MA, VetMB, MSc, MRCVS, DipECZM (Wildlife Population Health)

Session #2732

Affiliation: From the Lindsay Wildlife Experience, 1931 First Avenue, Walnut Creek, CA, 94597, USA (Woo), Department of Medicine and Epidemiology and the California Raptor Center, One Health Institute, School of Veterinary Medicine, University of California, 1 Shields Ave, Davis, Davis, CA, 95616, USA (Hawkins), The Royal (Dick) School of Veterinary Studies (Woo, Beckmann) and the Roslin Institute (Beckmann), The University of Edinburgh, Easter Bush Campus, Midlothian, EH25 9RG, United Kingdom.

Note: Abstract format selected by author(s). See Front Matter for description of conference proceedings manuscripts submitted as scientific abstracts.

Abstract: Lead is a toxic substance that negatively impacts animal and human health. One of the most common sources of lead exposure in raptors is via lead projectile ingested from prey or offal. The critically endangered California condor (*Gymnogyps californianus*) is very susceptible to the effects of lead intoxication.[1] Due to this, California implemented prohibitions of the sale and use of lead-based ammunition that first affected condor territory in 2008 and later became state-wide in July 2019. The purpose of this retrospective study was to evaluate associations between implementation of the statewide ban and blood lead concentrations (BLC) of raptor sentinel species presented at three California rehabilitation clinics: (1) within the current condor range; (2) on the border of their range; and (3) outside of their range. Species intake BLC and rescue locations were compiled, median BLC for each species were compared pre- and post- ban, and rescue locations were geospatially mapped using QGIS.[2] Data were collected from 199 raptors (2013- 2021) and statistics were calculated using R.[3] Median BLC decreased over time at Clinic 3 and increased over time at Clinic 2 but no significant correlation was identified between location and BLC. Species-specific median BLCs were variable regardless of clinic, likely due to prey preferences and foraging strategies. The varied temporal trends among clinics may reflect differing concentrations of lead exposure in different geographical areas over time. Future studies based on species' prey preference and lead isotope analysis may further aid our understanding of raptors' lead exposure risk in California.

References

1. BirdLife International. 2020. *Gymnogyps californianus. The IUCN Red List of Threatened Species* 2020: e.T22697636A181151405. Available at: https://dx.doi.org/10.2305/IUCN.UK.2020-3.RLTS.T22697636A181151405.en. Accessed April 5, 2022.

2. QGIS Association. QGIS Geographic Information System. Available at: http://www.qgis.org. Accessed April 4, 2022.

3. R Core Team (2020). R: A language and environment for statistical computing. Available at: https://www.R-project.org/. Accessed April 4, 2022.

Reference Ranges for Blood Lactate Levels in Bald Eagles, Great Horned Owls, and Red-Tailed Hawks

Dana Franzen-Klein, DVM, MS, and
Kyra Knutson, DVM

Session #2605

Affiliation: From The Raptor Center, the University of Minnesota, 1920 Fitch Ave, St Paul, MN 55108, USA.

Note: Abstract format selected by author(s). See Front Matter for description of conference proceedings manuscripts submitted as scientific abstracts.

Abstract: To determine if blood lactate levels can be used to assess prognosis and response to treatment in avian species, we first need to determine if point of care lactate meters are accurate in various avian species, and what the normal resting blood lactate value is in the species of interest. The primary objective of this research was to assess the utility of a point of care lactate meter (Lactate Plus, Nova® biomedical), and to determine normal reference ranges for three raptor species: bald eagles (*Haliaeetus leucocephalus*), great horned owls (*Bubo virginianus*), and red-tailed hawks (*Buteo jamaicensis*). A small pilot study also was conducted with emaciated raptors to assess the potential utility of blood lactate levels for initial treatment of this condition. The Lactate Plus meter produced consistent results, with the precision testing coefficient of variation measuring below 6%; however, the linearity testing with the Lactate Plus meter produced inconsistent results and is being further assessed. Blood lactate levels were measured in 50 bald eagles, 37 great horned owls, and 41 red-tailed hawks. Preliminary data analysis provided the following mean blood lactate ± one standard deviation values by species: bald eagle 11.0 ± 3.8 mmol/L, great horned owl 7.3 ± 3.2 mmol/L, red-tailed hawk 8.1 ± 2.9 mmol/L. Blood lactate levels in emaciated raptors were lower than the pre-release samples and did not show a consistent trend in response to treatment. This study obtained a good sample size to characterize resting, normal lactate values in three species.

Section 12

Diagnostics 2

Stephanie Lamb, DVM, DABVP (Avian Practice);
Hannah Attarian, DVM
Session Chairs

Blood Gas Analysis Under the Conditions of Using Emergency Protocol Air Sac Perfusion Anesthesia (APA)

Ruediger T. Korbel, Prof Dr med vet habil, DipECZM (Avian), and
Kari O. Lutro, DVM

Session #2651

Affiliation: From Center for Clinical Veterinary Medicine, Clinic for Birds, Small Mammals, Reptiles and Ornamental Fish, LMU University of Munich, Oberschleissheim, Germany (Korbel), Klinikken for Fugler og Eksotiske Dyr & Fornebu Dyrenklinikk AS, Oslo, Norway (Lutro).

Air sac perfusion (AP) and air sac perfusion anesthesia (APA), a technique established between 1994 and 1996,[1] in principle reflects a unidirectional carrier gas and anesthetic gas perfusion of the avian respiratory tract via an air sac cannula preferentially within the left caudal thoracic air sac. It represents a technique that has been established as a long-term anesthesia for the specific needs of ophthalmologic (especially intraocular) and head area surgery and as an emergency ventilation protocol in case of avian asphyxia and apnoea.

Major benefits of AP and APA include reversible apnea (resulting from a so called CO_2-wash out effect, lowered CO_2 partial pressure and thus lack of stimulation of the respiratory center) with complete absence of body movements, free surgical access to the head area, lowering of intraocular pressure (IOP) beneficial for intraocular microsurgery, outstanding anesthesia and flow rate control for ventilation beneficial for long term diagnostic and surgical procedures (such as MRT, CT, intraocular surgery), minimization of so called anesthesia waste gases due to the reduced carrier gas flow rates needed to protect surgical staff (at factor 10 compared to conventional intratracheal application).

While practical application of AP and APA and handling, especially in emergency situations with asphyxia due to isolated syrinx mycosis, intratracheal foreign bodies, intratracheal hemorrhage etc. may be easy to perform, underlying aerodynamic and physiologic principles are complex. Inadequate carrier gas flows will consequently result in continuous respiratory movements interfering with the needs of (micro-) surgical procedures and hypoxia (i.e., too low carrier gas flow rates) or life-threatening situations including hypocapnia with respiratory alkalosis caused by an increased CO_2-wash out effect followed by cardiac arrhythmia and finally cardiac arrest.

To determine appropriate air sac perfusion (oxygen) flow rates and to assure physiologic blood gas and cardiovascular values, blood gas analysis and pulse oximetry values—and their effects on the cardiovascular system, including blood pH and blood pressure values using a comparative model with racing pigeons (*Columba livia*, n = 24) and common buzzards (*Buteo buteo*, n = 12) at both 0.3 and 0.4 L/min/kg BW carrier gas flow rates using oxygen were investigated.

Results showed that constant carrier gas flow at 0.3 L/min/kg BW at an isoflurane concentration of 2.7 ± 0.2 vol% in racing pigeons and 1.7 ± 0.3 vol % in common buzzards results in a reliable surgical depth of anesthesia status with no hypocapnia, almost physiologic blood pH values and no cardiac arrhythmias. In contrast, even slightly increased air sac perfusion flow rates at 0.4 L/min/kg BW resulted in decreased CO_2 partial pressure ($paCO_2$) within the expired gases and consecutively, hypocapnia. At t = 40 mins after inducing the APA, n = 4 common buzzards started to show temporary cardiac arrhythmias which were presumably related to the hypocapnia.

Blood pH values and blood gas analysis values gained are not published so far, but are indispensable for choosing appropriate carrier gas flow rates within use of AP in emergency situations of avian asphyxia/apnea situations and APA for ophthalmologic and head area surgery. Typical indications using AP and APA will be demonstrated in presenting practical cases including asphyxia in psittacine birds and MRT investigation protocols.

References

1. Korbel R. Augenheilkunde bei Vögeln: Augenkrankheiten, Luftsack-Perfusionsanästhesie und ophthalmologische Photographie. Vet Habil. University of Munich/Germany; 1996.

Comparison of Lactate Measurements in Hispaniolan Amazon Parrots (*Amazona ventralis*) via Three Analyzers

Alexandra K. Mason, DVM, MS, CertAqV,
Jordan Padlo, CVT,
Mark A. Mitchell, DVM, MS, PhD, DipECZM (Herpetology),
J. Jill Heatley, DVM, MS, DABVP (Avian Practice), Dipl ACZM,
Thomas N. Tully, Jr., DVM, MS, DABVP (Avian Practice), DipECZM (Avian)

Session #2649

Affiliation: From Louisiana State University, School of Veterinary Medicine, Department of Veterinary Clinical Sciences, Skip Bertman Drive, Baton Rouge, LA 70803, USA (Mason, Padlo, Mitchell, Tully), Texas A&M, College of Veterinary Medicine & Biomedical Sciences, Veterinary & Biomedical Education Complex, 660 Raymond Stotzer Pkwy, College Station, TX 77843, USA (Heatley).

Abstract: Lactate is an important biochemistry analyte utilized in human and veterinary medicine to determine tissue perfusion and can be used as a prognostic indicator for certain disease conditions. While lactate is commonly measured using patient-side handheld monitors, these monitors have not been validated for companion avian species. The purpose of this study was to determine the level of agreement between two commercially available point of care lactate meters and a laboratory benchtop blood analyzer in Hispanolian Amazon parrots (*Amazona ventralis*). Our hypothesis was that good agreement would exist between the assays. Blood samples were collected from twenty adult parrots at Louisiana State University by drawing 1.5 ml of blood, placing one drop of whole blood on the Lactate Plus analyzer (Nova Biomedical, Waltham, MA, USA), and the remainder transferred into a lithium heparin microtainer. From the blood in the microtainer, 0.2 ml whole blood was introduced onto the EPOC Blood Analysis System (Siemens Medical Solutions USA, Inc., Malvern, PA, USA), while the remaining sample was centrifuged to obtain plasma that was immediately frozen and submitted to the Texas A&M clinical pathology laboratory for analysis. Bland Altman agreement plots measured the level of agreement between the lactate meters. Results indicate that resting lactate concentrations in the Amazon parrots had poor agreement with mean % differences between all devices being > 22% (EPOC and stat strip: 33.6% [95% CI: 27-40], EPOC and Texas Lab: 55% [95% CI:52-58], Texas Lab and stat strip: 22% [95% CI:16-28]). More research is needed to determine the value of lactate testing in psittacine birds.

A Comparison of Amazon Parrot (*Amazona ventralis*) Blood Gas Values on the i-STAT and EPOC

Jena Dicharry, BS,
Shannon Dehghanpir, DVM, MS, DAVCP,
Mark A. Mitchell, DVM, MS, PhD, DipECZM (Herpetology),
Thomas N. Tully, Jr., DVM, MS, DABVP (Avian Practice), DipECZM (Avian)

Session #2591

Affiliation: From the Department of Veterinary Clinical Sciences, School of Veterinary Medicine, Louisiana State University, Baton Rouge LA 70803, USA.

Note: Abstract format selected by author(s). See Front Matter for description of conference proceedings manuscripts submitted as scientific abstracts.

Abstract: This study was designed to establish blood-gas reference intervals for Hispaniolan Amazon parrots (*Amazona ventralis*) on the EPOC (Siemens Healthineers, Ottawa, ON, Canada) and i-STAT (Abbott Labs, Abbott Park, IL, USA) and to determine the agreement between analyzers. Twenty-six healthy parrots were used for this study. Exclusion criteria was a history of recent illness or abnormalities on physical exam. Jugular venipuncture was performed using manual restraint and blood samples were collected via syringe without additives, for immediate analysis. Data for venous partial pressure of oxygen (pO_2), partial pressure of carbon dioxide (pCO_2), pH, and bicarbonate were collected. Except for the EPOC data for pO_2, all analytes demonstrated Gaussian distribution. Reference intervals were established in accordance with the American Society for Veterinary Clinical Pathology for pCO_2, pH and bicarbonate on both devices.[1] Intervals were not established for pO_2 due to result variation and limited clinical application. Passing-Bablok regression and Bland-Altman plots were used to evaluate correlation and agreement between analyzers. All analytes were highly correlated ($r > 0.9$). Limits of agreement were set using Clinical Laboratory Improvement Amendments (CLIA) guidelines.[2] Agreement was good if all variances were within the proposed limits, agreement was fair if 80% of variances were within the limits, and poor if less than 80% of variances were within the limits. The data for pO_2 were in good agreement and pCO_2 was in fair agreement. The data for pH were in poor agreement. CLIA did not have established guidelines for bicarbonate; however, all variances were less than 15%.

References

1. Friedrichs KR, Harr KE, Freeman KP, et al. ASCVP reference interval guidelines: determination of de novo reference intervals in veterinary species and other related topics. *Vet Clin Pathol.* 2012;41(4):441-453.

2. Department of Health and Human Services. Clinical Laboratory Improvement Amendments of 1988 (CLIA) Proficiency Testing Regulations Related to Analytes and Acceptable Performance. Federal Register. 2019;84:1536-1566.

Section 13

Potpourri

Anna Osofsky, DVM, DABVP (Avian Practice)
Session Chair

Low Stress Parrot Examinations: The Silver Standard

Marcie Logsdon, DVM

Session #2676

Affiliation: From Washington State University Veterinary Teaching Hospital, 100 Ott Rd, Pullman, WA 99164, USA.

Abstract: Parrot training to decrease stress during handling and examinations has become the gold standard for avian practitioners and owners; however, sometimes the gold standard is not always achievable. In those situations, there are still a number of stress reduction techniques that should not be overlooked. From popcorn to tongue depressors, this lecture will discuss a variety of stress-reduction handling tips and techniques that every practice can use for examinations and medication administration, no matter the experience level of the handler or owner.

Introduction

Decreasing or eliminating the amount of stress a patient experiences in the exam room is beneficial not only for the patient, but also for the owner and veterinary staff. Clients are less likely to bring their bird back if they had a negative experience in the exam room, and veterinary team members are more likely to injure, or be injured by, a stressed parrot. Parrots have also been shown to exhibit a stress leukogram in response to handling and restraint.[1] While it can be difficult, or in some cases impossible to completely eliminate the stress experienced by a parrot in a clinical setting, every little bit of stress reduction counts. Below are a few examples of low input strategies to reduce stress during parrot exams and treatments.

Preparation is Key

Have everything ready first. This seems very straight forward but should never be overlooked. You want to have the bird in hand for as little time as possible. Things should be smooth and swift, but not fast or rushed. Have everything you might need readily available and within easy reach before restraining your bird.

Limit Number of Restraint Times

The fewer times a patient has to be restrained the better. Sometimes this may mean delaying the initial hands-on physical exam. If a patient presents for a condition that you know you will want further diagnostics for, it is ok to hold off on doing a physical exam in the exam room. The trick with this is clear communication with the owner. You want to make sure that they understand: (1) exactly why you are doing this (to limit the amount of stress on their pet) and (2) that their pet will receive a full physical exam before you perform the diagnostics and/or any anesthesia and that if you find anything at all that changes your mind, you will talk with them again before proceeding. Remember, what happens in the treatment area of any clinic is a complete black box to owners, and you don't want them filling up those blanks with their imagination. Similarly, take advantage of anytime the patient is anesthetized. If you anesthetize for radiographs, get owner permission to draw blood at the same time, even if you are unsure as to whether or not you will need to run it.

Tongue Depressors

Give your patients something to do during their exam. For medium to large size parrots, tongue depressors are perfectly size destruction-distractions. For most birds, chewing on a tongue depressor gives them something to take their mind off of the exam. You may need to have them pause for a minute during auscultation, but for the rest the exam, chew away! Just be careful to ensure that no one is hit by tiny wooden splinters and that the splinters are falling out of the bird's mouth, not into it. Fortunately, most tongue depressors are made of birch wood, which is non-toxic to birds.

Popcorn

This is the best medication delivery device since the invention of the syringe. There are wonderful methods available for positive-reinforcement training to get a parrot to willingly take medications from a syringe and far and away this is the ideal.[2,3] But some owners are simply not up to that level of training and time commitment. Providing the medication on a treat can be a great alternative to restraining and hand medicating a parrot on a daily or twice daily basis. The wonderful thing about popcorn is that it is cheap, readily available, easily soaks up medications, and does not crumble or fall apart the way that crackers or bread-type treats do. Of course, this only works for small to moderate volumes of medications that taste good, and you have to make sure that if the patient likes to dunk its food that it is given the popcorn well away from the water bowl. This trick has made the difference between treatment or not in many older patients that otherwise would not be receiving their daily arthritis medication.

The Weigh In

For patients that don't step up for new people or won't stand on a perch in an exam room, the easiest way to get a weight during the exam is by zeroing the entire travel crate on a scale and then reading the bird's weight in the negative once it is out of the crate. (If your scale doesn't read into the negative, get a new one.) If you miss this opportunity, you have a second chance when you go to put the bird back into its crate after the exam is finished.

Handle the Stressed Bird First

It's time for the annual exam on the Smiths' three birds. You see them once a year and once a year only. The owners mean well enough but don't do anything to desensitize them to their annual visit. As usual, two of the three birds are sitting in their travel enclosures relatively calmly while one is climbing circles in its crate getting more and more worked up. This is the bird that you want to examine first. It is amazing how often the nervous bird will calm down after it is done with its exam, especially if it has been there before and knows that after its exam, it is done. If you examine the other two birds first, the nervous one just has more time to get worked up.

Initial Restraint

Acclimating a parrot to being restrained in a towel is the gold standard and is something that can be accomplished by owners at home or with careful desensitization in the clinic; however, not having time to perform this type of desensitization is not an excuse to ignore other little things that can be done to decrease the stress of toweling.

If the patient must be grabbed directly out of the crate, move slowly and steadily. Do not pounce on the bird. Similarly, the age-old lights out trick (where you move into position, then quickly turn the lights off to prevent the bird from moving away while grabbing it) should only be used when the alternative is going to be more stressful (e.g., a flighted, non-hand tame budgerigar in a cage with multiple others).

References

1. McRee AE, Tully Jr TN, Nevarez JG, et al. Effect of routine handling and transportation on blood leukocyte concentrations and plasma corticosterone in captive Hispaniolan Amazon parrots (*Amazona ventralis*). *J Zoo Wildl Med.* 2018;49(2):396-403.

2. Heidenreich B. Train your parrot to accept oral medication tutorial. Available at: https://www.youtube.com/watch?v=FwtHWbnZexI. Accessed May 22, 2022.

3. McLaughlin A, Heidenreich B. Giving medication to birds without fear and stress-for birds or Humans! Fear Free Happy Homes. Available at: https://www.fearfreehappyhomes.com/giving-medication-to-birds-without-fear-and-stress-for-birds-or-humans/. Accessed April 18, 2022.

Standardized Sampling Technique for Blood Glucose Analysis in Apparently Healthy Chickens

Sarah R. Camlic, DVM, and
Brynn V. McCleery, DVM, DABVP (Avian Practice)

Session #2652

Affiliation: From Red Bank Veterinary Hospital, Tinton Falls, NJ 07724, USA.

Note: Abstract format selected by author(s). See Front Matter for description of conference proceedings manuscripts submitted as scientific abstracts.

Abstract: Point-of-care (POC) glucometers have been investigated in avian species including domestic pigeons, Hispaniolan amazons, and several wildlife species.[1-6] Certain glucometers have also been used to evaluate serum and plasma samples in domestic chickens, and there has been a study on the effect of refrigerated storage on chicken serum and plasma glucose.[5,6] Investigating short-term changes in whole blood glucose at room temperature using POC and countertop analyzer has clinical interest for many settings, especially for mobile veterinarians or clinics with in-house labs. This study evaluated for changes to blood glucose over time in whole blood from domestic chickens (*Gallus domesticus*), as well as sought to validate or invalidate the AlphaTRAK glucometer for use in this species. Blood was sampled from 12 apparently healthy hens of various breeds and immediately divided into aliquots in sodium heparin tubes. The first sample was immediately run as whole blood on an AlphaTRAK glucometer, and the remainder spun down and serum glucose evaluated using an Idexx Catalyst within 60 seconds. The remaining aliquots were kept at room temperature and blood glucose measurements were repeated on each sample at 5, 15, 30, and 60 minutes. Vital parameters at time of draw, packed cell volume/total solids, and serum character were also noted. The AlphaTRAK overestimated blood glucose values compared to the Catalyst. There was no significant decrease in glucose readings over time for either the AlphaTRAK or the Catalyst.

Acknowledgments: Funding was provided by a grant from Zoetis. The authors thank Kat Fairlie for her assistance in sample analysis.

References

1. Acierno MJ, Mitchell MA, Schuster PJ, et al. Evaluation of the agreement among three handheld blood glucose meters and a laboratory blood analyzer for measurement of blood glucose concentration in Hispaniolan Amazon parrots (*Amazona ventralis*). *Am J Vet Res.* 2009;70(2):172-175.

2. Acierno MJ, Schnellbacher R, Tully TN Jr. Measuring the level of agreement between a veterinary and a human point-of-care glucometer and a laboratory blood analyzer in Hispaniolan Amazon parrots (*Amazona ventralis*). *J Avian Med Surg.* 2012;26(4):221-224.

3. Lieske CL, Ziccardi MH, Mazet JA, Newman SH, Gardner IA. Evaluation of 4 handheld blood glucose monitors for use in seabird rehabilitation. *J Avian Med Surg.* 2002;16(4):277-285.

4. Mohsenzadeh MS, Zaeemi M, Razmyar J, Azizzadeh M. Comparison of a point-of-care glucometer and a laboratory autoanalyser for measurement of blood glucose concentrations in domestic pigeons (*Columba livia domestica*). *J Avian Med Surg*. 2015; 29:181-186.

5. Okorie-kanu CO, Igbokwe EC, Nwagbara, ND, Egeonu, MO. Evaluation of point-of-care glucometers for blood glucose determination in layer chickens. *J Sustain Sci Technol*. 2018;2:244-250.

6. Kawasaki T, Iwasaki T, Ohya I, Hasegawa Y, Noguchi M, Watanabe T. Effects of sampling and storage method on chicken blood glucose measurement. *Poult Sci J*. 2020;57(3):241-245.

Skin Microbiota of Monk Parakeets (*Myiopsitta monachus*) with Normal Feathering or Feather Loss via Next-Generation Sequencing Technology

Janina A. Krumbeck, PhD,
Debra D. Turner, MS,
Alison Diesel, Dipl ACVD,
Aline Rodrigues Hoffman, DMV, PhD, DACVP,
J. Jill Heatley, DVM, MS, DABVP (Avian Practice), DABVP (Reptile and Amphibian Practice) Dipl ACZM

Session #2517

Affiliation: From MiDOG LLC, Irvine, CA 92614, USA (Krumbeck), Schubot Center for Avian Health, College of Veterinary Medicine & Biomedical Sciences, Texas A&M University, College Station, TX 77843, USA (Turner), Department of Small Animal Clinical Sciences, College of Veterinary Medicine & Biomedical Sciences, Texas A&M University, College Station, TX 77843, USA (Diesel, Heatley), Department of Pathobiology, College of Veterinary Medicine & Biomedical Sciences, Texas A&M University, College Station, TX 77843, USA (Hoffman).

Note: Abstract format selected by author(s). See Front Matter for description of conference proceedings manuscripts submitted as scientific abstracts.

Abstract: Feather damaging behavior (feather picking) is a common malady of captive parrots. Diagnosis is challenging as a primary disadvantage to clinicians faced with diagnosis of parrot skin disease remains the lack of normative data for the healthy parrot epidermal microbiology. The objective of this pilot study was to establish bacterial and fungal cutaneous microbiota baselines for monk parakeets with and without feather loss. In healthy monk parakeets (MP; *Myiopsitta monachus* n = 12) with intact feathers (IF-MP), and to compare the IF-MP group to 12 MP with evidence of feather loss (FL-MP) via next-generation DNA sequencing. Of 351 bacterial and 97 fungal species identified, no statistical differences occurred in bacteria or fungal alpha- or beta-diversity based on age, sex, or feather loss; however, enclosure affected beta-diversity. The most abundant bacterial species for IF-MP were *Exiguobacterium indicum*, *Sphingomonas yabuuchiae*, and *Corynebacterium* spp., and *Fragaria/Populus euphratica*, *S. yabuuchiae*, and *E. indicum* for FL-MP group. *Lactobacillales*, *Acinetobacter*, *Kocuria* sp., and *Streptococcus epidermidis* were significantly enriched in IF-MP, while *Streptococcaceae*, *Methylobacterium*, *Clostridiales*, *Nocardioidaceae*, *Nesterenkonia*, and *Actinomycetospora* were significantly enriched in FL-MP. Both groups had *Cladosporium halotolerans/sphaerospermum* as the predominant fungus, while *Capnodiales* was significantly enriched in FL-MP. Use of next generation microbial sequencing as a preliminary diagnostic tool of bacterial and fungal communities of captive birds allows veterinarians to confirm cytologic diagnoses and circumvent the limitations of culture testing. Diagnostic implications include the potential transmission of pathogenic bacteria and/or fungi and pathogen resistance between cohoused birds.

Gap in pagination due to formatting issues.

Pages 167-442

Section 35

Hybrid: Mixed Topics

Kimberly Boykin, DVM, MS;
Caitlin Cunningham, DVM;
Lauren Gawel, DVM;
Kaitlyn Nikirk, DVM Candidate (Class of 2024)
Session Chairs

On-Call Exotic Emergency Services: One-Year Retrospective Caseload Analysis

Anna-Marie Ruoff, DVM,
Elizabeth Hyde, DVM,
Jay Johnson, DVM

Session #2534

Affiliation: From Arizona Exotic Animal Hospital, Mesa, AZ 85201, USA (Ruoff, Hyde, Johnson), The Center for Bird and Exotic Animal Medicine, Bothell, WA 98011, USA (Hyde).

Abstract: Exotic animal owners want options for emergency care for their pets, but few 24-hour small animal emergency hospitals are willing or able to handle exotic animal emergencies. We will discuss caseload and outcome data from the first year of an on-call, after-hours, exotic pet emergency service recently established by a high-volume, exotics-exclusive hospital in a large metropolitan area. 624 after-hours calls were fielded over 361 nights, averaging 1.73 calls per night (0-11 calls). Established clients accounted for 28.2% of calls, non-clients 64.1%, unknown client status 6.1%, and consults from other veterinarians 1.6%. Calls referencing small companion mammal patients encompassed 51.4% of calls, birds 18.9%, reptiles 14.9%, large mammals 4.3%, native wildlife 1.6%, amphibians 0.3%, fish 0.3%, and unknown species 8.2%. An after-hours emergency appointment was recommended by the on-call doctor for 59.0% of calls; of those, 34.8% came in for the appointment, 63.3% declined the appointment, and 1.9% were deceased on arrival. There was no significant difference between established clients and non-clients for compliance in making an after-hours appointment when one was recommended (42.1% and 32.9%, respectively). This information may guide other exotic animal practitioners to evaluate potential caseload, outcomes, and viability of establishing their own after-hours emergency services.

Introduction

Arizona Exotic Animal Hospital (AZEAH) is an exotic animal exclusive hospital with offices in Mesa, Phoenix, and Tucson, AZ. In October 2019, AZEAH established an after-hours on-call emergency service for its clients in the greater Phoenix metropolitan area. People who called the hospital after business hours were given the option to contact an on-call doctor if their pet was experiencing a medical emergency. These self-selected phone calls were forwarded through the Jive mobile VoIP application to the on-call doctor's cellular phone. Based on the signalment, presenting complaint, medical history, and/or client request, the on-call doctor could recommend and arrange for an after-hours emergency appointment to deliver care. Data from the phone calls were collected to better evaluate the needs of clients and the hospital. Data collected included date, species of the pet, presenting complaint and history, whether the owner was an established client of the hospital, the on-call doctor's recommendations, and whether the call resulted in an after-hours emergency appointment.

Caseload Analysis

In its first year, the after-hours emergency service received 624 calls over 361 nights. Data were not recorded for three nights (0.82% of 364 nights) due to recorder error or application malfunction; these were excluded from analysis. The average call volume was 1.73 calls per night, ranging from 0 calls to 11 calls. No calls were received on 109 nights (30.2%).

At the initiation of the service, business hours ended at 1800 hours (6:00 pm). After 7 months of the on-call service, the data were internally evaluated. The business hours of the hospital changed to accommodate the heavy call load immediately after closing; for the remaining 133 nights of analysis, business hours ended at 2200 hours (10:00 pm).

There were 481 calls during the 231 nights before the business hour change. Data were not recorded for three nights (1.3% of 231) due to recorder error or application malfunction; these were excluded from further analysis. The average call volume was 2.11 calls per night. No calls were received on 49 nights (21.5%).

There were 143 calls during the 133 nights after the business hour change. The average call volume was 1.08 calls per night. No calls were received on 60 nights (45.1%). The number of nights without calls differed significantly before and after the business hour change, X^2 (1, n = 361) = 22.24, $p < 0.05$.

Of the 624 total calls, 176 were from established clients (28.2%), 400 were from non-clients (64.1%), 10 were consults from other veterinarians (1.6%), and 38 were unknown status (6.1%, typically due to an unreturned voicemail message). Before the hours change, 140 were from established clients (29.1%), 315 were from non-clients (65.5%), 7 were consults from other veterinarians (1.5%), and 19 were unknown status (4.0%). After the hours change, 36 were from established clients (25.2%), 85 were from non-clients (59.4%), 3 were consults from other veterinarians (2.1%), and 19 were unknown status (13.2%). The clientele utilizing the service differed significantly before and after the business hour change, X^2 (3, n = 624) = 17.30, $p < 0.05$.

Clients were asked what species their pet was during most phone calls; however, the level of client specificity varied (e.g., jenday conure versus bird). The authors chose to categorize species data to allow for descriptive analysis. The categories were small companion mammals (including rabbits, guinea pigs, rats, hedgehogs, ferrets, sugar gliders, etc.), large mammals (including miniature pigs, small ruminants, camelids, macropods, etc.), birds, reptiles, amphibians, fish, native wildlife (which were excluded from previous categories), and unknown species (typically from unreturned voicemail messages). Calls referencing small companion mammal patients encompassed 51.4% of calls, birds 18.9%, reptiles 14.9%, large mammals 4.3%, native wildlife 1.6%, amphibians 0.3%, fish 0.3%, and unknown species 8.2%.

The resolution of each call was recorded by the on-call doctor who received it. These resolutions were then categorized for descriptive analysis. The on-call doctor recommended an after-hours emergency appointment for 368 calls (59.0%). Recommendations to monitor the pet's condition and/or make a daytime appointment were made during 167 calls (26.8%). For clinically unstable patients such as those in acute respiratory distress, the on-call doctor advised the caller to take the patient to the nearest available emergency facility (21 calls, 3.4%). Occasionally, the owner called while their pet was being treated at another veterinary facility or had already elected to go to another veterinary facility (19 calls, 3.0%). During 12 calls (1.9%), the pet had already died; condolences, necropsy, and cremation services were offered. Calls about native wildlife were referred to a local wildlife rehabilitation organization (10 calls, 1.6%). Veterinarians from other hospitals made 10 calls (1.6%) for consultation about exotic animal patients. Pets too large to transport to the hospital for care were advised to contact large animal mobile veterinarians as emergency house calls were not offered by this service (6 calls, 1.0%). Five callers (0.8%) were calling from out of state and recommended to consult a local veterinarian. Calls about laboratory results, hospital hours, or other non-patient-related questions were requested to call back during regular business hours (6 calls, 1.0%).

Of the 368 calls for which an after-hours emergency appointment was recommended, 128 resulted in an appointment (34.8%), 233 declined the appointment (63.3%), and 7 were deceased on arrival (1.9%).

Established clients were advised to come in for an after-hours emergency appointment for 102 of 176 calls (58.0%). Of those, 43 came in for an appointment (42.1%), 56 declined an appointment (54.9%), and 3 pets were deceased on arrival (2.9%). Non-clients were advised to come in for an after-hours emergency appointment for 258 of 400 calls (64.5%). Of those, 85 came in for an appointment (32.9%), 169 declined an appointment (65.5%), and 4 pets were deceased on arrival (1.6%). There were 38 calls from people of unknown client status, but these were typically short calls without much patient information, blocked numbers that did not leave identifying information for a return call, or a caller not answering or returning the doctor's call(s). An after-hours emergency visit was recommended in 8 (21%) of these cases, typically based on the severity of the presenting complaint, but no visits occurred (0%).

The authors hypothesized that established clients would be more likely to heed the on-call doctor's recommendation for an after-hours emergency appointment than non-clients; however, there was no significant difference between established clients and non-clients for compliance with recommended after-hours emergency appointments, X^2 (1, n = 353) = 3.06, p = 0.08. Phone calls with unknown client status were excluded from this analysis.

Conclusions

During the first year of the after-hours on-call service, 624 phone calls resulted in 128 after-hours emergency appointments. The service benefited both established clients and non-clients. Established clients utilized the service frequently and were able to get advice about or care for their pet during extended business hours or after-hours emergency appointments. Each phone call with a non-client was a point of contact for a potential new client, including if they came in for an after-hours emergency appointment or chose to schedule a daytime appointment. Other benefits to the practice included maintaining relationships with local veterinarians by offering after-hours consultations for exotic pet emergencies.

The service has progressed beyond its first year with continued success. In addition to the business hours change previously mentioned, other changes to staffing, on-call doctor scheduling, after-hours call record-keeping, and on-call software were made based on call data analysis and on-call doctor feedback. Further analysis, including continued call data collection and financial analysis will continue to guide business decisions regarding the service. The authors hope that these data can also serve to guide the creation of after-hours on-call emergency services at other veterinary facilities to better serve exotic pets and their owners.

Acknowledgments: The authors acknowledge the veterinarians, veterinary technicians, and paraprofessional staff of Arizona Exotic Animal Hospital for their support through the first year of the on-call emergency service.

Pandemic-Era Exotic Medicine: A Review of Changes in Medical Practices, Infrastructure, and Technology to Accommodate Unprecedented Circumstances

Alicia McLaughlin, DVM, CertAqVet, and
Meera Music, BVM&S, MRCVS

Session #2593

Affiliation: From The Center for Bird and Exotic Animal Medicine, 11401 NE 195th St Bothell, WA 98011, USA.

Abstract: The worldwide SARS-CoV-2 (COVID-19) pandemic that began affecting the United States in early 2020 resulted in major uncertainty and upheaval across society, including in veterinary clinics. The Center for Bird and Exotic Animal Medicine (CBEAM) is an exotics-exclusive hospital in Bothell, WA, that is staffed 24/7 and provides both routine and emergency (ER)/critical care to exotic pets. Since the start of the pandemic, the hospital experienced a drastic (50%) increase in patient numbers and high demand for emergency services, including a tripled average daily emergency caseload. This unexpectedly rapid growth resulted in a variety of necessary changes to patient intake procedures, hospital infrastructure, and technology to accommodate the needs of staff members and the community. The authors hope that sharing their experiences will help other clinics dealing with similar challenges.

Pandemic Medicine

The worldwide SARS-CoV-2 (COVID-19) pandemic that began impacting the United States in early 2020 resulted in major uncertainty and upheaval across society, including in veterinary clinics. To follow recommendations to protect staff and continue to operate, clinics had to rapidly change how they practiced medicine (with protocols changing as often as week-to-week to meet recommendations from the Centers for Disease Control and Prevention). Many practices switched to curbside medicine, where clients were not permitted to enter the veterinary hospital with their pets, to minimize exposure time between staff and clients. This new appointment structure resulted in decreased efficiency, decreased appointment availability, and longer wait times for clients. It also posed an additional challenge to build rapport with new clients due to a lack of face-to-face interaction with clients.

In addition to the significant biosecurity concerns regarding exposure of staff to clients and to other staff members, many issues arose surrounding use of biomedical equipment, including surgical gloves and masks. There were shortages of equipment (e.g., gloves, masks, medications) and mixed public health messages surrounding the benefits of biosecurity measures. In some cases, government mandates impacted how hospitals could operate (e.g., limiting routine surgical procedures, for example).

Staff availability was significantly affected by changes in childcare availability, changes in procedures connected to staff illness (e.g., quarantine measures, etc.), concerns about personal health, and burnout.

These issues were further compounded by a huge boom in new pet ownership.[1] People developed stronger bonds with their pets and noticed changes in behavior and medical conditions that might have been missed in the past due to increased time at home during lockdowns. The limited numbers of veterinarians graduating from college was insufficient to meet this increased demand for care. In addition, veterinarians were prone to the same life challenges everyone else experienced during the pandemic, including vulnerable family members, unexpected deaths, and changing access to childcare.

The decreased access to veterinary services associated with all of these factors was very stressful to both veterinarians and to pet owners. Medical care was impacted for many animals, sometimes to the point of loss of life. There was also a significant increase in inappropriate behavior of clients towards veterinary staff, including both physical and verbal threats and online campaigns against individual clinics. Veterinary staff were overworked and exhausted, increasing the potential for mistakes to occur (either communication or medical), and burnout in the field was rampant. Emergency veterinary clinics were utilized inappropriately for non-urgent medical concerns due to lack of routine appointment availability, leading to even longer wait times and frustration on both the part of staff and clients. Preventative care was delayed for many animals, which led to development of other illnesses.

One Clinic's Journey

The Center for Bird and Exotic Animal Medicine (CBEAM) is an exotics-exclusive hospital in Bothell, WA, that is staffed 24/7 and provides both routine and emergency/critical care to exotic pets. CBEAM is the only exotics-exclusive hospital with round-the-clock emergency (ER) services and staffing in the Pacific Northwest at this time. Since the start of the pandemic, the hospital experienced a drastic (50%) increase in patient numbers and high demand for emergency services, including a tripled average daily emergency caseload. This unexpectedly rapid growth resulted in the hospital leadership team making a variety of necessary changes to patient intake procedures, hospital infrastructure, and technology to accommodate the needs of staff members and the community.

Triage system

One of the hardest changes for CBEAM staff to make was to accept the fact that they could not continue to provide care for every single client that sought veterinary care for their exotic pet. The hospital did not have enough staffing at the start of the pandemic to accommodate the number of emergency patients demanding services on top of hospitalized, critically ill patients, and emergency surgeries, much less routine wellness care. The clinic significantly decreased appointment availability for preventative care to be able to redirect veterinarians as emergency clinicians while actively hiring additional clinicians and support staff to accommodate the extreme growth curve.

Even with this shift in resources, the ER service would routinely be overwhelmed. As a result, the clinic took a multi-pronged approach to ER scheduling.

1. Walk-ins were actively discouraged, though if a walk- in occurred, triage would occur on-site, and wait times would be appropriately estimated.

2. Emergency patient intake numbers were limited to three per hour, permitting 20 minutes per emergency patient for examination, stabilization treatments, and communication with clients to establish a plan for the day.

3. A color-coded triage system was implemented to guide emergency patient scheduling. If the emergency service was overwhelmed, a color code for which level of emergency patient would be accepted that day was initiated.

4. Specific time frames within the schedule were blocked off to prevent further emergencies from arriving during these times to permit care of hospitalized patients and diagnostic workups/procedures of previously admitted patients.

The goal for these policies was to prevent staff burnout and decrease the risk of medical errors while improving patient care for the hospitalized pets.

Space utilization

Boarding services were discontinued to allow staff to focus on caring for sick pets, and this facility space was repurposed. Previously used office space was transitioned into a triage area for ER patients. The leadership team developed new services and restructured others to mitigate the space limitations of the parking lot. The goal was to minimize the amount of time clients were present at the clinic to free up parking space for incoming ER clients. ER patients were dropped off shortly after arrival (assuming the client was reachable by phone). A new urgent care service helped accommodate scheduling of less critical emergencies to prevent the ER service from being overwhelmed. Rechecks of sick patients were scheduled as drop off appointments.

Staffing and scheduling changes

The leadership team increased support staff wages across the board to minimize turnover and address the needs of a rapidly changing economy. Priority was placed on hiring additional veterinarians and support staff as soon as possible to address the needs of the hospital. Members of the team that had not previously served in a leadership capacity were given new responsibilities to help improve cohesiveness of a rapidly growing team. A new training program was developed, since the old method of training staff on-the-job was not feasible with the fast pace of the clinic floor. A new staff shift (swing shift) was added to improve functionality of shift work and to prevent staff from feeling the need to work overtime to meet the needs of patients, further reducing incidence of burnout.

Technology upgrades

New technology was added to the hospital to address the shortcomings of client communication while practicing curbside medicine, including the use of a texting service (AllyDVM), a new interdepartmental communication process (Avimark Whiteboard), automated forms for history taking (Jotforms), and increased phone lines and updated computer systems. A new Client Communicator role was created to facilitate communication between the clients, reception team, and medical team, ultimately improving the client experience.

Client management

A no-tolerance policy for client abuse was instituted.

Summary

There is a new coalition being put together by Penn Vet to address the national crisis in pet care caused by the pandemic.[2] It will be interesting to see what solutions the coalition comes up with, and how veterinary clinics continue to evolve to meet the needs of their communities. The presenters hope that the description of their changing procedures to meet their community's needs will be helpful to other veterinarians facing similar challenges.

Acknowledgments: The authors thank Team CBEAM for their ongoing support, hard work, and dedication to provide the best care possible for exotic pets.

References

1. American Veterinary Medical Association. Pet population still on the rise, with fewer pets per household. Available at: https://www.avma.org/javma-news/2021-12-01/pet-population-still-rise-fewer-pets-household. Accessed March 1, 2022.

2. Columbia Broadcasting System. New Coalition Put Together By Penn Vet Aiming To Improve 'National Crisis' in Pet Care Caused By Pandemic. Available at: https://philadelphia.cbslocal.com/2021/09/23/penn-vet-coalition-pet-care-national-crisis-pandemic/ Accessed March 1, 2022.

The Impact of Dermal Characteristics on Low-Level Laser Power Measurement in Post-Mortem Zoological Species

Faith Burgos, DVM,
Michelle Borsdorf, DVM, Dipl ACZM,
John Ladner, LVT, MLSO,
Anne White, DVM, PhD, CVA,
Tara M. Harrison, DVM, MPVM, Dipl ACZM, Dipl ACVPM, DipECZM (Zoo Health Management), CVA

Session #2585

Affiliation: From North Carolina State University College of Veterinary Medicine, Department of Clinical Sciences, Raleigh, NC 27607, USA, Exotic Species Cancer Research Alliance, North Carolina State University College of Veterinary Medicine, Department of Clinical Sciences, Raleigh, NC 27607, USA (Harrison).

Note: Abstract format selected by author(s). See Front Matter for description of conference proceedings manuscripts submitted as scientific abstracts.

Abstract: The purpose of this study was to evaluate the power of a low-level laser after penetration through dermal tissue (~1 mm thickness) using a variety of post-mortem animal tissue with an ActivetPro low level laser similar to a previous study done in rats.[1] Specifically, this study evaluated the impact of fur, feathers, scales, and different pigments on penetration of the laser light. Frozen and thawed skin tissue samples were placed inside a light restricted box and exposed to a pre-programmed laser level from a Multiradiance ActivetPro photobiomudulation (PMB) device, and measured with a power meter to evaluate the light penetration through the tissue samples. Light penetration measurements were recorded at seven timepoints (range, 1-150 seconds). Notably, integument color appeared to impact power readings. In rabbits (*Oryctolagus cuniculus*), white hair had the highest power readings (14.98 Hz), followed by fawn/grey colored (0.83 Hz), then black colored (-0.05 Hz). Similarly in iguanas (*Iguana iguana*) and bullfrogs (*Lithobates catesbeianus*), lighter colored/pigmented specimens had higher power readings than darker specimens (16.42 Hz, 20.56 Hz respectively). In sea turtles, Kemp's Ridley (*Lepidochelys kempii*) had higher power readings than green sea turtles (*Chelonia mydas*) (36.88 Hz, 23.16 Hz respectively). In hawks (*Buteo jamaicensis*), covert and down feathers on the legs inhibited laser power measurements to minimal to none (3.15 Hz); however, when the feathers were damp, which allowed the skin tissue to have direct contact with the laser, power readings were measured (21.14 Hz). This is the first study, to the authors' knowledge, that investigates the effectiveness of low-level laser penetration through various species' tissues.

References

1. Joensen J, Ovsthus K, Reed RK, et al. Skin penetration time-profiles for continuous 810 nm and superpulsed 904 nm lasers in a rat model. *Photomed Laser Surg.* 2012;30(12):688-694.

2. Archibald KE, Harrison TM, Troan B, et al. Effect of multiradiance low-level laser therapy and topical silver sulfadiazine on healing characteristics of dermal wounds in marine toads (*Rhinella marina*). *Vet Med Int.* 2020;8888328.

3. Shaver SL, Robinson NG, Wright BD, et al. A multimodal approach to management of suspected neuropathic pain in a prairie falcon (*Falco mexicanus*). *J Avian Med Surg.* 2009;23(3):209-213.

4. Hewitt PG. *Conceptual Physics.* 9th Ed. Boston, MA: Addison Wesley; 2001.

Computed Tomography Basics and Artifacts

M. Scott Echols, DVM, DABVP (Avian Practice)

Session #2577

Affiliation: From Echols Veterinary Services, Salt Lake City, UT 84121, USA, Parrish Creek Veterinary Clinic, 86 N 70 W St, Centerville, UT 84014, USA, Medical Center for Birds, 3805 Main St, Oakley, CA 94561, USA.

Note: Abstract format selected by author(s). See Front Matter for description of conference proceedings manuscripts submitted as scientific abstracts.

Abstract: Computed tomography (CT) is becoming commonplace in avian and exotic animal medicine. Over the past 30 years, CT technology has rapidly advanced creating a diverse array of products and capabilities. Simply stated, not all CT scanners are equal or best suited for small avian and exotic animal species commonly encountered in practice. In particular, the newer high-definition volumetric cone beam scanners are providing significantly improved resolution and contrast over the more traditional slice based and older cone beam CT units. Contrast studies can dramatically improve visualization of soft tissue structures and better define normal versus abnormal anatomy not possible with plain CT. Newer iodine-based contrast protocols include higher doses (4-6 ml/kg IV in birds and reptiles and 3-4 ml/kg IV in exotic mammals) and longer administration times (full dose delivered over 2 minutes) compared to previously published protocols (generally consisting of 1-3 ml/kg IV delivered as a rapid bolus). Combined with pre, during and post-contrast CT scans, these new protocols are providing functional and significantly more patient data compared to older protocols. Additionally, CT technology is complex, and artifacts are common which can adversely affect interpretation. Artifacts can be divided into several types including physics based, patient based, scanner based and more. This presentation covers currently available CT technology, newer contrast protocols, and common artifact presentation and resolution.

The Role of the Veterinarian in the Newly Emerging Insects-as-Feed-and-Food Industry

Kimberly L. Boykin, DVM, MS, and
Mark A. Mitchell, DVM, MS, PhD, DipECZM (Herpetology)

Session #2654

Affiliation: From Louisiana State University School of Veterinary Medicine, Skip Bertman Drive, Baton Rouge, LA 70803, USA.

Note: Abstract format selected by author(s). See Front Matter for description of conference proceedings manuscripts submitted as scientific abstracts.

Abstract: Insects require substantially less resources to produce than traditional sources of animal protein and could help to feed an ever-growing human population with less effects on the environment and climate. Over the last decade, the insects-as-feed-and-food industry has rapidly emerged. As it continues to grow, the need for government regulation increases and veterinarians will likely be called upon to assist in assessing animal health, implementing biosecurity measures, and formulating/prescribing treatment protocols similar to what we have seen with the honeybee industry and the institution of veterinary feed directives (VFDs). Veterinarians will need to become knowledgeable regarding the diseases afflicting the feeder insect industry and how best to diagnosis pathogens of concern. Similar to other agricultural markets, high animal densities and management practices put insects at high risk for infectious diseases. In this presentation, we will focus on the cricket industry which has been particularly hard hit from viral diseases, but will also discuss the role that bacterial, fungal, and parasitic diseases could play in overall cricket health as well as the potential for crossover to humans and other animals. If the insects-as-feed-and-food industry is going to be viable, veterinary involvement will be essential to ensure that insects can be used as a safe source of food for all.

Comparison of Next-Generation DNA Sequencing as an Infectious Disease Diagnostic Tool to Traditional Culture and Sensitivity Testing

Janina A. Krumbeck, PhD

Session #2580

Affiliation: From MiDOG LLC, 14762 Bentley Circle, Tustin, CA 92780, USA, Zymo Research Corp., 17062 Murphy Ave, Irvine, CA 92614, USA.

Note: Abstract format selected by author(s). See Front Matter for description of conference proceedings manuscripts submitted as scientific abstracts.

Abstract: Successful infection treatments, require accurate pathogen identification.[1] Culture and sensitivity testing are considered the gold standard for infectious disease diagnosis for both human and pet patients; however, based on DNA/RNA sequencing research, it is estimated that only 1% of the world's microbes are culturable in the laboratory.[2] Especially for exotic patients, culturing of clinical samples has been challenging and diagnosis often relies on PCR testing to overcome culturing limitations. While PCR testing is fast and provides a high degree of sensitivity and specificity, it is limited to the target panel and provides limited clinical value when the panel returns with negative results. Next-generation DNA sequencing (NGS) has the same benefits as PCR testing, but also offers the de novo identification and quantification of all bacteria and fungi present in a sample, including novel pathogen discovery.[3] In this study, clinical samples from 30 exotic pet patients pre-diagnosed with bacterial and/or fungal infections are tested both with an NGS-based diagnostic tool provided by a commercial NGS-testing lab[4-10] and traditional culture and sensitivity testing. The results are compared for the microorganisms detected, cell numbers, and antimicrobial resistance profile(s). No treatment decisions were considered for this study. Preliminary results highlight a large bacterial and fungal species diversity in the study cohort and showcased the lack of sensitivity of culture testing. Specifically, fungal pathogens and anaerobic bacteria were often missed by culture. NGS has outstanding advantages in the diagnosis of bacterial and fungal infections and optimizes the treatment of infectious diseases.

References

1. Koch L. Metagenomics: pathogen diagnostics for the masses. *Nat Rev Genet.* 2016;17(7):378.

2. Barcina I, Arana I. The viable but nonculturable phenotype: a crossroads in the life-cycle of non-differentiating bacteria? *Rev Environ Sci Biotechnol.* 2009;8:254-255.

3. Tang S, Prem A, Tjokrosurjo J, et al. The canine skin and ear microbiome: a comprehensive survey of pathogens implicated in canine skin and ear infections using a novel next-generation-sequencing-based assay. *Vet Microbiol.* 2020;247:108764.

4. Melgarejo T, Oakley BB, Krumbeck JA, et al. Assessment of bacterial and fungal populations in urine from clinically healthy dogs using next-generation sequencing. *J Vet Intern Med.* 2021;35(3):1416-1426.

5. Krumbeck JA, Reiter AM, Pohl JB, et al. Characterization of oral microbiota in cats: novel insights on the potential role of fungi in feline chronic gingivostomatitis. *Pathogens.* 2021;10(7):904.

6. Niemiec BA, Gawor J, Tang S, et al. The bacteriome of the oral cavity in healthy dogs and dogs with periodontal disease. *Am J Vet Res.* 2022;83(1):50-58.

7. Rexo A, Hansen B, Clarsund M, et al. Effect of topical medication on the nasomaxillary skin-fold microbiome in French bulldogs. *Vet Dermatol.* 2022;33(1):10-e5.

8. Tsang W, Linde A, Krumbeck JA, et al. Occurrence of antimicrobial resistance genes in the oral cavity of cats with chronic gingivostomatitis. *Animals.* 2021;11(12):3589.

9. LafeberVet. The importance of next-generation sequencing in avian veterinary medicine. Available at: https://lafeber.com/vet/the-importance-of-next-generation-sequencing-in-avian-veterinary-medicine/. Accessed May 27, 2022.

10. Niemiec BA, Gawor J, Tang S, et al. The mycobiome of the oral cavity in healthy dogs and dogs with periodontal disease. *Am J Vet Res.* 2021;83(1):42-49.

Hematology and Biochemistry Reference Interval Dynamics in Captive Cockatoos and Bearded Dragons

Thanh Le

Session #2648

Affiliation: From Moichor Inc., 555 De Haro St Suite 220, San Francisco, CA 94107, USA.

Abstract: Reference intervals (RIs) present a reliable framework for interpreting hematology and clinical biochemistry results in avian and reptile patients. Specifically, RIs provide evidence-based support for veterinarians to make clinical decisions when comparing patients with the central 95% of reference sample populations; however, while RIs derived from a single, sizable healthy population serve a general interpretation framework, understanding how reference intervals may vary by major physiologic factors including sex, life stages, and seasonality enables more appropriate clinical decisions in captive exotic species. In this study, we examined these variations in hematology and clinical biochemistry RIs using wellness patient populations relegated to cockatoos and bearded dragons.

Introduction

The interpretation of quantitative laboratory results such as hematology and clinical biochemistry relies on establishing clinical reference intervals. Veterinarians can reasonably discriminate between healthy and abnormal states in novel subjects by comparing the patient's clinical values to a range of low and high reference limits derived from the central 95% of a known healthy population.[4]

The construction of RIs for an exotic species using the traditional direct approach follows three major steps. First, a reference population is selected using predefined wellness criteria, then measurements are conducted. Finally, reference intervals are derived using parametric and non-parametric statistical methods.[11] The selected population is often condition-free, but RIs can be established for any well-defined population. Ideally, a minimum of 120 samples are required to create robust clinical RIs. This stringent standard reduces the effects a small number of patients with sub-clinical diseases and outliers may have on RIs.[1,9]

There are unique challenges to provisioning RIs in exotic species. Predefining and selecting a sufficiently healthy population of any unique exotic species is difficult given the diversity in exotics.[4] Moreover, different study conditions such as the time of year or geographic locations may induce biases. In all, while RIs derived from a single population of healthy animals provide a general interpretation framework, different physiologic and lifestyle factors are known to mediate the baseline RIs for many species.[12,13,16,18,19]

This study applied an indirect non-parametric RI framework to examine the dynamics of hematology and clinical biochemistry RIs in two captive exotic populations: cockatoos (family Cacatuidae) and bearded dragons (species *Pogona vitticeps*). The wellness populations were stratified by sex and modeled smoothly through stages of life and time of year. We demonstrated that 1) the indirect RI derivation approach produced similar reference limits previously reported for cockatoos and bearded dragons 2) there may be significant differences in RIs depending on the factors used for population stratification and 3) while the difference in RIs through life stages and seasons can be discretized, the result shows that the change in RIs may be understood as continuous dynamics. Henceforth in this work, clinical variables and analytes were used interchangeably.

Materials and Methods

Sample collection

This study focused on RIs for cockatoos (family Cacatuidae) and bearded dragons (species *Pogona vitticeps*). We utilized an indirect RI approach to overcome the difficulty of sampling condition-free reference exotic populations. The main difference in the indirect RI derivation approach is that the reference intervals are determined a posteriori by statistical methods based on identifying a distribution in the midst of the data. This is contrary to the traditional approach which requires the assessment of individual results belonging to the pre-selected reference population.[11]

Routine wellness blood samples for cockatoos and bearded dragons from Moichor's partner clinics were drawn; hematology and clinical biochemistry results were processed and aggregated. Furthermore, Moichor hematology system and pathology findings by boarded pathologists confirmed the wellness status of the sample. Table 1 presents the distribution of samples by sex.

Table 1. Sample distribution by species and sex.

Animal	n (total)	n (male)	n (female)
Cockatoo	550	238	312
Bearded dragon	251	109	142

Hematology

For each collected sample, a blood smear was created and stained with modified Wright-Giemsa stain and processed using Moichor's Automated Hematology System.[14] The system operates by automatically classifying and quantifying leukocyte subtypes (WBC, Heterophils, Lymphocytes, Monocytes, Basophils, Eosinophils, Azurophils) under 400x magnification; subsequently, absolute leukocyte count and differential were produced and recorded.

Clinical chemistry

Plasma samples for clinical chemistry were collected in plasma separating tubes with light green tops. After the collection of whole blood samples at Moichor's partner clinics, samples were centrifuged to separate plasma from whole blood. A standard biochemistry profile was determined, including the concentration of total protein, albumin, AST, total bile acid, creatinine, uric acid, phosphorus, glucose, calcium, creatine kinase, cholesterol, and total triglycerides. The selected analytes were determined using colorimetric methods as measured by the Beckman Coulter AU680 Chemistry Analyzer (Beckman Coulter Inc, Brea, CA USA).

Statistical analyses

Statistical analyses were performed per individual hematology and biochemistry analytes. Because outliers artificially broaden the calculated reference intervals,[9] the outliers were removed from the aggregate data using the Interquartile Range Rule (IQR Method).[2] The remaining data for each analyte was aggregated by patient's sex; the patient's age was calculated based on the declared birth date, and the time of year or seasonality was determined by the number of days since the first day of the calendar years that a given sample was collected.

The reference interval for each analyte was calculated in aggregate based on how the clinical values are distributed.[3] For most clinical variables, an RI could be determined nonparametrically between 2.5th and 97.5th percentiles of the underlying distribution of values. On the other hand, when an analyte of interest or its transformation follows a Gaussian distribution, the parametric method can be applied, and the reference interval is to be within 1.96 SD around the distribution mean. Conceptually, both nonparametric and parametric methods of RI derivation attempt to determine the central 95% of clinical value distribution for any given analytes. While the nonparametric method is far simpler than estimating the RI based on the observed data and is generally preferred, the parametric method allowed for interrogating the dynamics of RIs in different analytes.

While most investigations into the seasonal or age dynamics in RI discretized the data into measurement groups for ease of experimentation and sufficient sample size.[12,13,16,18,19] it is likely that these changes take place in a more transient manner. Gaussian Process Regression (GPR),[17] a flexible Bayesian nonparametric model, could be used to model and visualize the smooth reference range dynamic for each analyte. For the purpose of examining the dynamics of RIs based on age and time of year, the clinical variables were observed and assumed to be approximately Gaussian within smaller epochs. GPR had several benefits: First, it worked well on limited data. Second, GPR gave the ability to provide uncertainty measures in terms of variance; in the context of RI, this corresponded to the definition of reference interval for Gaussian distributed variables. Third, when indexed by time, GPR is autoregressive, thus enabling forecasting or estimating of reference ranges over sparser sampling locations. In this study, GPR was fit to individual analyte datasets through stages of life and time of year. The kernel used in the regression model is a composite kernel[8]: the sum of Exponentiated Squared Kernel (RBF) and White Kernel.

Results

Static reference intervals are in agreement with previously reported RIs. For each of the two patient populations, static RIs were reported for males, females, and in aggregate. Furthermore, the distribution of each clinical value was also reported. In both captive populations, the static reference ranges were similar for most analytes (Fig 1). Furthermore, these RIs overlap with previously reported reference limits for these species.[7,15]

Static male and female RIs are in agreement. Without examining the RI dynamics, distributional characteristics for clinical values are similar between male and female patient groups for each clinical variable. (Figure 2).

Cockatoo	RI Male	RI Aggregate	RI Female
WBC (/uL)	4945 - 16383	4994 - 16810	5034 - 17327
Heterophil (/uL)	2514 - 11788	2475 - 11530	2445 - 11429
Lymphocyte (/uL)	798 - 5941	878 - 5902	905 - 5888
Monocyte (/uL)	0 - 788	0 - 759	0 - 755
Basophil (/uL)	0 - 538	0 - 535	0 - 517
Eosinophil (/uL)	0 - 389	0 - 422	0 - 437
Heterophil Diff (%)	0.38 - 0.85	0.39 - 0.85	0.39 - 0.84
Lymphocyte Diff (%)	0.11 - 0.54	0.11 - 0.55	0.11 - 0.55
Monocyte Diff (%)	0.00 - 0.09	0.00 - 0.09	0.00 - 0.09
Basophil Diff (%)	0.00 - 0.07	0.00 - 0.06	0.00 - 0.06
Eosinophil Diff (%)	0.00 - 0.04	0.00 - 0.05	0.00 - 0.05
Total Protein (g/dL)	3.0 - 4.2	2.9 - 4.1	2.9 - 4.1
Albumin (g/dL)	1.8 - 2.8	1.8 - 3.0	1.8 - 3.0
Asparatate Aminotransferase (U/L)	99 - 214	98 - 226	97 - 230
Total Bile Acid (umol/L)	9.2 - 68.4	10.3 - 76.5	10.6 - 77.8
Creatinine (mg/dL)	0.20 - 0.42	0.20 - 0.44	0.19 - 0.44
Uric Acid (mg/dL)	2.1 - 10.9	2.2 - 10.7	2.3 - 10.7
Phosphorus (mg/dL)	1.59 - 5.58	1.62 - 5.69	1.69 - 5.81
Glucose (mg/dL)	205 - 311	208 - 309	211 - 305
Calcium (mg/dL)	8.15 - 10.36	8.17 - 10.38	8.20 - 10.44
Creatine Kinase (U/L)	125 - 962	127 - 905	128 - 885
Cholesterol (mg/dL)	176 - 359	177 - 359	180 - 356
Total Triglycerides (mg/dL)	71 - 208	70 - 209	70 - 211

Bearded Dragon	RI Male	RI Aggregate	RI Female
WBC (/uL)	5395 - 27830	6149 - 29095	6353 - 29187
Heterophil (/uL)	1014 - 6388	979 - 9641	924 - 8051
Lymphocyte (/uL)	2300 - 18939	2182 - 18978	2184 - 18952
Monocyte (/uL)	0 - 577	0 - 582	0 - 594
Basophil (/uL)	182 - 1538	169 - 1565	170 - 1583
Eosinophil (/uL)	0 - 245	0 - 287	0 - 285
Azurophil (/uL)	91 - 1622	109 - 1769	132 - 1900
Heterophil Diff (%)	0.08 - 0.52	0.09 - 0.52	0.09 - 0.51
Lymphocyte Diff (%)	0.31 - 0.86	0.31 - 0.85	0.32 - 0.83
Monocyte Diff (%)	0.00 - 0.03	0.00 - 0.05	0.00 - 0.05
Basophil Diff (%)	0.01 - 0.13	0.01 - 0.13	0.01 - 0.13
Eosinophil Diff (%)	0.00 - 0.03	0.00 - 0.03	0.00 - 0.03
Azurophil Diff (%)	0.01 - 0.11	0.01 - 0.12	0.01 - 0.13
Total Protein (g/dL)	3.9 - 6.8	3.7 - 6.8	3.7 - 6.8
Albumin (g/dL)	2.8 - 5.6	2.7 - 5.6	2.7 - 5.7
Asparatate Aminotransferase (U/L)	13 - 64	13 - 80	13 - 88
Total Bile Acid (umol/L)	2.4 - 41.2	1.7 - 10.5	1.5 - 35.0
Uric Acid (mg/dL)	2.1 - 11.3	146.5 - 268.7	1.7 - 9.7
Phosphorus (mg/dL)	2.69 - 8.03	10.79 - 21.86	2.96 - 11.25
Glucose (mg/dL)	153 - 251	142 - 172	143 - 285
Calcium (mg/dL)	10.96 - 23.82	10.79 - 21.86	10.77 - 20.94
Creatine Kinase (U/L)	80 - 4354	70 - 4136	70 - 3262
Cholesterol (mg/dL)	156 - 456	158 - 451	163 - 443
Total Triglycerides (mg/dL)	60 - 520	51 - 527	49 - 521

Figure 1. Hematology and clinical biochemistry RIs calculated based on aggregated wellness cockatoo samples (left) and bearded dragon (right).

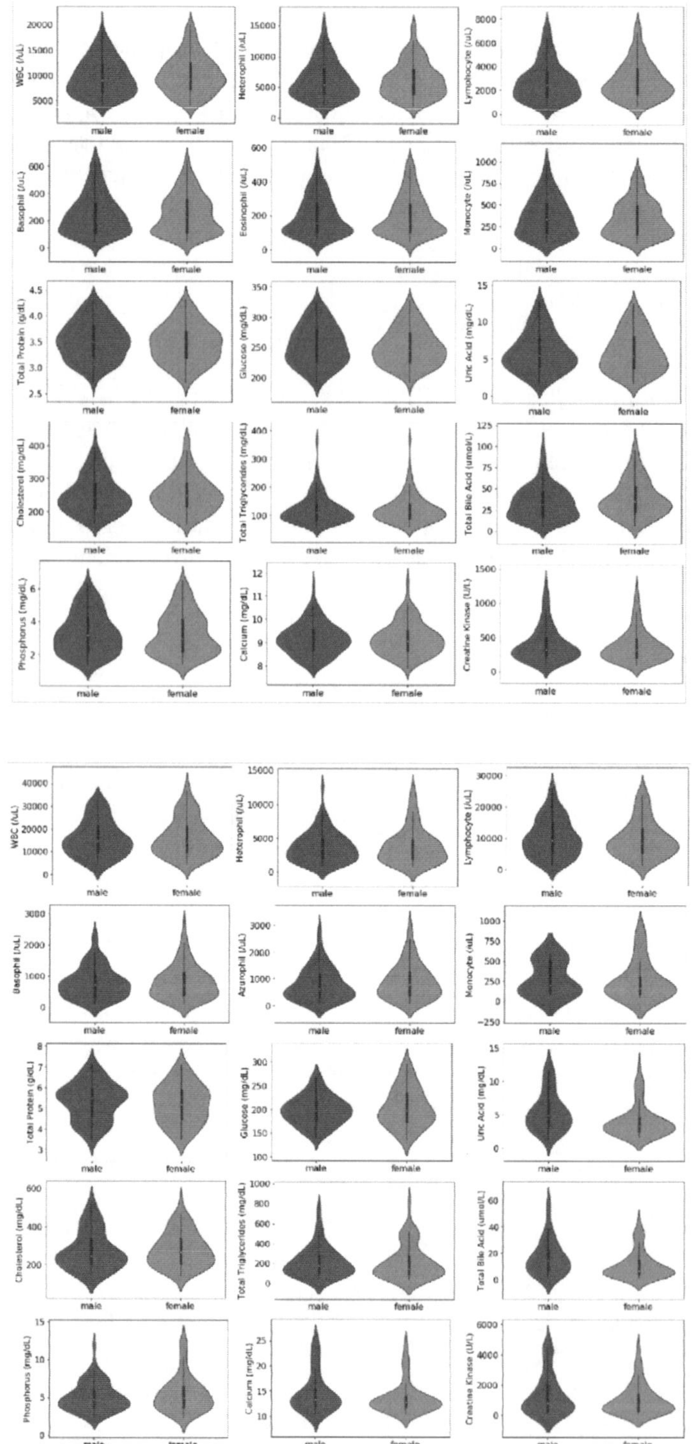

Figure 2. Distribution of hematology and clinical biochemistry values in male vs female wellness cockatoo samples (top panels) and bearded dragons (bottom panels). Each sub-figure represents a unique analyte.

Smooth dynamic RIs better reflected known seasonal trends and events. Like other parrots, captive cockatoos molt in the Spring and Fall during a process that may last 8-12 weeks.[5] During molting periods, higher WBC, heterophils, and lymphocyte counts were observed (Fig 3). This finding was consistent with previous studies on cockatoos and other avian species during their molting periods. This modulation in immune response was thought to be shaped by a trade-off between the availability of resources and the probability of encountering pathogens.[5,6] Moreover, the change in serum total protein level along with glucose and calcium (Fig 4) was also consistent with prior studies. Compared to the discretized reference intervals, the smooth dynamical RIs demonstrated the transient nature of cockatoo physiologic events. While we observed similar seasonal dynamics in bearded dragons (Figs 5 and 6), growing data would continue to highlight major seasonal trends.

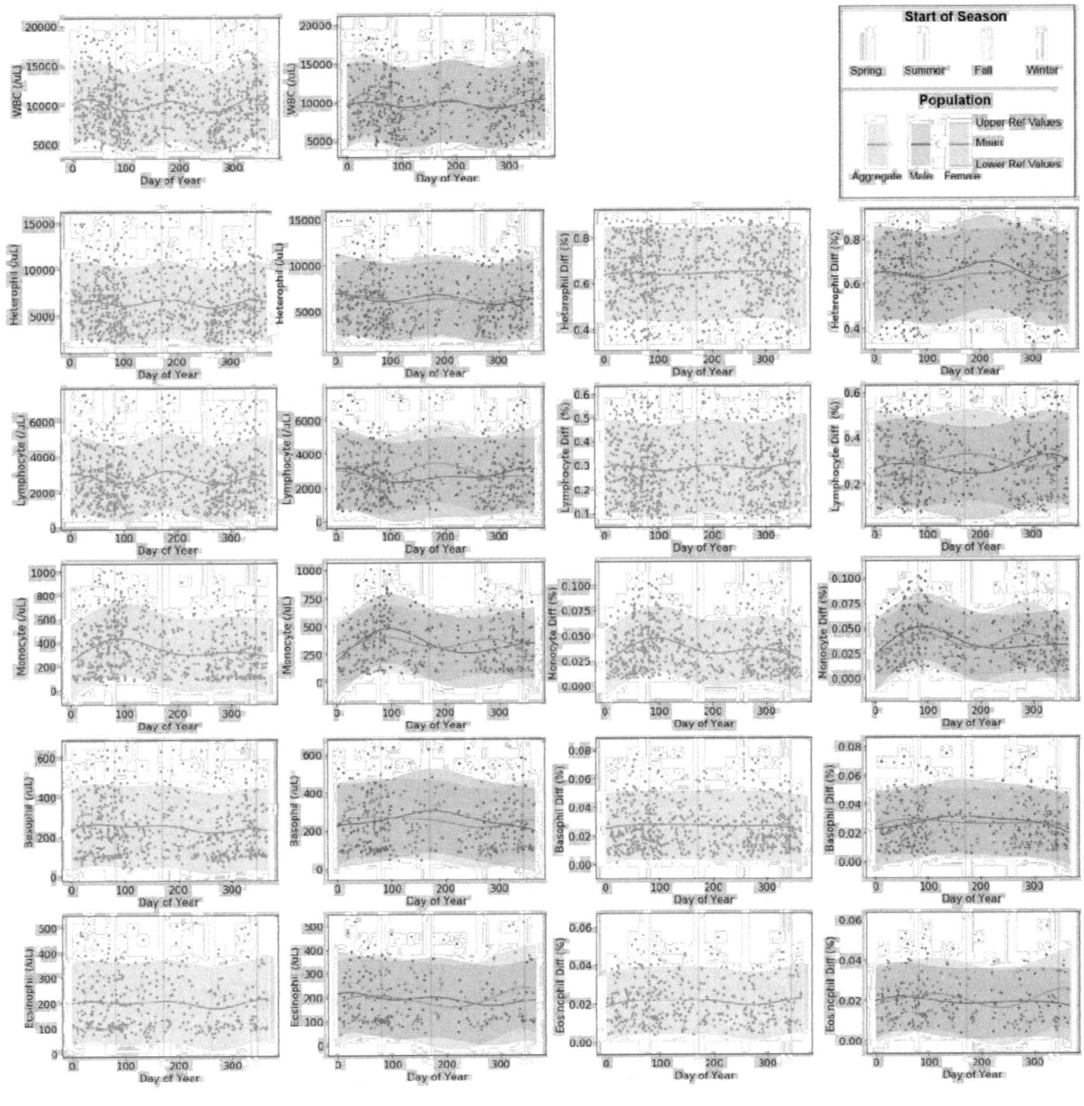

Figure 3. GPR Mean and 95% reference range of hematology variables in cockatoos through the seasons. Each variable consists of two panels representing regression on days since the beginning of the year using aggregate data (left) and data stratified by sex (right). Vertical lines in each sub-figure represent the beginning of each meteorologic season in the United States: Spring, Summer, Fall, and Winter.

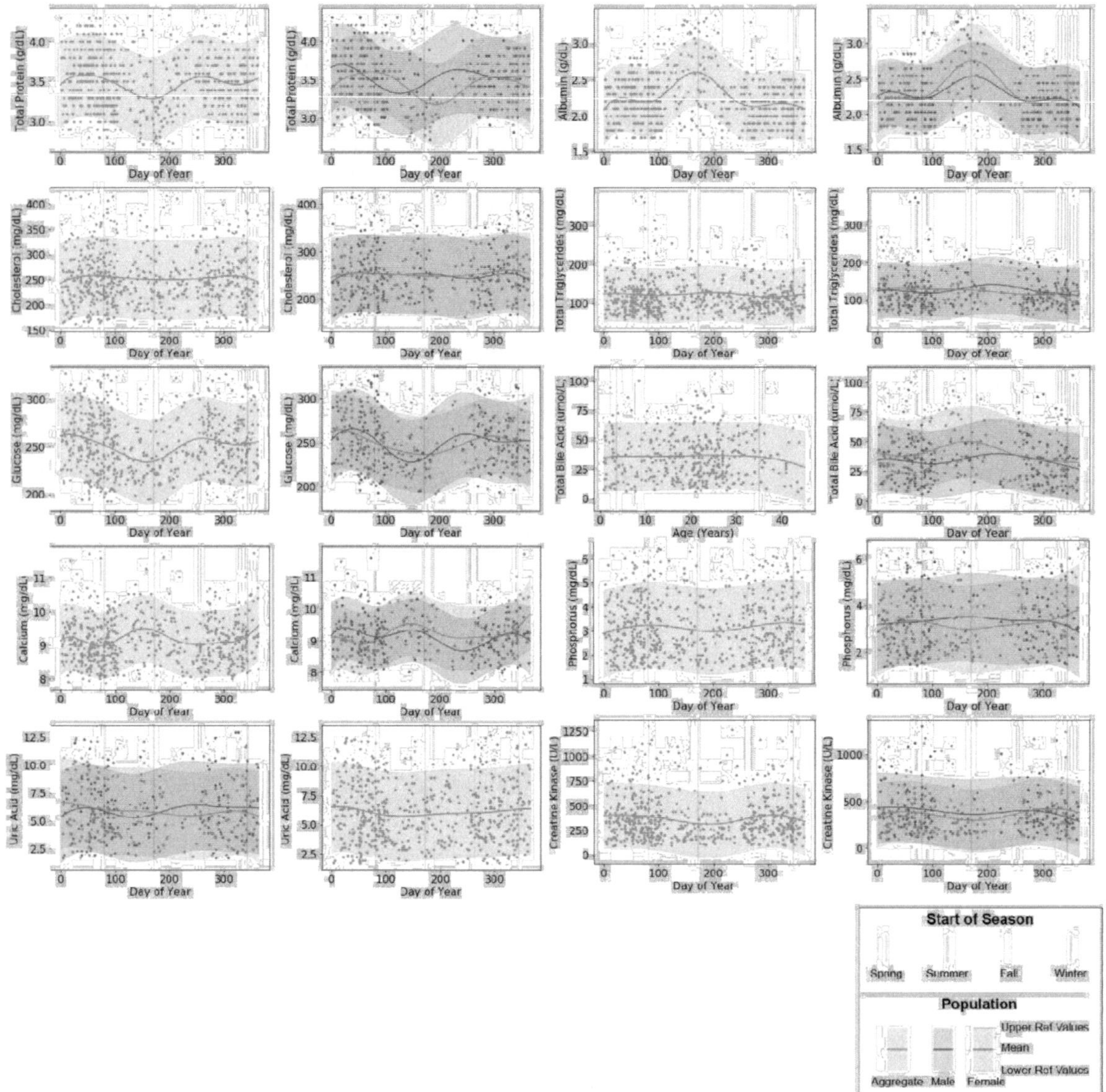

Figure 4. GPR Mean and 95% reference range of clinical biochemistry variables in cockatoos through seasons. Each variable consists of two panels representing regression on the day since the beginning of the year using aggregate data (left) and data stratified by sex (right). Vertical lines in each sub-figure represent the beginning of each meteorologic season in the United States: Spring, Summer, Fall, and Winter.

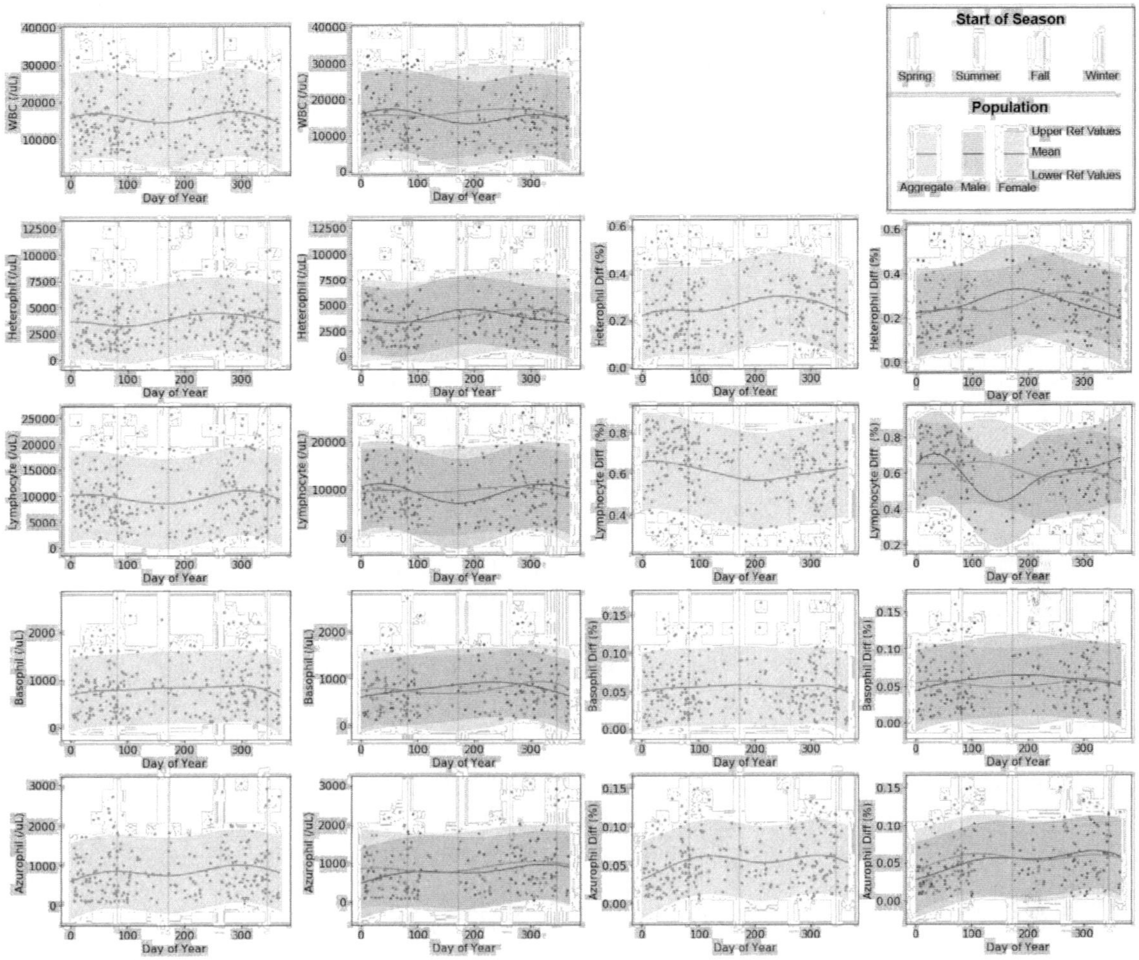

Figure 5. GPR Mean and 95% reference range of hematology variables in bearded dragons throughout the seasons. Each variable consists of two panels representing regression on the day since the beginning of the year using aggregate data (left) and data stratified by sex (right). Vertical lines in each sub-figure represent the beginning of each meteorologic season in the United States: Spring, Summer, Fall, and Winter.

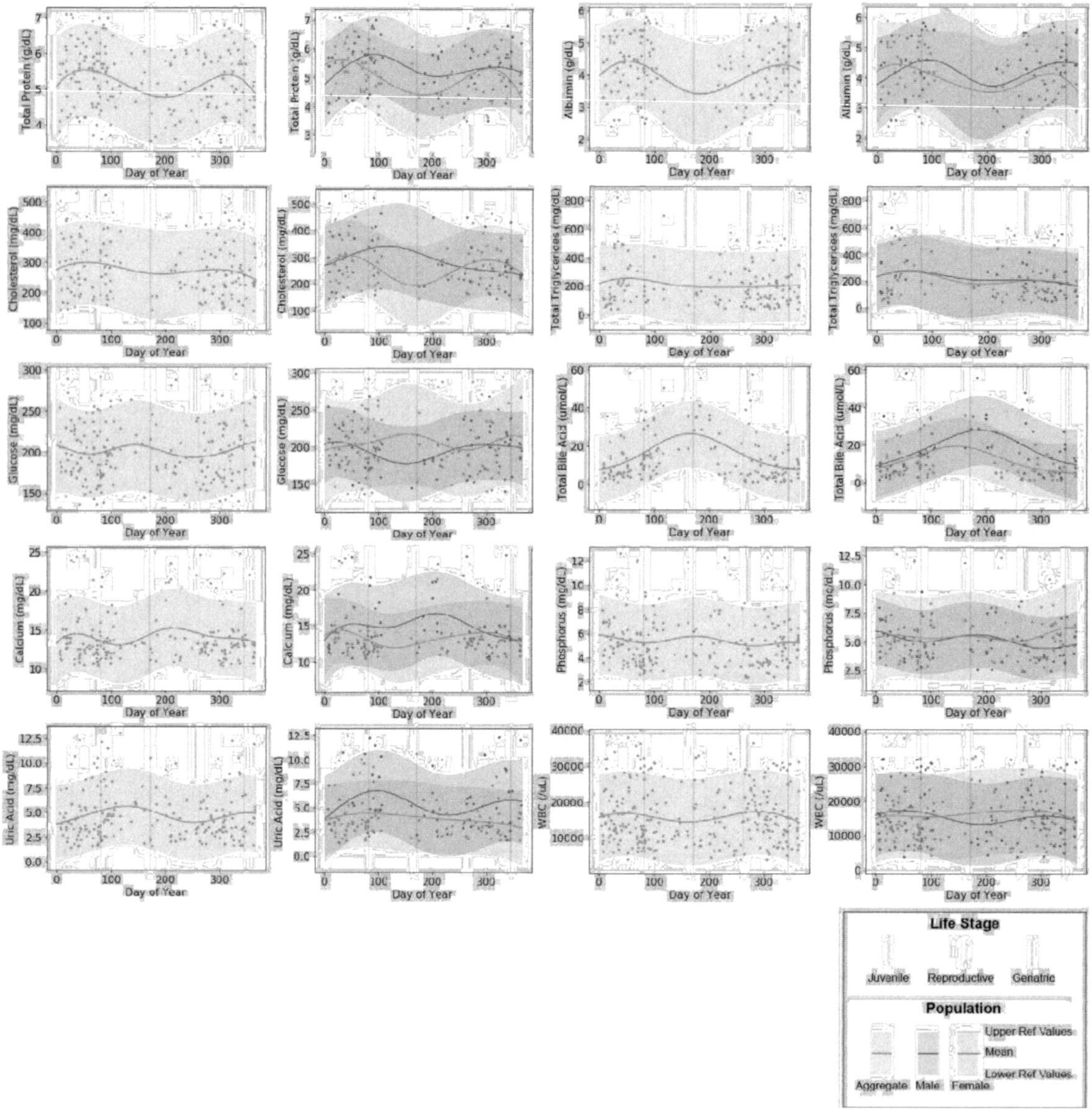

Figure 6. GPR Mean and 95% reference range of clinical biochemistry variables in bearded dragons throughout the seasons. Each variable consists of two panels representing regression on the day since the beginning of the year using aggregate data (left) and data stratified by sex (right). Vertical lines in each sub-figure represent the beginning of each meteorologic season in the United States: Spring, Summer, Fall, and Winter.

Smooth dynamic RIs showed marked changes in aging individuals. While the RI data suggest a relatively stable hematology and chemistry RIs for both adult cockatoos and bearded dragons, there was a more drastic shift in the baseline values for patients entering the geriatric stage of life in both reference populations (Figs 7-10).

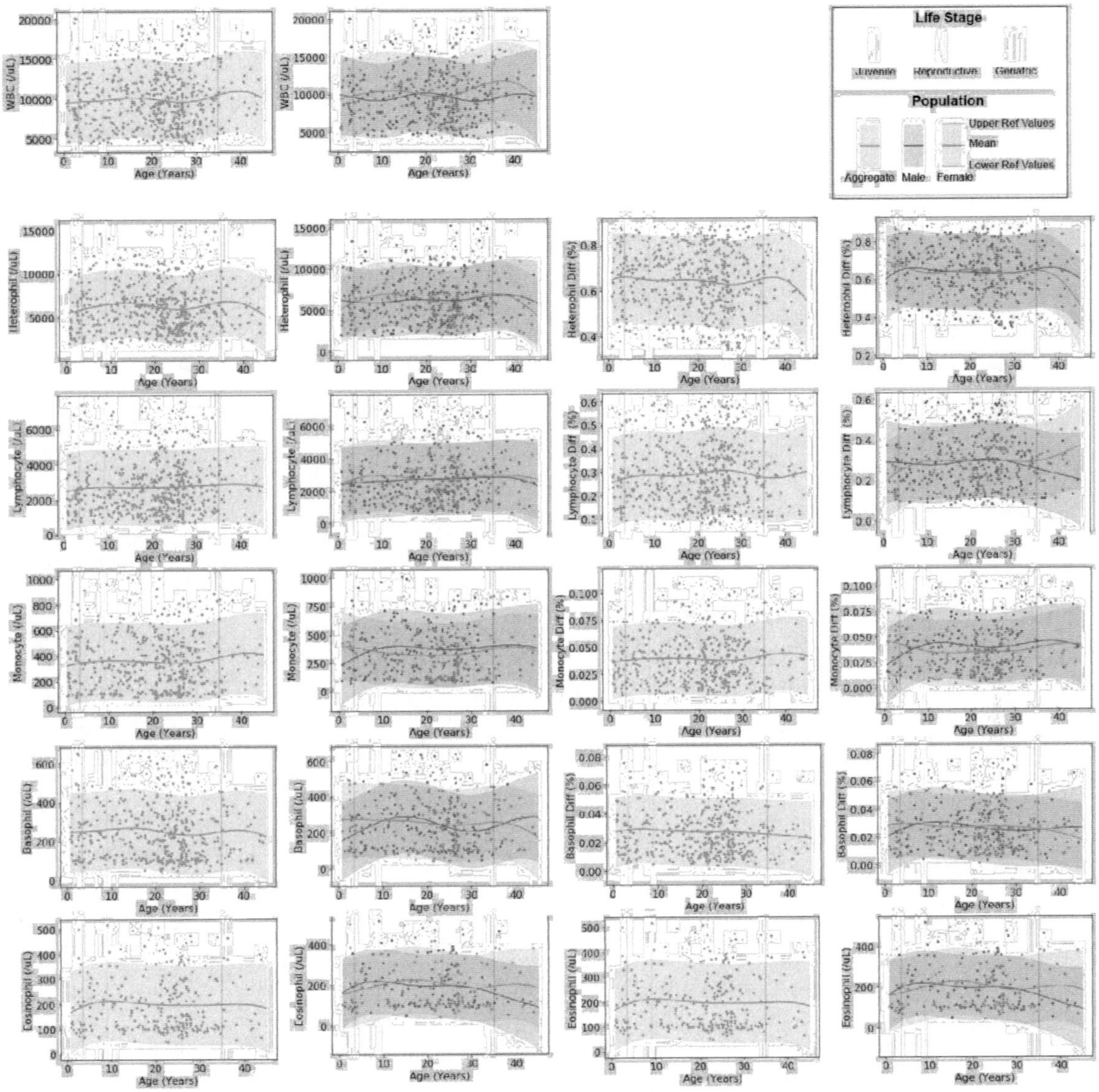

Figure 7. GPR Mean and 95% reference range of hematology variables in cockatoos throughout the stages of life. Each variable consists of two panels representing regression on age using aggregate data (left) and data stratified by sex (right). Vertical lines in each sub-figure represent the beginning of different stages of life: juvenile, reproductive, and geriatric.

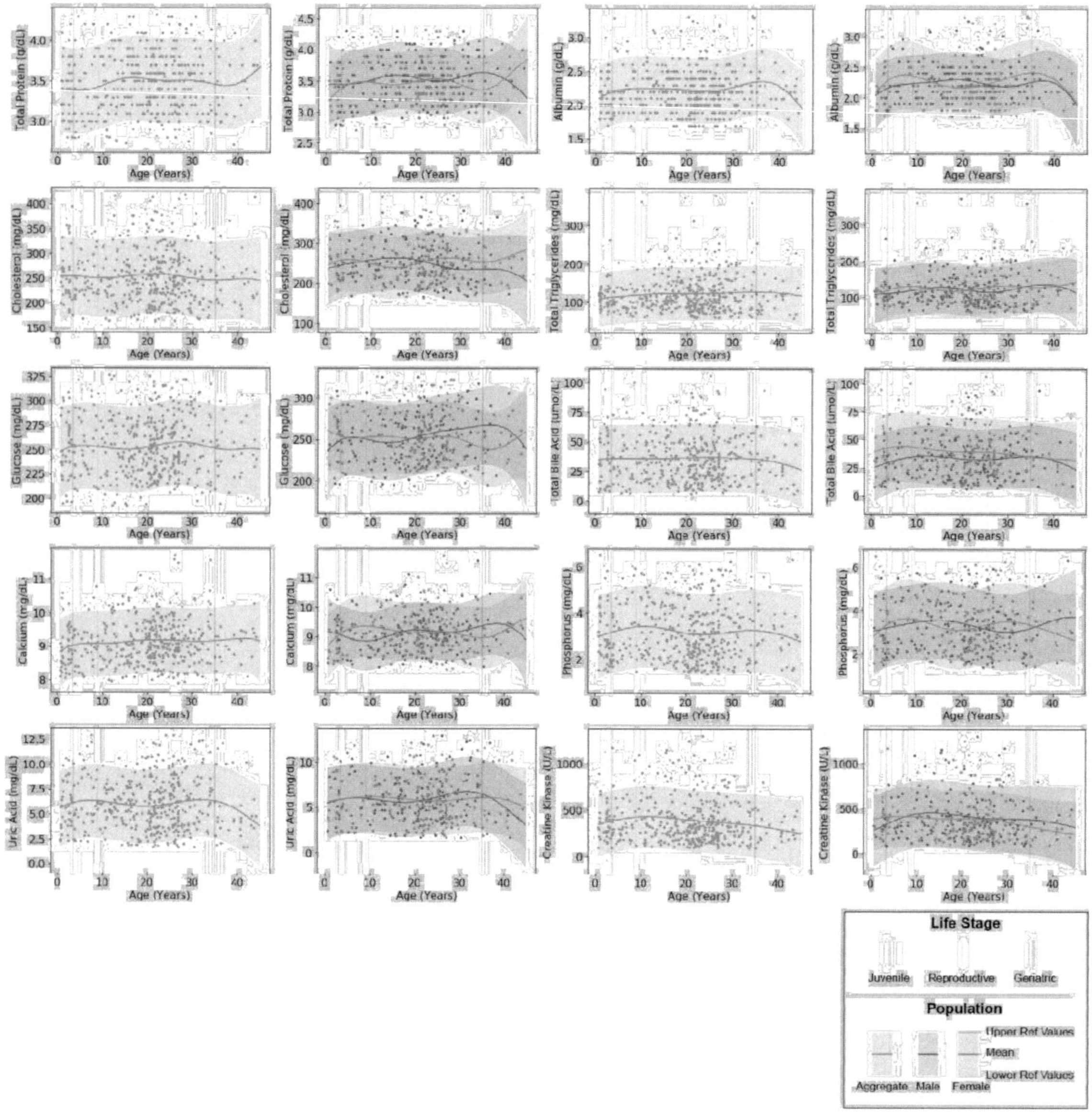

Figure 8. GPR Mean and 95% reference range of clinical biochemistry variables in cockatoos through the stages of life. Each variable consists of two panels representing regression on age using aggregate data (left) and data stratified by sex (right). Vertical lines in each sub-figure represent the beginning of different stages of life: juvenile, reproductive, and geriatric.

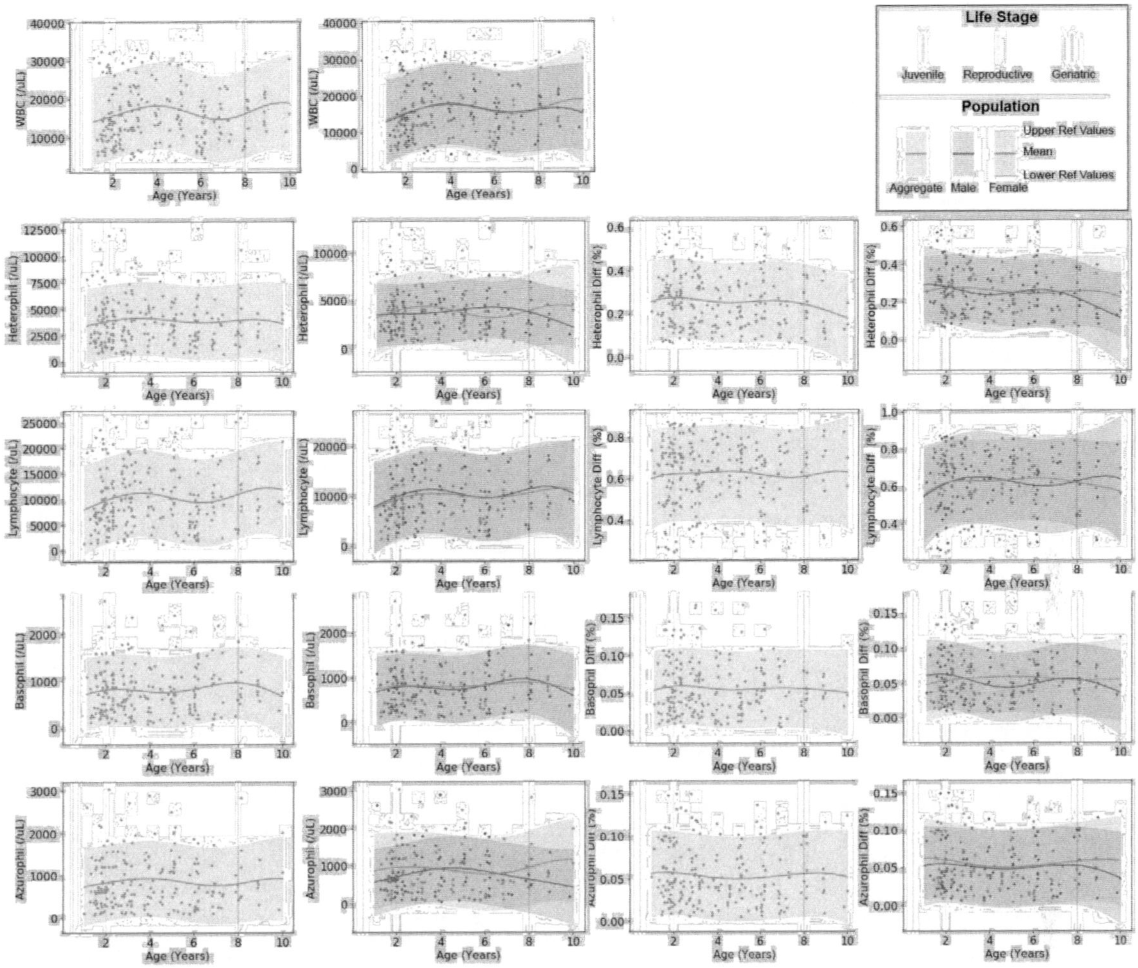

Figure 9. GPR Mean and 95% reference range of hematology variables in bearded dragons through stages of life. Each variable consists of two panels representing regression on age using aggregate data (left) and data stratified by sex (right). Vertical lines in each sub-figure represent the beginning of different stages of life: juvenile, reproductive, and geriatric.

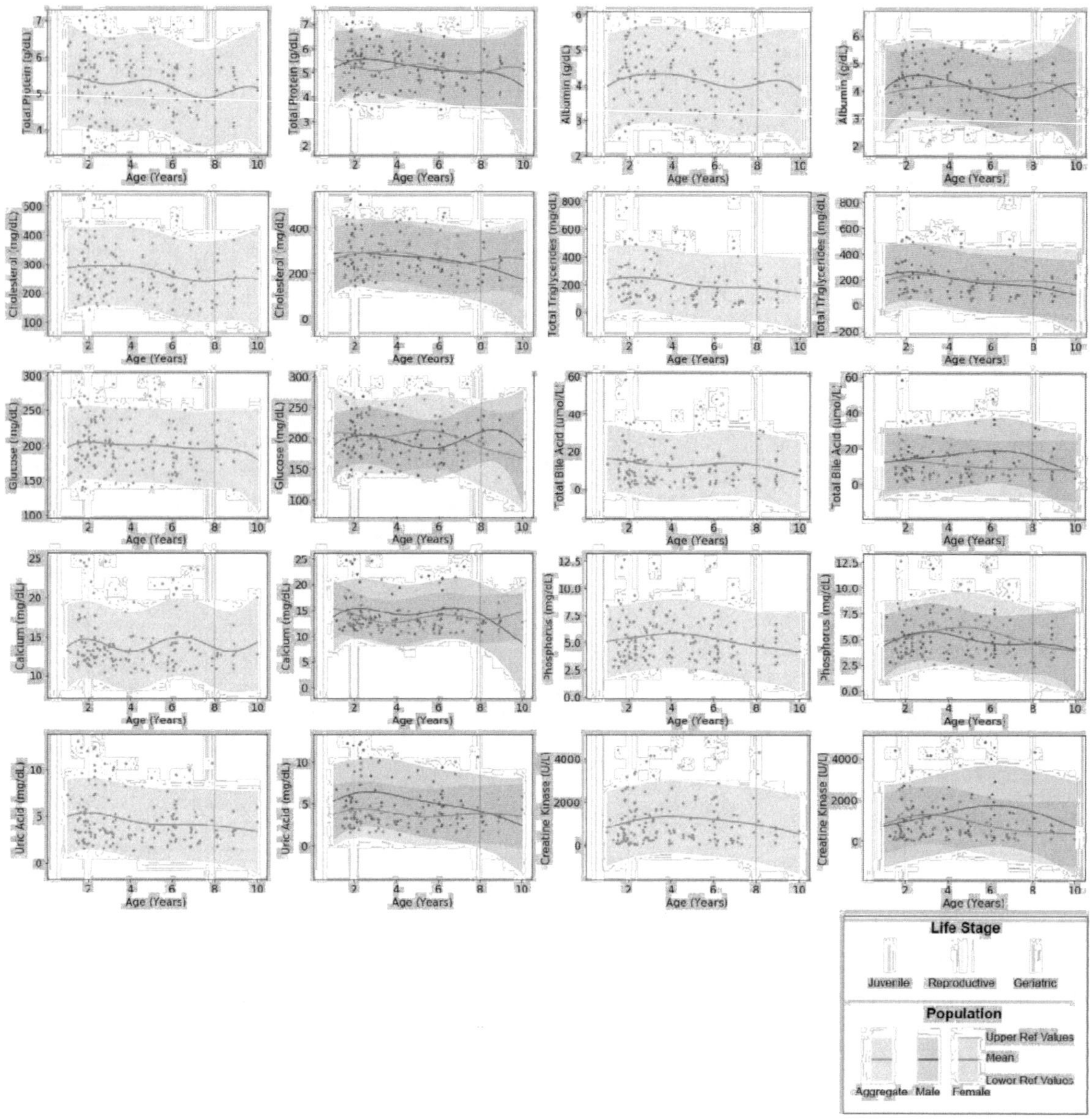

Figure 10. GPR Mean and 95% reference range of clinical biochemistry variables in bearded dragons through stages of life. Each variable consists of two panels representing regression on age using aggregate data (left) and data stratified by sex (right). Vertical lines in each sub-figure represent the beginning of different stages of life: juvenile, reproductive, and geriatric.

Discussion

In this study, we demonstrated the use of an indirect approach to construct dynamic reference intervals for cockatoos and bearded dragons. The resulting aggregate RIs significantly overlapped with the previously established reference limits using the traditional approach. Furthermore, we showed that while discretely constructed reference intervals are appropriate for many population stratifications such as by sex, seasonal, and life stages, RIs might be better conceptualized and understood as continuous dynamics.

The study populations were primarily companion animals; thus, these patients were kept in relatively controlled environments. Even though the population was captive, we could capture intrinsic seasonal and life stage changes through the construction of dynamic reference intervals. It would be informative to compare the differences in dynamic RIs between wild and captive/companion populations.

In addition, the study population originated from 41 states across the United States, resulting in a wide range of geographic sampling locations for the indirect reference populations. While we do not expect dramatic effects on reference intervals due to the geographic distribution of samples, we intend to explore these differences in RIs in the future.

It is crucial to note the assumption that clinical variables are approximately Gaussian in smaller time epochs. Although this assumption allowed us to efficiently model the dynamics of baseline values in most analytes for this study, in clinical practice, it would be essential to apply appropriate mathematical transformations to ensure the integrity of RIs built on this Gaussian assumption.[3,10]

Perhaps, the most important implication of this study is how the proposed dynamic RIs would affect veterinarians' understanding of patients' state of wellness. Future studies will focus on how each patient's intervention threshold would be affected due to differential diagnoses resulting from shifts in reference intervals.

References

1. American Society for Veterinary Clinical Pathology. Quality Assurance and Laboratory Standards Guidelines. Guidelines for the determination of reference intervals (RI) in veterinary species. 2022. Available at: https://cdn.ymaws.com/www.asvcp.org/resource/resmgr/QALS/Other_Publications/RI_Guidelines_For_ASVCP_webs.pdf. Accessed June 27, 2022.

2. Chandola V, Banerjee A, Kumar V. Outlier detection: a survey. ACM Computing Surveys, 14:15. 2007. Available at; https://www.researchgate.net/profile/Vipin-Kumar-54/publication/242403027_Outlier_Detection_A_Survey/links/0deec52d83600bb86c000000/Outlier-Detection-A-Survey.pdf. Accessed June 27, 2022.

3. Clinical and Laboratory Standards Institute. *Defining, Establishing, and Verifying Reference Intervals in the Clinical Laboratory; Approved Guideline.* 3rd Ed. Wayne, PA: CLSI; 2010.

4. Cray C. Reference intervals in avian and exotic hematology. *Vet Clin North Am Exot Anim Pract.* 2015;18(1):105-116.

5. Dawson A. Avian molting. In: Scanes CG, ed. *Sturkie's Avian Physiology.* St. Louis, MO: Elsevier; 2015: 907-917.

6. Demina I, Tsvey A. Total white blood cell counts but not HL ratio changes in the transition from post-juvenile molt to autumn migration in the first-year Eurasian blackcap (*Sylvia atricapilla*). *J Exp Zool A Ecol Integr Physiol.* 2022;337(2):159-170.

7. Doneley B, Monks D, Johnson R, Carmel B. Reference intervals for commonly kept reptile species. In: Doneley B, Monks D, Johnson R, Carmel B, eds. *Reptile Medicine and Surgery in Clinical Practice.* Ames, IA: Wiley-Blackwell; 2018:473-479.

8. Duvenaud D, Lloyd JR, Grosse R, et al. Structure discovery in nonparametric regression through compositional kernel search. *Proc Int Conf Machine Learn.* 2013;28(3):2203-2211.

9. Horn PS, Feng L, Li Y, Pesce AJ. Effect of outliers and nonhealthy individuals on reference interval estimation. *Clin Chem.* 2001;47(12):2137-2145.

10. Horn PS, Pesce AJ, Copeland BE. A robust approach to reference interval estimation and evaluation. *Clin Chem.* 1998;44(3):622-631.

11. Jones GR, Haeckel R, Loh TP, et al. Indirect methods for reference interval determination-review and recommendations. *Clin Chem Lab Med.* 2018;57(1):20-29.

12. Kasprzak M, Hetmanski T, Kulczykowska E. Changes in hematological parameters in free-living pigeons (*Columba livia* f. *urbana*) during the breeding cycle. *J Ornithol.* 2006;147(4):599-604.

13. LaGrange SM, Kimble SJ, MacGowan BJ, Williams RN. Seasonal variation in hematology and blood plasma chemistry values of the timber rattlesnake (*Crotalus horridus*). *J Wildl Dis.* 2014;50(4):990-993.

14. Le T, Lazo A, Powers L. Automation of avian and reptile complete blood count using artificial intelligence powered microscopy. *Proc ExoticsCon.* 2021;517-518.

15. Ritchie BW, Harrison GJ, Harrison LR, eds. *Avian Medicine: Principles and Application.* Lake Worth, FL: Wingers Publishing;, 1994.

16. Pages T, Peinado VI, Viscor G. Seasonal changes in hematology and blood chemistry of the freshwater turtle *Mauremys caspica leprosa. Comp Biochem Physiol A.* 1992;103(2):275-278.

17. Rasmussen CE, Williams CKI. Gaussian Process for Machine Learning. 2006. Available at: https://gaussianprocess.org/gpml/. Accessed June 27, 2022.

18. Schmidt EMDS, Paulillo AC, Santin E, et al. Hematological and serum chemistry values for the ring-necked pheasant (*Phasianus colchicus*): variation with sex and age. *Int J Poult Sci.* 2007;6(2):137-139.

19. Silva LFN, Riani-Costa CCM, Ramos PRR, Takahira RK. Seasonal influence on biochemical profile and serum protein electrophoresis for boa constrictor amarali in captivity. *Braz J Biol.* 2011;71(2):517-520.

Comparison of Ear Microbiome in Rabbits with and without Otitis Externa using Next-Generation DNA Sequencing

Gina Vecere, DVM,
Shahar Malka, DVM, DABVP (Avian Practice),
Natasha Holden, MSc,
Shuiquan Tang, PhD,
Janina A. Krumbeck, PhD

Session #2507

Affiliation: From the Long Island Bird and Exotics Veterinary Clinic, 333 Great Neck Road, Great Neck, NY 11021, USA (Vecere, Malka), MiDOG LLC, 17062 Murphy Avenue, Irvine, CA 92614, USA (Holden, Tang, Krumbeck).

Note: Abstract format selected by author(s). See Front Matter for description of conference proceedings manuscripts submitted as scientific abstracts.

Abstract: Next-generation DNA sequencing (NGS) technologies have been increasingly explored in veterinary medicine in recent years, providing new and practical opportunities for clinical case intervention.[1,2] In this study we aimed to characterize the normal external ear canal microbiome of clinically healthy rabbits compared to that of rabbits presenting with otitis externa. Swabs from the external ear canal of 34 clinically healthy rabbits and 16 rabbits diagnosed with otitis externa were collected. Next-generation DNA sequencing was used to analyze samples for their respective bacterial and fungal microbiomes. A variety of previously unrecognized organisms were noted in analyses of both the clinically healthy and otitis externa groups. These organisms did not align with those traditionally considered to be the primary organisms in the microbiome of the external ear canal of rabbits, for clinically healthy or disease state ears.[3-5] In addition, there was a significantly higher number of different bacterial and fungal species in the clinically healthy group compared to the otitis group ($P = 0.002$ and $P < 0.001$, respectively). This study provided evidence that the rabbit aural microbiome profile is distinctly different between a clinically healthy and an otitis state. It also highlighted new bacterial and fungal organisms of note in rabbits diagnosed with otitis externa compared to those previously thought to be the primary disease-causing organisms. Understanding the microbial population dynamics in the rabbit aural microbiome is particularly critical for helping clinicians recognize, prevent, or revert the progression of otitis.

References

1. Tang S, Prem A, Tjokrosurjo J, et al. The canine skin and ear microbiome: A comprehensive survey of pathogens implicated in canine skin and ear infections using a novel next-generation-sequencing-based assay. *Vet Microbiol*. 2020;247:108764.

2. Krumbeck J, Reiter A, Pohl J, et al. Characterization of oral microbiota in cats: novel insights on the potential role of fungi in feline chronic gingivostomatitis. *Pathogens*. 2021;10(7):904.

3. Mancinelli E, Lennox A. Management of otitis in rabbits. *J Exot Pet Med.* 2017;26(1):63-73.

4. Oglesbee BL. Otitis externa and media. In: Oglesbee BL, ed. *The 5-Minute Veterinary Consult: Ferret and Rabbit*. Ames, IA: Wiley-Blackwell; 2006:308-311.

5. Quesenberry KE, Carpenter JW. Neurological and musculoskeletal diseases. In: Quesenberry KE, Carpenter JW, eds. *Ferrets, Rabbits, and Rodents: Clinical Medicine and Surgery.* 3rd Ed. St. Louis, MO: Elsevier Saunders; 2012:245-255.

Pododermatitis in Small Mammals and Birds

Nico Schoemaker, DVM, PhD, DipECZM (Small Mammal), DipECZM (Avian), and
Yvonne van Zeeland, DVM, MVR, PhD, DipECZM (Avian), DipECZM (Small Mammal), CPBC

Session #2657

Affiliation: From the Division of Zoological Medicine, Department of Clinical Sciences, Faculty of Veterinary Medicine, Utrecht University, Utrecht, 3584 CM, The Netherlands.

Previously published in Schoemaker N, van Zeeland YRA. Pododermatitis in birds and small mammals. Proc 8th World Congr Vet Dermatol Contin Educ. 2016;403-409.

Abstract: Pododermatitis is a common condition in small mammals and birds, characterized by lesions in areas where soft tissue structures overly the bony structures of the feet. Upon cross-comparison, many striking similarities can be identified. For example, lack of exercise, inadequate nutrition, overweight and improper perching or floor substrate are considered portent contributing factors in most species. As such, addressing the underlying cause is an important cornerstone of treatment, in addition to treating pain, secondary infections and managing the wound. Staging of the lesions is deemed helpful to aid in adequate decision-making on treatment and prognosis. While many cases can be managed with the proper care, wound healing may take a considerable time, with advanced cases often responding poorly to therapy. Educating clients on prevention of pododermatitis is therefore considered key to minimize the incidence of this condition in practice.

Introduction

Pododermatitis, also referred to as bumblefoot (in birds) or sore hocks (in small mammals), is a condition characterized by the presence of infection, inflammation and/or degenerative changes of the ventral surface of the feet. The condition is commonly seen in both small mammals and birds. Without appropriate therapy, the disease can quickly progress to not only involve the skin and connective tissues of the foot, but also the deeper tissues (tendons, joints and skeletal structures), thereby resulting in significantly decreased or loss of limb function.

Etiology and Pathogenesis

In both birds and small mammals, pododermatitis has been linked to continuous or uneven pressure applied to the plantar surface of the feet. This pressure may subsequently result in disruption of the epithelial surfaces and ischemia and pressure necrosis of the soft tissue structures overlying the bony structures of the feet. As a sequela to the ischemia and necrosis, inflammatory mediators are released, which, together with the oxidative damage caused during the intermittent reperfusion of the area, result in further tissue damage and vascular thrombosis. At this stage, chronic inflammation, hyperkeratosis, erosions and ulcerations may be seen. Invasion of bacterial organisms, in particular *Staphylococcus aureus* (all species), *Escherichia coli* (in raptors, parrots) and *Pasteurella multocida* (in rabbits), may subsequently result in secondary bacterial infections of the skin and underlying tissues. In severe cases, infection may spread to the underlying bones, tendons and joints, resulting in osteomyelitis, tenosynovitis and arthritis, thereby complicating the disease process and reducing the chances of the limb function returning to normal, especially if permanent damage to the tendons (e.g., displacement, rupture) is present. Occasionally, ulcerative lesions may also disrupt the integrity of the blood vessels of the foot, thereby resulting in significant hemorrhage and anemia. In advanced cases, pododermatitis may furthermore lead to systemic spread of the infection with subsequent endocarditis or polyarthritis.

Predisposing Factors

Many factors may play a role in the development of pododermatitis. In general, any management-related, medical or behavioral condition that results in an increased pressure of the weight bearing surfaces, alterations in the weight distribution or stance of the animal and/or increased friction and irritation of the skin may predispose an animal to developing pododermatitis (Table 1). Management-related factors that have been found to increase the risk for pododermatitis include lack of exercise, confinement to a small space, improper substrate or perches, poor or unsanitary housing and malnutrition. In addition, excess body weight, trauma and other physical or behavioral conditions that lead to an altered stance, activity and/or weight bearing (e.g., spondylosis, arthritis, neurologic disease, nervousness, conformational deformities) may increase the risk of the animal developing pododermatitis.

Table 1. Predisposing risk factors for pododermatitis in birds and small mammals.

Risk factors
• Obesity or, in rabbits, emaciation
• Pregnancy (in small mammals)
• Inadequate nutrition (e.g., hypovitaminosis A [psittacine birds], biotin deficiency [turkeys], hypovitaminosis C [guinea pigs])
• Lack of exercise
• Improper perches or substrate (e.g., inappropriately sized perches; abrasive surfaces such as carpets, sandpaper; hard floors such as tiles or cement; wire bottom cages)
• Moist and unhygienic conditions, leading to urinary and/or faecal soiling of the feet
• Poor husbandry (e.g., small enclosures, overcrowding)
• Unilateral lameness or pain, causing increased weight bearing to the contralateral limb (e.g., arthritis, fractures)
• Conditions that lead to reduced mobility, paresis, paralysis, or altered gait (e.g., spondylosis, neurologic disease, limb deformities)
• Trauma and abrasions to the feet (e.g., overgrown talons [raptors], puncture/bite wounds, trap injury, thermal and electrical burns/wounds, frostbite)
• Infections (e.g., *Staphylococcus aureus*; *Pasteurella multocida* [rabbits], poxvirus [raptors, poultry], *Knemodikoptes* [budgerigars]) – may both be the primary cause as well as secondary to erosion/ulceration of the plantar surface, thereby complicating disease
• Loss of protective hair covering (e.g., iatrogenic clipping of hair) in rabbits
• Behavioural: nervousness, excessive thumping (rabbits); bating of jesses (raptors)

Although pododermatitis may develop in any bird or small mammal, certain species or breeds have been found to be more susceptible. For example, large and giant rabbit breeds (weighing >5 kg), raptors (e.g., falcons, eagles, osprey, red-tailed hawks), penguins, larger ducks and swans, and heavy-bodied psittacine birds (e.g., Amazon parrots, budgerigars, cockatiels) appear to be predisposed. Moreover, pododermatitis appears to occur particularly in rabbit breeds that have fine, sparse hair on the metatarsus (e.g., the Rex), thereby offering little protection against local injury, ischemia and pressure necrosis. Similarly, guinea pigs and rats are susceptible to sustaining injury to the plantar and palmar surfaces of their feet because of their relatively thin skin (similar to rabbits) and lack of hair on their feet and metatarsi. A study concerning risk factors of pododermatitis in rabbits from the UK revealed age and gender as predisposing factors, whereby they found pododermatitis to be more common in female, older rabbits. Moreover, neutering appeared to predispose the animals to pododermatitis; however, it cannot be excluded that this effect is the result from bias resulting from these animals also being more likely to be overweight. In other small mammals and birds, no age or sex predilection has been identified.

Clinical Presentation and Staging

Pododermatitis encompasses a range of clinical presentations varying from mild erythema, swelling and/or superficial ulcerations to deep ulcerations with concurrent osteomyelitis, arthritis and/or tenosynovitis with loss of limb function. In birds, the plantar metatarsal pad or plantar digital pads are most commonly affected, whereas in rabbits and rodents the lesions are usually located on the caudal aspect of the tarsus and metatarsus and, less frequently, the metacarpus and phalangeal regions of the front limbs. In the initial stages, lesions in small mammals may be limited to focal hair loss (in rabbits), erythema, scaling and thinning of the skin of the plantar surface over the bony prominence of the metatarsus. In addition, erosion, dry crusty scabs or callous-like swellings, combined with mild to moderate inflammation may be noted in the earlier stages of the disease. In more advanced cases, inflammation may be more severe and also involve the deeper structures, with obvious ulceration and tissue necrosis. Secondary infections may occur as well and result in suppurative dermatitis, abscessation, tenosynovitis, osteomyelitis and/or arthritis. At this stage, loss of pedal function may also be observed, (e.g., due to rupture or displacement of a tendon). Dependent on the severity of the podermatitis, mild to non-weight-bearing lameness of the affected limb may be seen. If pododermatitis is present bilaterally, animals may also show reluctance to move.

In both raptors and rabbits, classification schemes have been designed which help to grade pododermatitis depending on the extent, severity, and prognosis (Tables 2 and 3).

Table 2. Grading system for pododermatitis in raptors.[1]

Grade	Description
Grade I	Early lesion of the plantar surface involving integument only, with no secondary infection of deeper tissues. A loss of papilla, erythema and/or hyperkeratosis may be seen. Prognosis is excellent.
Grade II	Infection of the subcutaneous tissues with no gross swelling of affected feet. Localized signs of inflammation/infection and/or ischemic necrosis. Prognosis is good.
Grade III	Infected, swollen, painful feet without apparent damage to deep vital structures. Serous or caseous fluid draining from fibrotic lesion. Prognosis is good to guarded.
Grade IV	Infection of deep vital structures such as tendons and bones, though pedal function is retained. Patients may have concurrent tenosynovitis, arthritis, and/or osteomyelitis. Prognosis is guarded to poor.
Grade V	End-stage disease with osteomyelitis, crippling deformity, and loss of pedal function. Prognosis is grave and euthanasia is often performed.

Table 3. Grading system for pododermatitis in rabbits.[2]

Grade	Description
Grade 0	No lesions.
Grade 1	A small, circular area on the plantar aspect of the metatarsal bone-calcaneus (mono or bilateral lesions), with minimal alopecia, minimal epidermal hyperaemia and/or hyperkeratosis of the skin, but with no evidence of infection or bleeding of underlying tissues.
Grade 2	Circumscribed area of varying size localised at the caudal plantar aspect of the metatarsal-calcaneal area or extending linearly along the plantar aspect of the cranial metatarsal area with alopecia, erythema and scaling of surrounding tissues.
Grade 3	Area of varying size focally ulcerated and with varying degree of keratinisation abnormalities. Infection of subcutaneous tissue present.
Grade 4	Full-thickness skin loss with swelling and necrotic debris may be present with infection of underlying tissues. Purulent exudates may be adherent to the lesions.
Grade 5	Severe infections with involvement of deep structures including bones and tendons with tenosynovitis, osteomyelitis and arthritis.
Grade 6	End-stage disease with loss of pedal function.

Diagnostic Approach

A definite diagnosis of pododermatitis can usually be made following the history and physical examination. Moreover, the history may reveal information on the potential risk factors such as nutrition, housing, bedding, and exercise. Similarly, a full and systematic physical examination is needed to discover concurrent diseases or predisposing factors for the pododermatitis. For this purpose, the patient should initially be observed in its own cage or enclosure, whereby any changes in gait or abnormalities in resting or perching posture are recorded. A body weight and body condition score should also be obtained. Upon physical examination, special attention should be paid to the plantar (and palmar) surfaces of the feet to evaluate the presence of alopecia (in rabbits), erythema, erosions, ulcerations, necrosis, swelling and/or exudate. In addition, an orthopedic and/or neurologic examination may be performed to assess the limb function and presence of (other conditions resulting in) pain and reluctance to move.

Diagnostic work-up may further include a complete blood cell count, serum chemistry panel, and urinalysis (if applicable) to identify concurrent diseases or predisposing factors. Moreover, impression smears and a culture and sensitivity of the affected area may help to determine involvement of bacteria and guide the choice of antibiotic therapy. Preferably, cultures are performed from swabs are taken from deep within the lesion or pieces of infected tissue. If the lesion has extended into the deeper layers, imaging is recommended to identify involvement of the underlying bone or joints. Radiographs will usually suffice for this purpose (and may also help to assess presence of underlying limb and/or spinal disorders). However, Computed Tomography or Magnetic Resonance Imaging may be considered to detect more subtle abnormalities of the bone and soft tissues.

Management and Prognosis

Management of pododermatitis often requires a multimodal approach, which, dependent on the severity of the lesions, may include a combination of management, medical and/or surgical interventions. In any patient, treatment should be aimed at correction of the underlying causes and concurrent disease conditions. In addition appropriate wound care and bandaging, topical medication, systemic antibiotics, analgesics and/or surgical intervention may be indicated.

Correction of underlying causes

Any underlying causes should be addressed appropriately. This generally involves adjustments to the animal's housing, flooring, bedding and/or nutrition. For small mammals, cage floors should be adequately padded and/or bedded to reduce pressure on the feet (e.g., using foam rubber, thick towels, hay or fiber bedding). Bedding should be cleaned regularly to prevent risk of contamination and subsequent infection of the pedal lesions. In perching birds, perches of various sizes and types (e.g., wood, rope, rubber) should be provided. In addition, perches may be wrapped in safe, soft materials (e.g., cotton padding or bandaging tape) to provide extra protection. For raptors, flat perches covered with artificial turf (for falcons) or round/elliptical perches made of hemp or sisal rope (for eagles, hawks, owls) may be used. As for small mammals, surfaces should be kept clean and dry, and changed frequently. In waterfowl, access to swimming water (preferably a pool with gradual slopes) is important. Moreover, soft flooring or mats should be provided to reduce the pressure on the animal's feet. Enclosures should furthermore be designed in such a way that movement, exercise and activity are stimulated. Enrichment (e.g., feeding stations, toys, foraging enrichment, tunnels, solid wheels) may also help to achieve this goal.

Aside from housing, correction of the diet is also important, especially in overweight animals. Moreover, dietary corrections may be aimed at resolving any potential nutritional deficiencies (e.g. hypovitaminosis C in guinea pigs, hypovitaminosis A in parrots) that may contribute to the problem.

Wound care

In general, pododermatitis lesions require open wound management and healing by secondary intent. It is important to keep the ventral foot surface clean and dry, especially if ulcerations are present. For small mammals, hair surrounding the lesions may be trimmed (though not down to the skin!) to prevent matting and contamination. In addition, the wound may be thoroughly cleaned using chlorhexidine or povidone iodine solutions (in the initial phases of treatment), or by soaking it in warm saline or lactated Ringer solution (in the later stages of treatment). Following wound irrigation, necrotic tissue may be removed.

Bandaging

Bandaging is usually necessary, except in mild cases. Their function is to prevent tissue desiccation and protect the wound against further pressure, trauma and contamination. Generally, bandages are composed of three layers: 1) a primary layer consisting of a dressing that is in direct contact with the lesion and provides protection as well as encourages drainage of exudate, delivery of medication and debridement of necrotic tissue; 2) a secondary layer that is composed of an absorptive material such as cast padding or cotton padding; and 3) a tertiary layer that keeps the bandage in place (e.g., Vetrap or similar product). In severe cases, additional protection may be applied in the form of interdigitating bandages, dental acrylic or polypropylene foam shoes, styrene plastic polymer foot casts, ball bandages and/or doughnut-shaped bandages. These bandages may all help to relieve pressure on prominent weight-bearing surfaces and redistribute pressure onto other surfaces. The type of bandage chosen usually depends on the stage of healing, For example, wet-to-dry bandages are often used in the initial drainage stage, whereas ball bandages, polypropylene shoes and interdigitating bandages may be used in the granulation stage, early and later stages of healing, respectively. Regardless of the type of dressing and bandage chosen, bandages should be changed regularly, starting with daily changes until the condition improves, following which the time between bandage changes can be prolonged to once every 2-3 days up to a week until the condition resolves. Preferably, both feet are bandaged to avoid development of pododermatitis of the other foot as a result of the shift in weight-bearing.

Topical medication

Topical medication may be used alone or in conjunction with a bandage. Examples of medications that can be used topically to treat pododermatitis include silver sulfadiazine, fusidic acid, mupirocin calcium, Manuka honey, antibiotic ointments (e.g., piperacillin, rifampicin, lincomycin), dimethyl sulfoxide (DMSO) and/or products containing proteolytic enzymes. Care must be taken when using topical antibiotics, because ingestion or absorption can result in toxicity in many species, whereas oil-based products may affect feather quality and impair thermoregulation in birds. Topical corticosteroids are generally not recommended because of their immunosuppressive effects and delayed wound healing. Footpad strengthening products may be applied during advanced stages of healing.

Systemic antibiotics

Systemic antibiotics may be indicated in case of persistent or deep infections (grade 2 dermatitis or higher). Selection is preferably based on a culture and sensitivity, but should this not be available, a broad-spectrum antibiotic with action against gram-positive and gram-negative bacteria (e.g., trimethoprim-sulfamethoxazole) may be chosen. Antibiotic-impregnated beads (containing e.g., gentamycin, tobramycin or amikacin) also offer an effective method of antibiotic delivery to an infected ischemic area, especially in birds that underwent aggressive surgical debridement

Analgesia

Analgesics comprise an important part of the treatment regimen. Generally, analgesia is provided through administration of nonsteroidal anti-inflammatory drugs (e.g., meloxicam, carprofen) and/or opioids (e.g., buprenorphine, butorphanol).

Surgical intervention

More advanced cases of pododermatitis (grade 3 or higher) often require aggressive surgical debridement to remove infected or necrotic tissue, promote drainage (especially in case of an abscess), stimulate formation of granulation tissue and improve circulation to the area. The wound may then be flushed with warm saline after which antibiotic impregnated beads may be left behind in the wound. In addition, bandages mag be placed to protect the wound and reduce pressure on the weight-bearing surface. Generally, wounds are left open to heal by secondary intent, but in some patients (primary) closure of the wound may be attempted after a granulation bed has developed. In severe cases (e.g., pododermatitis in conjunction with tendon displacement, severe osteomyelitis, septic arthritis and/or loss of limb function), limb amputation may be considered as a salvage procedure (Note: not advisable for wild birds!).

In addition to surgical intervention, low-level laser therapy or therapeutic laser has also been found to improve wound healing by promoting fibroblast development, collagen production, and epithelialization; enhancing leukocyte infiltration and macrophage activity; promoting angiogenesis and stimulating vasodilatation; and increasing lymphatic drainage of the region.

Follow-Up and Prognosis

Pododermatitis can be challenging and difficult to treat. Particularly chronic pododermatitis will take several months to heal, depending on the severity and response to treatment. Prior to initiating therapy, owners should always be properly informed about the probable course and outcome. For mild to moderate cases of pododermatitis (grade 1-3) prognosis for recovery is generally considered good, especially if the underlying causes can be corrected. For grade 4-5 pododermatitis, the prognosis is considered guarded to poor, especially if osteomyelitis, septic arthritis and/or tenosynovitis are present. Euthanasia may be warranted in animals that suffer from severe, chronic pododermatitis, especially if more than one limb is affected, concurrent disease is present and/or the animal responds poorly to therapy and/or is not able to stand without severe pain.

Throughout the course of treatment, regular monitoring is advised whereby daily to weekly or biweekly rechecks are advised, dependent on the severity and stage of the disease process. Measuring the size of the lesion and documentation of the progress with the help of photographs are useful to evaluate the success or failure of the treatment plan. Aside from evaluating the lesion, assessment of the patient's general condition, as well as the efficacy of the used protocol for pain relief, is considered essential.

Prevention

Proper husbandry, diet and client education are important tools to help prevent the development of pododermatitis. This includes the implementation of correct husbandry practices (e.g., using various perch types and materials, suitable flooring and soft, dry bedding, sufficient space), proper sanitation, offering the animal environmental enrichment and exercise (including flight in birds of prey) as well as a nutritiously balanced diet for the species in question, and preventing the animal from becoming overweight. In addition, periodic evaluation of the animal's feet may help to detect pedal lesions in an early stage, thereby increasing the chances of successful intervention.

Selected References

1. Remple JD. Raptor bumblefoot: a new treatment technique. In: Redig PT, Cooper JE, Remple JD, et al, eds. *Raptor Biomedicine.* Minneapolis, MN: University of Minnesota Press; 1993:154-160.

2. Mancinelli E, Keeble E, Richardson J, Hedley J. Husbandry risk factors associated with hock pododermatitis in UK pet rabbits (*Oryctolagus cuniculus*). *Vet Rec.* 2013;174(17):429-429.

3. Blair J. Bumblefoot: a comparison of clinical presentation and treatment of pododermatitis in rabbits, rodents, and birds. *Vet Clin North Am Exot Anim Pract.* 2013;16(3):715-735.

4. Brown C, Donnelly T. Treatment of pododermatitis in the guinea pig. *Lab Anim.* 2008;37(4):156-157.

5. De Jong IC, Reimert H, Rommers JM. Effect of floor type on footpad injuries in does: a pilot study. *Proc 9th World Rabbit Congr.* 2008.

6. Harcourt-Brown, NH. Foot and leg problems. In: Benyon, PH, Forbes, NA, Harcourt-Brown, NH, eds. *Manual of Raptors, Pigeons and Waterfowl.* Cheltenham, UK: BSAVA; 1996:163-167.

7. Harkness JE, Turner PV, Vande Woude S, Wheler CL. Specific diseases and conditions: pododermatitis. In: Harkness JE, Wagner JE, eds. *Biology and Medicine of Rabbits and Rodents.* 5th Ed. Ames, IA: Wiley-Blackwell; 2010:249-396.

8. Pignon C, Mayer J. Guinea pigs: integumentary disorders. In: Quesenberry KE, Orcutt CJ, Mans C, Carpenter J, eds. *Ferrets, Rabbits, and Rodents: Clinical Medicine and Surgery.* 4th Ed. St Louis, MO: Elsevier; 2020:287-290.

9. Remple JD, Forbes NA. Antibiotic-impregnated polymethyl methacrylate beads in the treatment in the treatment of bumblefoot in raptors. In: Lumeij JT, Remple JD, Redig PT, et al, eds. *Raptor Biomedicine III.* Lake Worth, FL: Zoological Education Network; 2000:255-262.

10. Remple JD. A multifaceted approach to the treatment of bumblefoot in raptors. *J Exot Pet Med.* 2006;15: 49-55.

11. Sanchez-Migallon, D. Birds: Pododermatitis. In: Mayer J, Donnelly TM, eds. *Clinical Veterinary Advisor, Birds and Exotic Pets.* St Louis, MO: Elsevier Health Sciences; 2012:219-221.

12. Varga M. Skin diseases. In: Varga M, ed. *Textbook of Rabbit Medicine.* 2nd Ed. Oxford, UK: Butterworth-Heinemann; 2014:233-240.

13. Varga M, Paterson S. Dermatologic diseases of rabbits. In: Quesenberry KE, Orcutt CJ, Mans C, Carpenter J, eds. *Ferrets, Rabbits, and Rodents: Clinical Medicine and Surgery.* 4th Ed. St Louis, MO: Elsevier; 2020:220-232.

14. Vella D. Rabbits: Pododermatitis. In: Mayer J, Donnelly TM, eds. *Clinical Veterinary Advisor, Birds and Exotic Pets.* St Louis, MO: Elsevier Health Sciences; 2012:407-409.

Echocardiographic Evaluation in Koi Carp (*Cyprinus carpio*) Under Manual Restraint Compared to Anesthesia with Isoeugenol

Megan Partyka, DVM,
Michael Perlini, DVM, Dipl ACVR,
Carl Toborowsky, MS, MA, VMD,
Alvin Camus, DVM, PhD,
Joerg Mayer, DVM, MS, DABVP (Exotic Companion Mammal Practice),
DipECZM (Small Mammal), Dipl ACZM

Session #2635

Affiliation: From Department of Small Animal Medicine and Surgery (Partyka, Toborowsky, Mayer), Department of Veterinary Biosciences and Diagnostic Imaging (Perlini), and Department of Pathology (Camus), College of Veterinary Medicine, University of Georgia, Athens, GA 30602, USA.

Note: Abstract format selected by author(s). See Front Matter for description of conference proceedings manuscripts submitted as scientific abstracts.

Abstract: Fish are affected by a wide range of cardiovascular conditions including infectious (e.g., viral, bacterial, fungal, and parasitic) and non-infectious (e.g., congenital, degenerative, environmental, idiopathic) diseases. There is an overall a lack of knowledge and information in the literature regarding both normal cardiac parameters and the cardiovascular effects of anesthetic agents, such as isoeugenol, on fish including koi carp (*Cyprinus carpio*). Koi carp are popular ornamental fish found worldwide as display animals in private collections, zoological institution and aquariums. A cross-over study evaluated echocardiography in 40 adult koi carp and determined ultrasound to be a non-invasive method for cardiac evaluation under both manual restraint and anesthesia with isoeugenol (AQUI-S) at 50 ppm. Mean heart rate was significantly higher in anesthetized fish compared to manually restrained fish. Respiratory rate decreased significantly in anesthetized fish compared to manually restrained fish. Mean ejection fraction, stroke volume, and cardiac output were significantly lower in anesthetized fish than in manually restrained fish. Overall poor measurement agreement and correlation were found between veterinarians measuring the same echocardiography study for each fish. This study provides the first description of echocardiography in koi carp and demonstrates significant differences in cardiopulmonary parameters under manual restraint compared to anesthesia with isoeugenol.

Clinical Findings and Surgical Repair of Acquired Coelomic Hernias with Polypropylene and Porcine Mucosa Meshes in a White Spotted Bamboo Shark (*Chiloscyllium plagiosum*)

Julie Pujol, DMV,
Claire Vergneau-Grosset, DMV, IPSAV, CES, Dipl ACZM,
Marion Jalenques, DMV,
Juliette Raulic, DMV, IPSAV,
Marie-Pierre Lessard,
Wahiba Ait Youcef,
Noémie Summa, DMV, IPSAV, Dipl ACZM

Session #2711

Affiliation: From the Faculté de médecine vétérinaire, Université de Montréal, 3200 rue Sicotte, Saint-Hyacinthe, QC J2S 2M2, Canada (Pujol, Vergneau-Grosset, Jalenques, Raulic, Summa), The Aquarium du Québec, Sépaq, 1675 avenue des hôtels, QC G1W 4S3, Canada (Lessard, Ait Youcef).

Note: Abstract format selected by author(s). See Front Matter for description of conference proceedings manuscripts submitted as scientific abstracts.

Abstract: A captive adult male white spotted bamboo shark exhibited a sudden onset of depressible distension of the cranial third of the coelom. A discontinuity of the coelomic muscle wall associated with moderate coelomic effusion was noted on coelomic ultrasound. Coelomic hernia and focal coelomitis were suspected and treatment was initiated (ceftiofur 6.6 mg/kg and florfenicol 40 mg/kg, IM once). Hematology and biochemistry revealed an increase in urea and total proteins.[1,2] Upon exploratory coeliotomy under anesthesia (MS-222 50 mg/L and propofol 2.5 mg/kg IV[3]), a double coelomic hernia of the ventral coelomic wall without entrapment of vital structures was confirmed. The larger hernia of the transversus abdominis[4] muscle was reduced with a polypropylene mesh (Prolene mesh, Ethicon, Inc.) and a porcine submucosa mesh implant (Surgisis, Cook Medical, Ind.). The second hernia was closed with an interrupted cruciate pattern. Bacterial culture of the muscular wall revealed a multidrug resistant *Pseudomonas* sp. Postoperative treatment included gentamycin 6 mg/kg IM q24h × 7 days based on culture sensitivity, meloxicam 0.5 mg/kg once and nutritional support. Postoperative complications included skin erythema and subcutaneous edema adjacent to the wound for up to 1 month post-surgery. Inflammatory nodules developed on the scar four months after the herniorrhaphy. No wound dehiscence or hernia recurrence were noted on follow-up 9 months post surgery. The definitive etiology of the hernia remains unknown but it is suspected to be traumatic. To our knowledge, this is the first report of a successful treatment of coelomic hernia with polypropylene and porcine mucosa meshes in elasmobranchs.

References

1. Alexander AB, Parkinson LA, Grant KR, et al. The hemic response of white-spotted bamboo sharks (*Chiloscyllium plagiosum*) with inflammatory disease. *Zoo Biol.* 2016;35(3):251-259.

2. Parkinson LA, Alexander AB, Campbell TW. Variability in hematology of white-spotted bamboo sharks (*Chiloscyllium plagiosum*) in different living environments. *Zoo Biol.* 2017;36(4):284-288.

3. Miller SM, Mitchell MA, Heatley JJ, et al. Clinical and cardiorespiratory effects of propofol in the spotted bamboo shark (*Chylloscyllium plagiosum*). *J Zoo Wildl Med.* 2005;36(4):673-676.

4. Ashley LM, Chiasson RB. *Laboratory Anatomy of the Shark.* 5th Ed. Ontario, Canada: McGraw-Hill College; 1988.

Point-of-Care Ultrasonography in Exotic Animals

Cathy A. Johnson-Delaney, DVM

Session #2733

Affiliation: From Washington Ferret Rescue & Shelter, 12522 128th Lane NE, Kirkland, WA 98034, USA, NW Zoological Supply, 3614 Smith Ave, Everett, WA 98201, USA.

Abstract: Point-of-care ultrasound (POCUS) and AFAST (abdominal focused assessment with sonography for trauma), and TFAST (thoracic focused assessment with sonography for trauma) techniques can be done in exotic patients such the ferret, rabbit, guinea pig, parrot, chicken, bearded dragon, chameleon, chelonian, and snakes. This POCUS can quickly look for fluids in the body cavities and be used as screening for other diseases such as masses, organomegaly, and uroliths. Sonographic uses for aspirations, and needle biopsies will be discussed. POCUS to assess the heart can facilitate initiation of treatment, although it may not take the place of a detailed echocardiograph. Ultrasound is also useful to visualize gut motility. Because of differences in anatomy, scanning positions for different organs are used. In rabbits and guinea pigs, paralumbar scanning is useful for kidneys and the reproductive tract. Ferret positioning differs from those used in dogs and cats. Assessment in reptiles requires anatomic knowledge but is useful for coelomic enlargements. A standoff may be necessary for scanning reptiles and birds. The handheld scanner that projects onto a cellphone or tablet computer makes point-of-care scanning practical to add to any examination.

Introduction

Wouldn't we all love to be able to use a Star Trek medical tricorder and scanner right now? In some ways, use of a point-of-care ultrasound (POCUS) device, particularly one that is a handheld unit going wirelessly to a cell phone, tablet or computer gives us some of that ability right now! While doing a physical examination is still a starting point for the hands-on exam, results can be inconsistent if you encounter a tense body, even after administration of a sedative. In many, you cannot fully palpate all the organs because other organs are in the way, such as a cecum in a rabbit or guinea pig making palpation of the rest of the abdomen difficult. Being able to observe organs, fluids, and even function (think heart) while having the animal on your exam table or cageside, can expedite care timelines, improve outcomes, and even give clients peace of mind that you are really looking thoroughly at their pet. Ultrasound units are now in a price range and of improved technology that every practice can afford to have it available.

An animal needs to be in an appropriate position to be comfortable, while allowing you to scan through the abdomen's four quadrants. Consider the cranial quadrant containing the liver, spleen, stomach and associated lymph nodes. The left quadrant contains the left kidney, adrenal, retroperitoneal space. From the same probe position, you can scan the caudal abdomen including the urinary bladder and reproductive tract. The right quadrant contains the right kidney adrenal, pancreas, and associated lymph nodes of the mesentery. In rabbits and rodents with large cecums, you may need to move the probe ventrally to look at the bladder and reproductive tracts. But even so, within 3-5 minutes you can have some answers – this essentially is the abdominal focused assessment with sonography for trauma (AFAST) technique so you also will be able to tell if there is free fluid in the abdomen. You can apply the thoracic focused assessment with sonography for trauma (TFAST) technique to the thoracic area, and quickly look for free fluid, major cardiac dysfunction, even air or fluid within the lungs. Importantly for many species, the ability to scan the cranial thoracic area can rule out masses versus enlarged lymph nodes.[1-8]

Application for Use in Exotic Pets

POCUS techniques have been described in detail for dogs and cats, but unfortunately applications for use in exotic pets have not been published. It is possible to apply these techniques, but given the varied anatomy of the different mammals, birds and herpetologic species, the practitioner requires guidelines for the best positioning of the probe. Practitioners need to: (1) have knowledge of the normal anatomy of the different species and (2) practice! Do a lot of scans on normal, healthy individuals of the different species. In small animals, and particularly in those with little cutaneous fat, a stand-off may be needed. Herpetologic species usually need adjustment to account for the presence of scales, scanning from the ventrum where the scales may be thinner can help. In chelonians, you image through leg fossa. The size of the probe versus the size of the chelonian may limit what you can do, although using a standoff in the small leg fossae will allow you to image a fair amount of the internal body. These can be made using distilled water or ultrasound gel filled procedure glove. There are commercial standoffs available, but most are too large for the animals you need to scan. (3) Practice! Do scans on ill animals. Many of these should be following administration of an analgesic and/or a mild sedative. If an animal is being anesthetized for radiographs or other diagnostics, while the animal is being induced or recovered, you can quickly check some organs or body cavities. For example, every male ferret over the age of 3 years, quickly scan the bladder and prostate during the physical examination. POCUS can be incorporated into every geriatric animal examination. The examination fee can be adjusted to just include this 3- to 5-minute scan.

Remember that gas/air, bone, and hair can introduce artifacts. Uroliths or cecoliths or even calcified eggs, tumors or other structures will cause artifacts and difficulties imaging organs that are masked by these. This is why you can tailor POCUS to any animal and look at almost any organ. Birds, however, are still difficult due to air sacs, but by using an organ such as the liver as a window you can image the heart. The avian GI and reproductive system can be challenging, but if you locate the ventriculus, you can actually see the contractions, and through it you may be able to see gonads, the spleen, and even assess the proventriculus. You just need to picture the anatomy and move the probe where it can penetrate to the organ or body cavity to be viewed. Again, many of us learned ultrasound as a rather static imaging system with specified probe locations and angles, but this was in dogs and cats. Be creative with how you do imaging. It is also helpful to image an animal before going to surgery or necropsy, and then compare the ultrasonographic image to what is actually present. Again, practice!

It is advisable to have ready resources showing the normal anatomy of various species. There are some references that have enough ultrasound pictures of normal anatomy to aid in viewing any patient.[9,10] Examples of scans that can be of immediate help on the examination table or cageside include the following.

Birds

Abdominal area, midline: presence of egg, fluid, stasis of ventriculus, masses, liver disease. Angle the probe through the liver cranially to view the heart. Fluid in the pericardium and general status of the heart.

Lizards

Paralumbar, lumbar fossa area to view kidneys, from ventrum: bladder, liver, reproductive tract. Probe placement axillary area aiming towards shoulder girdle, imaging of the heart.

Snakes

Ventrum, using a standoff: view entire snake. Especially helpful to look for obstructions, masses, retained eggs, cecoliths.

Chelonians

All leg fossas: internal anatomy.

Ferrets

Usually scan with the ferret either in dorsal recumbency or scruffed and held up. Start at bladder, prostate. Doing palpation along with the ultrasound can visualize quickly almost all organs. Angling up cranially through the liver may show fluid in the thoracic cavity. The heart and cranial mediastinum are easily viewed.

Rabbits, guinea pigs, and others with large cecums

From para/lumbar fossa area view kidneys, and possibly some of the reproductive tract. From the caudal ventrum view the bladder directly and use the bladder as a standoff to view the reproductive tract. The liver is viewed moving the probe ventrally and tracing the liver silhouette. The heart is imaged from the lateral thoracic areas, realizing rib artifact. The cranial mediastinum can be imaged as well from the lateral body wall.

Other small animals

Hedgehogs: sedation is almost always needed to be able to image. All images come from under the mantle. In lateral recumbency, image positions are close to those described for a dog.

Rodents

The probe size may image almost all internal structures from the ventrum of the abdomen or the lumbar fossa areas. Hamster cheek pouches can be checked by imaging over the pouch.

Conclusions

Principles of POCUS can be used in exotic pet practice, although positioning of the probe may be different depending on species. The practitioner needs to be familiar with normal anatomy, as well as be able to recognize abnormalities. The development of handheld or portable ultrasound probes makes it easy for all exotic practitioners to do ultrasonography. It also has become economically feasible to own the equipment.

References

1. Boysen SR, Rozanski EA, Tidwell AS, et al. Evaluation of a focused assessment with sonography for trauma protocol to detect free abdominal fluid in dogs involved in motor vehicle accidents. *J Am Vet Med Assoc.* 2004;225(8):1198-1204.

2. FASTVET. Available at: http://www.FASTvet.com. Accessed April 2, 2022.

3. Lisciandro GR, Lisciandro SC. Global FAST for patient monitoring and staging in dogs and cats. *Vet Clin North Am Small Anim Pract.* 2021;51:1315-1333.

4. Lisciandro GR. AFAST target-organ approach and fluid scoring system in dogs and cats. *Vet Clin North Am Small Anim Pract.* 2021;51:1217-1231.

5. McMurray J, Boysen S, Chalhoub S. Focused assessment with sonography in nontraumatized dogs and cats in the emergency and critical care setting. *J Vet Emerg Crit Care.* 2016;26(1):64-73.

6. Lisciandro GR. Abdominal and thoracic focused assessment with sonography for trauma, triage and monitoring in small animals. *J Vet Emerg Crit Care.* 2011;21(2):104-22.

7. Boysen SR, Lisciandro GR. The use of ultrasound in the emergency room (AFAST and TFAST). *Vet Clin North Am Small Anim Pract.* 2013;43(4):773-797.

8. Lisciandro GR. *Point-of-Care Ultrasound Techniques for the Small Animal Practitioner.* 2nd Ed. Ames, IA: Wiley-Blackwell; 2021.

9. Krautwald-Junghanns ME, Pees M, Reese S, Tully R. *Diagnostic Imaging of Exotic Pets.* Hannover, Germany: Schlutersche Verlagsgesellschaft mbH & Co; 2011.

10. Johnson-Delaney CA. Ultrasonography in ferret practice. In: Lewington JH, ed. *Ferret Husbandry, Medicine and Surgery.* 2nd Ed. Sydney, NSW, Australia: Elsevier Saunders; 2007:417-429.

Exotic Emergency Triage, Medicine, and Surgery

Meera Music, BVM&S, MRCVS,
Alicia McLaughlin, DVM, CertAqVet,
Anneliese Strunk, DVM, DABVP (Avian Practice)

Session #2569

Affiliation: From The Center for Bird and Exotic Animal Medicine, 11401 NE 195th St. Bothell, WA 98011, USA.

Abstract: Exotic animals commonly present to veterinary clinics on an emergency basis. Both acute and chronic diseases may manifest with a sudden onset of symptoms due to the tendency of exotic species to hide signs of illness until it is advanced. Accurate diagnosis of the underlying cause for illness is essential to provide appropriate treatment, but in many cases, diagnostic testing must be staged to avoid inadvertent mortality of these fragile patients. In this presentation, we discuss how to optimize the clinical approach to the critically ill exotic animal, and review emergency triage, medical, and surgical management of exotic mammals, reptiles, and birds.

Introduction to Triage

The purpose of this presentation is to aid the emergency (ER) clinician in their ability to understand and utilize accurate triage to manage a hectic (typical) ER shift. Appropriate triage is one of the most valuable tools in any ER clinician's repertoire, and can improve efficiency, optimize staff and resources, and advance patient and client care.[1] Speed and accuracy during triage are essential on an ER shift, as the condition of patients can rapidly change.

Exotic pets frequently arrive at emergency veterinary clinics in critical condition and may be suffering from one or more concurrent chronic diseases resulting in acute presentation.[2] There is a large body of literature reviewing emergency medical techniques in exotic animals that clinicians should familiarize themselves with prior to treating these patients. Emergency techniques from dog and cat medicine can often be extrapolated to exotic patients, although anatomy and physiology will vary widely depending on species.

Triage is an essential skill for both the ER clinician and other veterinary support staff. Triage must occur numerous times throughout the intake process including the following:

1. Scheduling: reception staff communicates with the client and establishes whether the reported symptoms warrant an emergency visit or can be scheduled as an urgent care appointment or routine exam

2. Entrance of the animal to the emergency clinic: veterinary technicians or assistants acquire a brief history and perform an initial visual assessment

3. Physical exam: the veterinarian examines the patient and orders stabilization care (when/if appropriate)

4. Diagnostic sample collection/procedures: staff and resources must be organized to determine order of diagnostic sample collection, placement of intravenous (IV) or intraosseous (IO) catheters, setting up continuous rate infusions (CRIs), and prioritization/timing of indicated procedures

5. Diagnosis: interpretation of the onsite diagnostic results and communication of the findings with the client

6. Treatment plan: prioritizing order of additional treatments/procedures based on test results

Standardization of triage is important to consistently and accurately communicate the condition of the animal, both to the client and to fellow staff members. This also aids in scheduling and sign-posting an accurate timeline to clients. Managing client expectations is an important part of an ER day, and keeping communication open is key.

Triage lists can be an excellent way to standardize patient care (Table 1). They can assist receptionists in making scheduling recommendations and help medical team members accurately prioritize critical patients. Any time phone triage occurs, it is important to warn clients of likely wait times based on the current status of the ER caseload. Clients should also be cautioned that ultimately, an in-person exam combined with diagnostic evaluation is the only way to determine accurate triage status.

Clinical symptoms of exotic pets are often masked at home as part of their drive to hide illness.[3] Symptoms are often hidden even more effectively in clinical settings secondary to stress, especially on initial presentation. Allowing a patient time to adjust to a new environment may aid in clinical assessment; however, if an animal is in critical condition, rapid decompensation is possible anytime from presentation to during or after assessment, as the ultimate presentation of acute-on-chronic disease.[4]

On presentation at the clinic, the technician or assistant should gather a brief history at the door to help prioritize patient care and identify the particular client's concerns. Important information that must be covered includes the following:

1. Whether or not the owner wishes for cardiopulmonary resuscitation (CPR) efforts to be pursued in the event of cardiopulmonary arrest.

   This should be noted clearly in the medical chart and on the patient's enclosure with either CPR or DNR (do not resuscitate) to improve clinic response in the event of an emergency. All CPR patients should have dosages of emergency drugs calculated at the time of presentation and kept with the patient at all times to maximize the chances of a successful CPR. There are automatic drug calculators that can be used to easily and quickly make these calculations.[5,6]

2. Client permission to provide supportive care if the animal appears to need pain medication or fluid support.

   If the animal is actively bleeding or struggling to breathe, then the only conversation that should be had prior to bringing the animal to the treatment area should be CPR status to prevent delays in providing appropriate patient stabilization.

A more detailed medical history, including husbandry conditions, can be triaged to a later time. Alternatively, stock electronic questionnaires (such as JotForms.com) can be utilized so the client can fill this out while their animal is being initially assessed.

Table 1. Triage system.

Color: Urgency (time frame)	Description of animal condition
Red: Now	Severe lethargy, birds sitting on the bottom of enclosure and fluffed Any animals laying on side/unable to stand Seizure Severe trauma, even if the animal appears to be ok Uncontrolled bleeding Fractures with open wounds or protruding bone Difficult, rapid, or labored breathing Not eating for 24+ hours (not reptiles) Animals that have been scratched/bitten by a cat, mammal, or bird Painful postures, hunched over, possibly rigid, possibly grinding teeth Genital bleeding, exhibiting signs of a prolapse Very small or young bird or mammal not eating for 12+ hours.
Orange: Today	Diarrhea Vomiting Poor response to stimulation Mild lethargy Birds, not acting quite right Diminished stool production over last day or days Concerns about recent surgery Any open wounds Eye injuries
Yellow: Tomorrow	Sneezing Eye discharge Masses Ear infections
Green: Can be seen in the next few days	Reptiles that are inappetent or constipated Reptiles with bad sheds Mild skin problems with no other concerns Annual or boarding exams with no other concerns

Emergency patients will be seen and given immediate stabilization treatment in order of triage severity. Unfortunately, exotic pets tend to hide symptoms of illness until they are so sick they cannot do so any longer, and they can suddenly decompensation without warning. Triage status may change as more information is obtained and as exam findings change. Wait times vary depending on caseload and may exceed 6 hours.

After the patient arrives at the clinic, the doctor should perform an immediate visual examination of the animal. This can occur even if the vet is on the phone with another client. If the animal is in critical condition or in need of more direct support, it is appropriate for the veterinarian to tell the client on the phone that they must end the conversation to tend to an emergency that has just come through the door.

Visual examination should include evaluating mentation, posture, and respiratory rate/effort. If the animal is struggling to breathe on presentation, it should be immediately placed in an oxygen chamber prior to a hands-on physical exam to permit stabilization. Ideally, the reception team should inform the treatment team that a dyspneic animal emergency was scheduled, so that an O_2 chamber can be prepared ahead of presentation.[3,7]

A temperature should be taken from mammals on intake (when feasible - this may not be possible for species such as hedgehogs). Rectal temperatures are an objective measurement of stability of the patient. Prognosis for rabbits[8] and guinea pigs[9] is significantly reduced when these animals are hypothermic on presentation. As temperatures continue to decrease below intake rectal temperature, the odds of death significantly increase. Another immediate point-of-care analysis that should be available is blood glucose. This is particularly relevant in the classic case of a seizuring or apparently comatose ferret secondary to an insulinoma, but it is also helpful in neonatal or very small animals. These animals are often hypothermic on presentation, but providing heat alone without glucose supplementation and complete, balanced nutritional support could result in a secondary crash after the patient is normothermic. Be aware that most point-of-care glucometers are less accurate compared to i-stat or VetScan analyzers for exotic species.[10]

Exams on critical patients should be performed in a staged fashion to minimize the length of time the patient is experiencing stress.[2] For example, the most prudent approach to an exam of a dyspneic cockatiel would be to quickly assess the following parameters while transferring the animal to oxygen: a brief palpation of the keel to determine body condition, which can indicate chronicity of the issue; brief coelomic palpation, and brief auscultation, which can alert the clinician to presence of coelomic or other disease.[3] The patient should then be placed in an oxygen chamber while a conversation with the client occurs that outlines realistic possible outcomes at this stage, including death. A delicate patient may pass away even with a cursory exam, let alone one that requires further handling, and so warning the client of possible outcomes prior to embarking on a more thorough examination is the kindest, and most appropriate option.

Once a complete examination has been performed, a follow up conversation with the client must occur, outlining the physical exam findings in more detail, the differentials that could be involved in the possible disease processes, possible prognosis, and diagnostic testing recommendations. Discussion of euthanasia in a case where the animal is in critical condition (especially when the client has expressed financial constraints) is important. Though it can be difficult to discuss ending the life of a family member, it is also important not to be vague or ambiguous about costs of long-term care (financial and time/effort) so that the client is able to make a fully informed decision. Even if finances do not play a significant role in medical decisions for the client, trying to be realistic about prognosis is very important so clients are fully aware of likely outcomes is very important.

Once the decision to move forward with hospitalization is made between the client and veterinarian, and the risks, prognosis, and finances have been discussed, then it is time to decide how to stage the diagnostics and treatments. The client should understand that the treatment plan is based on the patient's response to therapy, and/or changes to repeat diagnostic testing, as this will directly impact the decision to further hospitalize, move to surgery, or consider euthanasia.

It is important to prioritize diagnostics in order of those that will change the treatment approach. For example, in the case of the dyspneic patient, both radiographs and a complete blood count (CBC) would be important diagnostics to consider, but if finances were a concern, or staging the diagnostic tests needed to occur due to patient instability, radiographs would give more immediately actionable clinical information if the animal has evidence of pneumonia versus evidence of perihilar edema and an enlarged heart which is indicative of heart disease. [7] The decision to place a dyspneic animal on antibiotics with evidence of pneumonia on radiographs does not require an elevated white blood cell count or relative heterophilia on a CBC to make that clinical decision; however, these additional diagnostic tests still have value, as they may reveal unexpected pathologic changes, provide additional prognostic information, and provide the clinician with the ability to monitor trends and response to therapy.[3]

One useful tool when embarking on an exam of a dyspneic, stressed, or painful patient that is at risk for decompensation is sedation. Many sedation protocols exist, though at the authors' hospital low doses of midazolam ± butorphanol or buprenorphine are commonly utilized to lower stress levels and relax the patient. This can often improve outcomes when needing to handle the patient for any period of time out of the comfort of an oxygen chamber.[4]

Keep in mind with dyspneic avian and reptile patients, positioning can affect the stress level of the patient. In birds, an upright posture is ideal to reduce pressure on the respiratory tract. In the event that radiographs, bloodwork, and ultrasound are all approved for an avian patient with coelomic distention and/or respiratory distress, the first diagnostic to perform should be the ultrasound. Coelomic ultrasound can provide important information about the level of coelomic disease present.[3] If fluid is identified on ultrasound, it should be aspirated prior to taking radiographs, as reduction of the fluid will help to improve radiographic coelomic detail and will also drastically improve the animal's comfort, ability to breathe, and stability.[11] Remember that unpositioned dorsoventral and horizontal beam lateral radiographs can be used so that the patient can be kept in a normal, anatomic, upright posture, rather than restraint that may increase dyspnea. The final step in the diagnostic workup of a dyspneic animal is often phlebotomy, and this may be done immediately after radiographs. If the patient appears stressed or declining due to the positioning needed to obtain the sample, then the animal should be placed back into oxygen for a break prior to phlebotomy.

In the event ultrasound had revealed a significant problem (large mass present, large volume of fluid), then prior to performing the rest of the approved diagnostics, it would be best to discuss this current finding with the client. It is important for the client to be included in the decisions of the animal's care – even if the mass type or source of fluid is unknown, what is known is that it is significant enough to affect breathing to the point that the animal cannot breathe comfortably out of oxygen. It would be important to have another conversation regarding prognosis, possible medical therapeutic options (for example, in the case of reproductive disease leading to fluid accumulation or tumors, GNRH agonist therapy may be an excellent therapeutic option), surgical options, ongoing oncology referral, or euthanasia.[12] In the event that surgery is desired, then determining what is necessary for the animal to stabilize to the point the surgery is feasible is the next consideration.

If the client is interested in continuing treatment, then the rest of the diagnostics should be gathered to obtain information regarding the animal's stability, organ function, among other things, because if there are concurrent pathologic changes, they need to be addressed prior to surgery. It can be tempting when an animal presents in an emergency to have a more fundamental approach to a mass: cut it out! But it is important to go in a stepwise manner through the possible outcomes, and be realistic and patient when needed. Although surgery is a vitally important tool in any ER vet's repertoire, time needs to be taken to determine if surgery is truly in the best interest of the patient and if it is, then the stage should be set to give the patient the best possible outcome during and after surgery.

Surgery

Surgical intervention is required in a number of exotic emergency cases;[13] however, stabilization and careful preparation is required to maximize the chance of successful outcomes. In many cases, administration of fluid therapy, analgesia (including constant rate infusions), oxygen support (if indicated), control of hemorrhage via short-term application of pressure wraps, and time will improve an emergency patient's chances of surviving an anesthetic procedure.

Select cases that require immediate surgical intervention

The following are examples of cases that require immediate surgical intervention:

- Uncontrollable hemorrhage (no response to direct pressure, inability to place a pressure wrap)
- High likelihood of internal hemorrhage or damage
  - Liver lobe torsions with free fluid, decreasing packed cell volume (PCV), or uncontrollable pain
  - Extensive prolapse of organs from the cloaca (birds or reptiles) or rectal prolapse in mammals
  - Penetrating injuries to the thorax, abdomen, or coelom (predator bite wounds, projectile injuries)
- Tracheal obstruction (requires tracheostomy in mammals or airsac cannulation in birds)[14]
- Egg retention with respiratory distress that does not respond to oxygen therapy
- Severe abdominal pain that does not respond to analgesic therapy (including CRI administration)
- Sudden temperature drop and/or change in attitude in mammals with GI signs
- Herniation

In an effort to improve successful surgical outcomes, preparation is critical, even in emergency situations. Staff should be trained to place IV catheters, initiate CRI administration, and perform blood transfusions (especially autotransfusions) quickly. Ensure that equipment and tools are labeled and available for easy access. Utilize procedure equipment checklists, drug calculators, and walkthroughs with staff to prevent key supplies from being forgotten. There are resources to help with setting up protocols for constant rate infusions.[15,16]

Common emergency procedures performed at CBEAM

The following are emergency procedures that are commonly performed at CBEAM:

- Trauma surgery-laceration repairs, wound exploration, enucleations, amputations, fracture repairs, etc.
- Occlusal adjustments when intraoral severe soft tissue trauma is present (exotic mammals)
- Avian tracheoscopy and airsac cannulation

- Ferret urethral catheterization (with possible stone retropulsion)

- Ferret GI obstruction
  - Gastrotomy
  - Enterotomy
  - Intestinal resection and anastomosis

- Rabbit liver lobectomy (with concurrent autotransfusion if hemorrhage present)

- Gastric decompression

- Rabbit abdominal exploration
  - Gastrotomy
  - Enterotomy

- Reptile coelomic exploratory surgery/foreign body removal

Surgical intervention is not without risk, and post-operative recovery can be complicated and expensive. For these reasons, detailed discussions with owners are required even in emergency cases, so they are prepared for possible negative outcomes, and for the long-term expectations for case management and cost.

For example, a bird in respiratory distress with a tracheal obstruction would improve with placement of an airsac cannula, but if an owner is unable or unwilling to proceed with the diagnostics to identify the problem or treatments to correct the problem (that might include surgical intervention or months of oral ± nebulized medication), euthanasia should be recommended as an option from the start. If an owner is not sure about long term commitment, it still may be indicated to place an airsac cannula to allow the owner time to process the situation. It should be made clear that placement of the tube is not an insignificant procedure and requires hospitalization while the tube is in place.

Some procedures have historically been considered too dangerous to perform in exotic species, either related to complications from the procedure or anesthesia. With appropriate anesthetic monitoring (by both equipment and staff), maintenance of an airway/ventilatory support, and fluid support with IV or IO catheterization, anesthetic safety has dramatically improved in exotic species. Rabbit/rodent abdominal exploration and GI surgery has been considered too likely to result in post-operative complications in the past; however, if an animal is showing clear indications for surgical intervention, especially related to an obstructive process in the GI tract, the best outcomes are achieved from addressing the problem sooner than later to prevent issues from necrosis the longer an obstruction persists. In addition, this will minimize the amount of suffering the patient experiences if it turns out that the problem is beyond surgical correction, and euthanasia on the table may be the most compassionate decision. Please see below for further discussion of rabbit/small mammal GI disease surgical management.

Specific case examples

Severe avian trauma

In the author's (AS) experience, birds that sustain a traumatic partial amputation from a predator attack tend to have a better outcome when stabilized for 24 to 48 hours before surgery. As long as there is no active bleeding and no coelomic penetration has occurred, birds should be bandaged and given time to recover from the shock of the traumatic event before surgery.

Small mammal gastric dilation

Initial stabilization of cases with moderate to severe gastric dilation should include intensive management with IV fluids, CRIs, and anxiolytics to improve patient comfort. This may help symptoms resolve without the need for further intervention (such as gastric decompression or exploratory surgery). Gastric decompression is a technique that should be performed promptly if there is no improvement with initiation of the intensive care, and should be performed before surgical intervention is recommended.[17] Decompression can rapidly decrease the amount of tension within the gastric wall (which can lead to necrosis), dramatically increase patient comfort, and alleviate pressure on sections of the intestinal tract that might be compressed by the dilated stomach.

Decompression should be performed under general anesthesia using a large red rubber catheter that has been fenestrated and lubricated prior to passing into the stomach.[18] Severely depressed patients may only need sedation for decompression. Risks of the procedure may include rupture of the cranial GI tract, aspiration, and destabilization secondary to anesthesia. If blood is seen on the red rubber catheter, then the procedure should not be continued or repeated. Blood in this instance does not necessarily indicate perforation of the GI tract but does suggest severe mucosal irritation or inflammation.

If the gastric dilation is easily alleviated with decompression, but recurs afterwards, more than one decompression may be indicated; however, this makes it more likely there is another problem in the distal GI tract. If the gastric dilation continues to recur rapidly after decompression, or if the stomach cannot be easily/effectively decompressed, an exploratory surgery/gastrotomy should be performed.

Liver lobe torsion

Liver lobe torsions (LLT) have been identified in rabbits presenting with a wide variety of clinical signs. This includes rabbit GI stasis syndrome (RGISS) or any other nonspecific clinical signs of illness.

Physical exam findings of LLT may include cranial abdominal pain, palpation of a firm liver lobe, decreased body temperature, and pale mucous membranes; however, many rabbits present with sudden onset of inappetence as the only symptom. The authors have not noted a clear breed or sex predisposition, although lops have been previously reported to be overrepresented.[19] Classic laboratory findings supportive of LLT include anemia (usually moderate, sometimes severe) and mild to severe elevations in alanine transaminase (ALT). The range of ALT values documented in confirmed LLT is 89 mg/dl to >1000 mg/dl. Radiographic findings are inconsistent.

Although LLTs can be diagnosed on computed tomography (CT),[20] ultrasound is the most useful and rapid diagnostic modality for identification of the affected lobe(s). It is vital to visualize all lobes and compare blood flow between each lobe to accurately identify an absence or relative decrease in blood flow. The caudate lobe is the most frequently affected, but other liver lobes can also torse, and more than one lobe may be affected in the same animal. To identify the caudate lobe most easily, the authors utilize the following approach. Place the ultrasound probe caudal to the right side of the ribcage aiming dorsally. Visualize the right kidney and sweep cranially to identify the caudate lobe, which is usually in contact with the kidney. Evaluate the echotexture in comparison to the other lobes and use Doppler to assess blood flow. If free fluid is present, be prepared for performing an autotransfusion during surgery.

Although some rabbits will survive with a torsed liver lobe, surgical excision of the affected lobe(s) results in the fastest resolution of clinical signs and ongoing risk. With practice and an assistant surgeon to help with retraction and visualization, this procedure can be performed in less than 20 minutes. If the PCV is less than 20%, blood transfusion should be considered regardless of whether surgery is elected. The authors have seen two cases of concurrent gastric torsion in rabbits with LLT.

Advanced Imaging in Emergency Settings

CT is an advanced imaging modality that can be used in an emergency setting to rapidly identify problems, better determine prognosis, and guide treatment recommendations. In many cases, quick scans can be performed with sedated patients securely wrapped in a towel and/or placed in a plastic container. More intensive scans with contrast can be considered depending on patient stability and vascular access.

Case presentations that would benefit from CT evaluation

Cases that benefit from CT evaluation include the following:

- Abdominal wall hernias
- Suspected LLT
- Penetrating injury/traumatic injury
- Dyspnea/lower airway disease
- Nasal hemorrhage/obstruction
- Coelomic/abdominal disease
- Appendicitis in rabbits[21]

Summary and Conclusions

Addressing and standardizing triage across every aspect of the intake procedure of a patient is an important part of effective ER management. By appropriate triage at the outset of patient presentation and reevaluating case status numerous times throughout the hospitalization process, an ER clinician can make the best medical, diagnostic, and surgical recommendations, significantly improving patient outcomes.

Examples of emergency cases seen by an exotic emergency veterinarian are provided for discussion in Appendix 1.

References

1. Mathews K. *Veterinary Emergency and Critical Care Manual.* 3rd Ed. Guelph, ON: LifeLearn; 2017.

2. Graham J, Doss G, Beaufrere H. *Exotic Animal Emergency and Critical Care Medicine.* Hoboken, NJ: Wiley-Blackwell; 2021.

3. Music M, Strunk A. Reptile critical care and common emergencies. *Vet Clin North Am Exot Anim Pract.* 2016;19(2):591-612.

4. McLaughlin A, Music M, Strunk A. Stayin' alive: current topics in exotic emergency medicine. *Proc ExoticsCon.* 2019;20-33.

5. Costello M. Principles of cardiopulmonary cerebral resuscitation in special species. *Sem Avian Exot Pet Med.* 2004;13(2):132-141.

6. Lichtenberger M. Shock and cardiopulmonary-cerebral resuscitation in small mammals and birds. *Vet Clin North Am Exot Anim Pract.* 2007;10:275-291.

7. McLaughlin A, Strunk A. Common emergencies in small rodents, hedgehogs, and sugar gliders. *Vet Clin North Am Exot Anim Pract.* 2016;19(2):465-499.

8. Di Girolamo N, et al. Rectal temperature is a prognostic indicator in client-owned guinea pigs. *J Small Anim Pract.* 2021;62(10):861-865.

9. Di Girolamo N, Toth G, Selleri P. Prognostic value of rectal temperature at hospital admission in rabbits. *J Am Vet Med Assoc.* 2016;248(3):288-297.

10. Petritz O, Antinoff N, Chen S, et al. Evaluation of portable blood glucose meters for measurement of blood glucose concentration in ferrets (*Mustela putorius furo*). *J Am Vet Med Assoc.* 2013;242(3):350-354.

11. LafeberVet. Abdominocentesis in birds. Available at: https://lafeber.com/vet/abdominocentesis-in-birds/. Accessed Feb 1, 2022.

12. LafeberVet. Reproductive emergencies in birds. Available at: https://lafeber.com/vet/reproductive-emergencies/. Accessed Feb 1, 2022.

13. Bennett RA, Pye GW. *Surgery of Exotic Animals.* Hoboken, NJ: Wiley-Blackwell; 2022.

14. LafeberVet. Air sac cannula placement. Available at: https://lafeber.com/vet/air-sac-cannula/. Accessed May 27, 2022.

15. Schnellbacher R, Comolli J. Constant rate infusions in exotic animals. *J Exot Pet Med.* 2020;35:50-57.

16. Veterinary Information Network. Constant rate infusion calculator. Available at: www.vin.com. Accessed Feb 1, 2022.

17. Lennox A. Updates in medical versus surgical management of rabbit GI stasis syndrome. *Proc ExoticsCon.* 2021.

18. Guerrera B. Successful decompression technique in rabbits presenting with gastric outflow obstructions. *Proc ExoticsCon.* 2019;423-426.

19. Graham JE, Orcutt CJ, Casale SA, et al. Liver lobe torsion in rabbits: 16 Cases (2007 to 2012). *J Exot Pet Med.* 2014;23(3):258-265.

20. Dagget A, Loeber S, Le Roux A, et al. Computed tomography with Hounsfield unit assessment is useful in the diagnosis of liver lobe torsion in pet rabbits (*Oryctolagus cuniculi*). *Vet Radiol Ultrasound.* 2021;62(2):210-217.

21. Di Girolamo N, Petrini D, Szabo Z, et al. Clinical, surgical, and pathological findings in client-owned rabbits with histologically confirmed appendicitis: 19 cases (2015-2019). *J Am Vet Med Assoc.* 2021;260(1):82-93.

Appendix 1. A day in the life of an exotic emergency veterinarian. Cases for discussion.

1. The case of the fluidy fowl
    - Owner brought chicken in for suspected egg binding and lethargic behavior
    - Physical exam shows severe ascites with no palpable egg(s)

2. The case of the steamed Senegal
    - Parrot landed in a pan of boiling water
    - Mild inflammation at burn site

3. The case of the flying ferret
    - Ferret jumped off a balcony
    - Femoral fracture palpable, very painful, hematoma at site

4. The case of the blah bunny
    - Normal temperature, mildly dehydrated, quiet, tense abdomen

5. The case of the lethargic lagomorph
    - Low body temperature (98 °F), mildly dehydrated, quiet, slightly pale mucous membranes

6. The case of the wryneck rabbit
    - Rabbit with sudden onset of head tilt (<30 minutes' duration) that is stable and not dehydrated
    - Owner is extremely worried

7. The case of the respiratory rat
    - Older rat, moderately obese, in respiratory distress
    - No upper respiratory congestion, other 5 rats in household asymptomatic

8. The case of the gasping galah
    - Open mouth breathing with neck extended, marked biphasic respiratory effort

9. The case of the painful pig
    - Intact female guinea pig that is straining to urinate
    - Patient's bladder is small and expressible, but she vocalizes on palpation and is straining
    - Urine is pink tinged

10. The case of the hemorrhagic hedgehog
    - Hedgehog comes in for second opinion, veterinarian 1 month ago saw it for a "urinary tract infection" and started antibiotics, bleeding went away but now it is back
    - Patient is producing massive amounts of blood clots but running around and acting normal

11. The case of the dying dragon
    - Bearded dragon that hasn't eaten in 1 year
    - Patient is emaciated and near death

Getting the Most Out of Our Journals: Finding the Evidence in a Peer-Reviewed Paper

Mark A. Mitchell, DVM, MS, PhD, DipECZM (Herpetology)

Session #2701

Affiliation: From Louisiana State University School of Veterinary Medicine, Skip Bertman Drive, Baton Rouge, LA 70803, USA.

Abstract: We all have a desire to use evidence-based medicine to manage the needs of our patients. Unfortunately, we are in a field where objective, evidence-based data are lacking. To counter this, it is important for all professionals in our field to contribute new knowledge. One of the reasons more research is not done has been attributed to a lack of understanding of study design and statistics. The author believes that these types of obstacles can be overcome if veterinarians approach these challenges as they do veterinary medicine. Veterinarians are trained to use hypothetico-deductive reasoning and decision tree analysis when approaching their cases. Patterns evolve out of these methods which help grow our confidence as veterinary scientists. Similar patterns exist with study design and statistics. By following these similar thought processes, it is possible for veterinarians to gain the necessary confidence they need to perform these important tasks and contribute to our body of evidence-based medicine. In addition, knowledge of these methods can also improve the veterinary scientist's ability in interpret the peer-reviewed literature. As study flaws are common, understanding the limitations of a peer-reviewed paper can minimize the likelihood of incorporating this flawed information into our knowledge base. The goal of this masterclass is to provide attendees with an introduction to the methods used by the author to perform these important tasks. By learning these patterns, attendees can increase their confidence and understanding of study design and statistics.

Introduction

Study design

Success with research starts with the design of a study. Many of the failures clinicians experience can be directly related to a lack of planning during the original design and development of a study. The types of studies performed in veterinary medicine are varied and may include basic bench top research to clinical trials. With such a broad range of topics and hypotheses to be tested, it should not be unexpected that the types of studies required to answer these scientific questions can vary. Developing a sound understanding of the different types of studies available and when/how to use them to answer specific research questions is essential to the success of an individual(s) planning a research project.

There are four basic study designs that will be discussed: cross-sectional studies, case-control studies, cohort studies, and experimental studies. To decide which study design is best, first consider the types of data to be collected and the specific hypotheses being proposed. Other considerations when planning a study include the feasibility of collecting the data, any temporal components associated with the study, and the resources (both financial and personnel) available. If you can answer these questions in the early stages of planning a study, it will help ultimately determine which study design will be best fit your needs.

Cross-sectional studies are an excellent first step for attempting to ascertain the occurrence of some outcome (e.g., disease process) and any associated independent variables in a population. This type of study is commonly referred to as a prevalence study. Cross-sectional studies evaluate a study population at a given point ("snapshot") in time. These types of studies are routinely pursued in veterinary clinical medicine because the data can be collected from patients as they present to determine the prevalence of some occurrence. These types of studies are well suited for those research questions attempting to identify how common or uncommon a disease might be or whether there are any inherent risk factors associated with a specific disease. A deficiency of these types of studies is that they do not provide information relative to the temporal nature of a disease in a case/population. For example, if you are interested in estimating the prevalence of *Ophidiomyces* in snakes by evaluating PCR results, you will be able to determine which snakes are positive or negative and which, if any, risk factors (e.g., species, sex, activity) predispose them to an outcome of being positive or negative; however, you will be unable to determine when the snakes were exposed to the fungus or for how long. Of course once you have completed a cross-sectional study and determined such a disease was prevalent, follow up studies (e.g., cohort studies) could be designed to further evaluate a temporal question.

Case-control studies are commonly done in veterinary clinical medicine because they are often designed to be retrospective in nature and can utilize historic (already available) information. These types of studies can provide a great deal of information regarding risk factors for a disease or outcome without needing a large amount of research dollars. With case-control studies the disease outcome is known and the relative effect different risk factors have on that outcome can be assessed by matching the patients to conspecifics that are presented for different outcomes. This type of study is well suited for preliminary investigations (pilot studies) that hope to collect sufficient results to pursue research grants for more developed studies or as a stand-alone study that can contribute to evidence based medicine.

Cohort studies are not commonly done in veterinary clinical medicine because of their longitudinal design and cost. Whereas a case-control study evaluates a known outcome and seeks to determine the effect of specific risk factors on that outcome, cohort studies seek to longitudinally evaluate the potential outcome(s) that occur as a result of various risk factors. This type of study should not be used for rare disease events, as it might not be possible to measure a specific outcome within a reasonable amount of time. These types of studies have contributed significantly to human medicine, and would likely do the same in veterinary medicine if the resources were available to pursue them.

Experimental or intervention studies are one of the most common types of studies pursued in veterinary clinical medicine. This type of study, when done properly, places the control of the research in the hands of the investigators. Experimental studies can range from being as simple as evaluating a simple outcome while controlling a single independent variable (e.g., one group receiving a treatment and another a placebo) to more complex cross-over studies evaluating multiple independent variables and outcomes.

Regardless of the study design being used, it is imperative to consider the sample size required for the study, the methods used for the randomization of the study subjects, and the potential for introducing and controlling different types of bias that could have negative effects on the results. Something that is expected for most study proposals through granting agencies and institutional animal care and use committees is the pre-determination of the sample size required to answer specific hypotheses. In the past, investigators might get by using examples of previous studies to estimate their sample size for a study; however, this is not considered sufficient and pre-determining sample size using a priori information specific to the sample population and the data collected are now required. Being able to have a sound estimate for a sample size for a study can go a long way in limiting problems later (e.g., introduction of a type II error). Random allocation of subjects into test groups should be done using appropriate methods, such as random number generators. Software to perform this is readily available, so it is now possible to be thorough and random when implementing a study. Finally, bias is a major contributor to study failure. Pre-evaluating a study for these types of error can help minimize the likelihood of them occurring, which will increase the overall power in the results.

There is always a great deal of work required to complete a research study. To ensure that the results of the study will have value and contribute to evidence-based medicine, it is important to design a study that will answer the specific hypotheses being proposed. Taking time to consider the potential pitfalls that might occur in a study during the initial planning stages will greatly improve an investigator's likelihood of success.

Parametric versus Non-Parametric Statistical Tests

The first step in selecting statistical tests is to determine which group of statistics to use. For simplicity, we will divide the statistics into two categories: parametric and non-parametric statistics. Certain assumptions must be met in the study design to properly perform and utilize a statistical test. Two of the most common assumptions, which are needed in both parametric and non-parametric tests are: (1) data collected for analysis should be obtained from study subjects randomly selected from the population represented in the study and (2) the individuals and their data should be independent of one another. These represent basic tenets to follow regardless of the statistics used. Note other assumptions must be met for each specific statistical test, the experimental design should reflect these assumptions. A major difference between the two types of statistics is related to the distribution of the data; parametric statistics should only be used on data that meets the assumption of normality, while non-parametric statistics do not have this restriction. Thus, if we simplify this to the types of data, it is obvious that categorical and ordinal data must always be analyzed using non-parametric statistics and that continuous data that doesn't meet the assumption of normality must also be analyzed using non-parametric statistics. The only data type that we would use parametric statistics for are those that meet the assumption of normality. Now, it is important to point out that continuous data that doesn't meet the assumption of normality can be transformed (e.g., log transformed) to potentially normalize the data. If this transformation is successful, then the more robust parametric statistics could be used to analyze the data; however, it would still be important to report the data by their median and percentiles as the transformed data would not be clinically relevant.

Statistical software

Statistical software is available that has simplified the process for analyzing data. These software programs come in a variety of packages, from those that require programming to those that don't require programming. The programs that require programming tend to be more robust in what they can accomplish. Examples of these types of programs include SAS (SAS Institute Inc., Cary, NC USA) and R (https://www.r-project.org/). R is free. MedCalc and SPSS (current version 24.0; IBM Statistics, Armonk, NY USA) do not require programming, although programming can be done with SPSS when more involved statistical methods are required. The author uses a combination of these programs. For those running very basic statistics, such as chi square tests and Fisher exact tests, free calculators are available on the Internet. The Centers for Disease Control and Prevention also have a free software program, EpiInfo (https://www.cdc.gov/epiinfo/index.html). Regardless of the software being used, it is important to understand when to select a test to ensure the analysis is correct, as these software programs will generate results for whatever test is selected, even if it is the wrong test.

Summary

The field of zoological medicine requires evidence-based knowledge to expand our understanding of these magnificent animals. Evidence-based knowledge requires hypothesis generated research that is validated using appropriate study design and statistical analysis. Every member of our profession can and should feel an obligation to contribute the generation of evidence-based science. For it is only through this new knowledge that we can achieve our potential as a true specialty.

Section 36

Hybrid: Emergency and Critical Care

In Honor of Marla Lichtenberger, DVM, DACVECC

Anna-Marie Ruoff, DVM;
Julianne Vickstrom, DVM
Session Chairs

Wildlife Rehabilitation: Rapid Diagnosis!

Renée Schott, DVM, CWR

Session #2519

Affiliation: From Wildlife Rehabilitation Center of Minnesota, 2530 Dale St. N, Roseville, MN 55113, USA.

Abstract: This fun and interactive class utilizes real cases and many pictures to review some of the most common presentations of wildlife and the diagnoses associated with those cases. Be prepared to participate, laugh, and maybe even learn something! Species covered include songbirds, waterfowl, waterbirds, turtles and more! Types of cases span those with simple diagnoses and quick turnarounds to complicated cases with involved treatment plans.

Recommended Reading

1. Hernandez SM, Barron H, Miller E, et al, eds. *Medical Management of Wildlife Species.* Hoboken, NJ: Wiley-Blackwell; 2019. doi:10.1002/9781119036708.

2. Mullineaux E, Keeble E, eds. *BSAVA Manual of Wildlife Casualties.* 2nd Ed. Gloucester, England: British Small Animal Veterinary Association; 2016.

3. Friend M, Franson JC, eds. *Field Manual of Wildlife Diseases: General Field Procedures and Diseases of Birds.* Madison, WI: US Geological Survey; 1999.

4. Ruth I. *Wildlife Care Basics for Veterinary Hospitals: Until the Rehabilitator Arrives.* Washington, DC: Humane Society of the United States; 2012.

5. Gage L. *Hand-Rearing Wild and Domestic Mammals.* Ames, IA: Blackwell Publishing; 2002.

6. Miller E, ed. *Minimum Standards for Wildlife Rehabilitation.* 4th Ed. St. Cloud, MN: National Wildlife Rehabilitators Association and International Wildlife Rehabilitation Council; 2012.

7. Stocker L. *Practical Wildlife Care.* 2nd Ed. Oxford, England: Blackwell Publishing; 2005. doi:10.1638/1042-7260(2000)031[0586:br]2.0.co;2.

8. Meredith A, Keeble E. *Self-Assessment Colour Review: Wildlife Medicine and Rehabilitation.* London, England: Manson Publishing; 2011. doi:10.1201/b15226.

Basic Wildlife Rehabilitation Triage

Renée Schott, DVM, CWR

Session #2518

Affiliation: From The Wildlife Rehabilitation Center of Minnesota, 2530 Dale St. N, Roseville, MN 55113, USA.

Abstract: It can be difficult for a practitioner to know which wild animals will survive the rehabilitation process to release and which will not. This master class uses data from The Wildlife Rehabilitation Center of Minnesota, one of the largest wildlife rehabilitation centers in the country, to help predict which patients will make it to release and which will not. The goal of this class is to help those in private practice know when to euthanize on admission and when to treat and transfer to a wildlife rehabilitator.

Adapted from Schott R. Wildlife rehabilitation: Until the rehabilitator arrives. Proc ExoticsCon. 2020.

Introduction

Private practitioners are often regarded as experts across all taxa, regardless of the focus of their practice. Clients and the general public will occasionally call with questions about injured or orphaned wildlife, or may simply show up in the lobby with the animal. It behooves the clinic to be prepared to deal with these scenarios so the finders have resources and options. It is important that clinics are prepared to handle and refer these calls appropriately for the sake of efficiency, the welfare of the wild animal, and the safety of the patients in the hospital. In addition, it is important for private practitioners to be armed with basic triage and euthanasia skills to humanely end the suffering of wildlife that cannot recover.

The Wildlife Rehabilitation Center of Minnesota (WRCMN) admits > 19,000 sick and injured wild animals each year. Asterisks (*) below indicate older data from WRCMN and updated data from WRCMN is used during the live class to help others make informed decisions. It is difficult to compare data from one rehabilitation center to another due to differences in location, resources, capacity, species, diseases, etc. For more details about how the data was compiled, please contact renee@wrcmn.org.

Preparation

Before wildlife is in the lobby or calls start coming in, clinics should prepare by following the steps below.

Find a wildlife rehabilitator or facility to work with

It is illegal to possess native wildlife without the proper permits. Veterinarians must be working with a permitted rehabilitator in order to possess wildlife they are treating. In addition, most states require rehabilitators to work with a licensed veterinarian to keep their permits active. Rehabilitators are familiar with natural history and veterinarians know veterinary medicine; working together ensures the best outcome for the animal. In addition, rehabilitators can determine when an animal needs to be rescued and when it should be left alone. Calls are best referred to rehabilitators or facilities set up to field these questions, as questions are complicated, and answers aren't always straight forward.

Compile resources

Where can you refer phone calls or injured patients to? Do you know the addresses or hours of facilities in the area? Who is a contact at your state natural resources or fish and game department and at US Fish and Wildlife? Making a resources sheet and SOPs for your reception will limit miscommunications.

Know the laws

Typically, each state regulates who can rehabilitate wildlife and how. Federal regulations exist for migratory birds as well. While it is usually legal for veterinarians to treat and/or euthanize wildlife, they must transfer all migratory birds to a licensed wildlife rehabilitator within 24 hours of stabilization[1] and they cannot possess other species for more than 24-48 hours unless they are already working with a permitted rehabilitator. There are often additional guidelines and regulations surrounding threated or endangered species. Keep in mind, species that have a hunting season in your area must be treated as food animals; withdrawal times and the list of prohibited medications apply.[2] More details about legal responsibilities and restrictions can be found in Duerr and Whittington, 2017.[3]

Euthanasia

For a wild animal to allow capture, it is usually very debilitated. Release proportions vary according to species, age and injuries, but range from 24%-85%.[4-7] Euthanasia is common, necessary and humane. In most areas, the number of animals that need rehabilitation are too numerous for the available resources. When considering the below cases, it is important to consider these resources, the anticipated number/species/age of animals needing rehabilitation, rehabilitator/facility limits, and welfare. The ability of the rehabilitator, facility, or clinic to accept animals for rehabilitation will be dependent on the above factors. If resources (time, space, money, people) are low, the overall ability of the caretakers will diminish which will lower welfare for everyone. Euthanizing the least likely to survive and basing the number of animals in care on resources available, will allow those with the best prognosis to receive the best care and ultimately the best chance at release. Because stretching resources results in a lower standard of care for all, euthanizing more animals on admission in a strategic manner results in an increase in the overall number of animals being released (*).

What types of orthopedic conditions should be euthanized?

Prognoses and treatment options are taxa specific.[8] Consider your ability, the ability of the wildlife rehabilitator, and the species. Below are generalizations that apply to most taxa unless specified otherwise.

Necrotic or old, compound fractures: Compound fractures are inherently more prone to healing complications owing to the increased likelihood of contamination and/or infection. Necrotic bone must be removed and in general, most species need to have almost perfect fracture alignment with < 10% loss of long bone length. Necrotic bone is almost always accompanied by necrotic soft tissue. Many wild animals (especially birds) have very little soft tissue over certain bones and the loss of that soft tissue means the bones underneath will not heal or will have complications.[9] Older fractures are also inherently prone to more complications; if the bone has stopped healing, then all bone ends must be freshened to be stimulated to heal again. In addition, muscles, tendons and ligaments often have contracted, making reduction difficult. Not only do the bones need to heal perfectly, but all joints need near perfect range of motion. The definition of "old fracture" varies with species, age and type of fracture. For instance, a two-day old compound, comminuted fracture in a nestling songbird may be considered old, whereas a 7-day old closed, simple, midshaft fracture in a fox may be fresh enough to fix.

Traumatic joint injuries: In the author's experience, wildlife that experience traumatic joint injuries, begin trying to reduce the abnormal movement in the joint by producing fibrous tissue around the joint. This reduces range of motion chronically which leaves the animal non-releasable. In the author's experience, luxations and sub-luxations (excluding coracoid luxations off the keel, which usually heal well) will almost always heal with reduced range of motion and should be euthanized. In addition, traumatic joint luxations frequently will cause osteoarthritis.[9,10]

Successful surgical management of elbow luxations in raptors[11] and a woodchuck[12] have been reported; however, in both papers the possibility of osteoarthritis was not discussed. Unsuccessful treatment of luxations in wildlife are also reported.[12,13] Articular fractures need perfect alignment and rigid fixation to prevent secondary degenerative joint disease.[14] Wildlife develop reduced range of motion and signs of osteoarthritis within weeks of sustaining an articular fracture, no matter how rigid the fixation and perfect the anatomic alignment. WRCMN euthanizes all traumatic joint injuries that result in intra-articular fractures or luxations/subluxations of joints (excluding coracoid luxations off the keel) because of past experiences (*).

Shoulder fractures in aerial insectivores: in North America, swallow species (including but not limited to Barn Swallows *Hirundo rustica* and Cliff Swallows *Petrochelidon pyrrhonota*), swift species (including but not limited to Chimney Swift *Chaetura pelagica* and Vaux's Swift *Chaetura vauxi*), nightjar species (including but not limited to Common Nighthawk *Chordeiles minor* and Eastern Whip-poor-will *Caprimulgus vociferus*) have very high metabolic demands; Vaux's swifts may need to eat >5000 arthropods each day![15] Flight must be perfect in order to accomplish this. A bird with even the slightest flight deficit may die of starvation after release. In addition, most of these species migrate to Central or South America for the winter which also requires sustained flight. To adequately assess pre-release flight ability, these species need a 30' (swallows/swifts) and 50' (nightjars) flight cage. Any deficit should be heavily scrutinized. In the author's and other's experiences (Veronica Bowers, Native Avian Care, personal communication, 2017) < 1 % of shoulder fractures in these species will recover to a releasable state. Several species do not eat in captivity and must either be trained to accept handfeeding, gavage or force fed. All are high stress species. Feather condition must be pristine. Husbandry for these species is complicated and best not undertaken by a novice.

All aerial insectivores with any shoulder girdle (coracoid, scapula, clavicle) fractures are now euthanized at WRCMN due to past experiences, other's experiences and the intense husbandry care these species require (*).

What type of traumatic eye injuries should be euthanized?

In general, any condition that will result in permanent vision loss should be considered for euthanasia, even if only in one eye. Prey animals will be left unable to watch for predators, predators may have a difficult time catching enough prey (no live prey testing in captivity is equitable to catching prey in the wild). This is a controversial topic and some wildlife rehabilitation veterinarians and rehabilitators have differing opinions. This author does not release one-eyed animals, with an exception being some aquatic turtle cases. Traumatic eye injuries that result in vision loss include luxated lens (posteriorly and anteriorly luxated), ruptured globes, complete retinal detachment with incomplete reattachment, and any other condition that results in permanent vision loss.

What types of other issues should be euthanized?

Chronic pain: Wild animals have evolved to hide signs of pain and to self-repair injuries as much as possible. If they did the opposite, they would likely become prey. Therefore, any condition which has the potential to result in chronic pain in a human, should be assumed to have the potential to result in a chronic pain for the animal, should be grounds for euthanasia of the wild animal for humane reasons. Wild animals cannot receive chronic pain medications or other palliative care in the wild.

Spinal trauma with no deep pain: Most rehabilitators and veterinarians recognize spinal trauma via pelvic limb paresis or paralysis. Because of the anatomy of the spinal cord, loss of deep pain perception indicates a grave prognosis for return to function. In addition, neuropathic pain has been reported in paralyzed companion animals.[16] Therefore, animals without deep pain should be humanely euthanized. To test for deep pain perception, use a toe on the limb that is paralyzed/paretic. Pinch the toe; start with a superficial pinch and then slowly pinch harder. Watch for head movement, vocalization or other indication that the pain sensation made it up the spinal cord and to the brain. Ignore all movement in the limb that is being pinched as the withdrawal reflex may be intact or hyper reflexive, depending on where the lesion is. One cannot discern between the withdrawal reflex and voluntary movement when pinching a toe, and wild animals can have very strong reflexes, so it is imperative to ignore movement in the leg and watch the head of the animal. In addition to spinal trauma, brachial plexus avulsion occasionally occurs. If deep pain is not present in one forelimb of a unilaterally forelimb paralyzed animal, brachial plexus avulsion should be considered. Brachial plexus avulsion will not heal to a releasable state and these animals should be euthanized. Animals with no deep pain in a limb should be euthanized as they will not recover to a releasable state.

Seizing rabies vector species: Rabies vector species (raccoon, coyote, fox, bat, skunk) with abnormal neurologic signs should be considered for euthanasia due to public health concerns. It is important to know your rabies vector species and rabies virus variants in your area and what your department of health requires you to do; they may want the animal tested for rabies or to have the contact of the finder to talk about exposure. Generally, a seizing animal bears a poor prognosis, regardless of species.

Distemper virus infection suspects: Distemper suspect species (i.e., raccoon, coyote, fox) with abnormal neurologic or severe respiratory clinical signs, with or without leukopenia should be considered for euthanasia. Consider the infectious nature of this pathogen and the domestic animals in your clinic which are susceptible. In addition, wild animals have a high probability of permanent neurologic issues even if they survive, which is rare.

Only 6.7% (25) of the 373 adult raccoons admitted to WRC from 2011-2017 were able to be released (*). Treatment was attempted with 68 and 25 (36.8%) of treated were released. While not all of these cases were distemper suspects, adult raccoons are difficult for the public to capture and transport; consider how debilitated this species must be to present for rehabilitation and this bears out in prognosis.

Infants: The younger the animal is, the lower it's chance of survival when being raised in captivity by humans. Reviewing data from previous years can help facilities find which variables are best correlated with survival to release.

WRCMN data (*):

- Eastern cottontail rabbits (*Sylvilagus floridanus*): WRCMN received 3133 young cottontails in 2017 alone. Historically, under 65 grams, healthy Eastern cottontail rabbits have < 30% chance of release. Under 75 grams, unhealthy Eastern cottontail rabbits have a < 30% chance of release. Definition of healthy: mild-moderate dehydration, thin (not emaciated), BAR, no injuries.

- Eastern grey squirrels (*Sciurus carolinensis*): WRCMN received 1285 young grey squirrels in 2017 alone. Historically, healthy grey squirrels under 40 grams have < 30% chance of release.

- Virginia opossums (*Didelphis virginiana*): WRCMN received 119 young Virginia opossums in 2017. Opossums 20-40g have a 50% chance of survival.

- Hatchling altricial birds (just out of shell) are euthanized as they have 0% chance of release at WRCMN.

Adult Eastern cottontails and Eastern grey squirrels with abnormal brain neurologic clinical signs
Adult Eastern cottontails and Eastern grey squirrels commonly present to WRCMN with abnormal brain neurologic clinical signs (e.g., head tilt, circling, head turn, ataxia). Necropsy results most commonly conclude the etiology to be *Baylisascaris procyonis* (or a nematode tract through the brain) or head trauma.

WRCMN received 233 adult cottontails with abnormal central nervous system signs in 2017 (*). From 2011-2017, 1727 total cottontail adults were admitted. Of these 641 had abnormal brain neurologic clinical signs. Of the 641, 3.3% were eventually released. Of the ones that were treated, 8% were released. This is compared to all other adult cottontails which had an overall release proportion of 5.5% and a treated released proportion of 19%. This author recommends euthanasia of all adult cottontails with abnormal brain neurologic clinical signs due to the poor release proportion.

WRCMN received 94 adult grey squirrels with abnormal central nervous system signs in 2017 (*). From 2011-10.22.2017, 1295 total grey squirrel adults were admitted. Of these 414 had abnormal brain neurologic clinical signs (such as head tilt, circling, head turn, ataxia, etc). Of the 414, 8.7% were eventually released. Of the ones that were treated, 16.1% were released. This is compared to all other adult grey squirrels which had an overall release proportion of 10% and a treated released proportion of 29.1%. This author recommends heavily scrutinizing any adult Eastern grey squirrel with abnormal brain neurologic signs before choosing to treat.

Conditions to Not Euthanize (if you have the resources)

Mycoplasma gallisepticum conjunctivitis in house finches

Mycoplasma gallisepticum (MG) originally spread from the production poultry industry to house finches (*Haemorhous mexicanus*) in the 1990s[17] and is now endemic in the house finch population.[18] When it first appeared, the poultry industry and biologist were afraid of MG recovered house finches becoming latent carriers and some antibiotics being ineffective;[19,20] however, it is very hard for wild house finches to actually transmit *M. gallisepticum* to domestic poultry[21,22] and there is a treatment, that when given appropriately can eliminate the carrier state.[23] Euthanizing one or even several hundred house finches because they have this disease will not affect the population or disease status. WRCMN treats and releases house finches with presumed MG conjunctivitis according to Mashima et al 1997.[23]

Severe mange in red fox

Juvenile red fox (*Vulpes vulpes)* occasionally are admitted in the fall with severe mange, presumably after having been forced out of maternal territory. While some of these fox are too sick or emaciated with secondary conditions to survive, many can be treated and released.[24] These patients require intensive critical care as they are usually severely emaciated, chronically anemic, and have a myriad of other health concerns that will develop during treatment (i.e., refeeding syndrome).

Animal missing digits

Certain species can survive well in the wild with missing digits; however, the location and number of missing digits that allow successful release are species dependent.[8]

How to Euthanize Various Wildlife Species

The techniques described below are used most frequently at WRCMN. They take into consideration the wild animal species, age, and physiology of the animal being euthanized as well as the safety of the handler; priority is given to decreasing apparent animal stress. There are other technique options. In the author's experience, the methods below have proven reliable and cause the least amount of stress to the animal.

Euthanasia Techniques

Isoflurane (or sevoflurane) via face mask

Restrain the patient and place head in mask through diaphragm or hook up vaporizer to an induction box with the animal inside. Start at maximal percentage until patient is fully anesthetized. Then administer appropriate amount of pentobarbital IV or IC.

Isoflurane box

- Liquid isoflurane is poured into a clear Rubbermaid-type container with a lid. Either do this in a vent hood or ensure container is on the ground/floor and you remain above to prevent accidental inhalation of isoflurane, which is heavier than room air. Ideally, a respirator would be worn to avoid inhalation of all fumes when not in the hood.

- If the animal is fractious, aggressive or high stress, ideally the container would be big enough for a live trap or kennel. Having several sizes of containers would be ideal for different sized animals.

- Use 1 tbsp liquid isoflurane for every 4 gallons container volume.

- The box should be mostly sealed (but not completely sealed as the gas expands).

- After isoflurane is poured into the container, the animal (or animal in a live trap) can be transferred to the container and the lid replaced.

- Monitor the patient to ensure it quickly lose consciousness.

- Leave patient in the tub for at least 10 minutes to ensure over dose of isoflurane has caused death. Overdose of isoflurane is an approved humane euthanasia method according to the AVMA.[25]
  - Pinkie mammals may need longer depending on their body temperature and age.

- Rock tub before opening to ensure no movement.

- Open tub outside (tub on ground, ensure you are above tub or are wearing a respirator) or in a vent hood.

- Use a stethoscope to ensure heartbeat of the patient has stopped.

- Leave tub outside or in vent hood for at least 10 minutes.

Table 1. Euthanasia choices for wildlife.[a]

Species/group	Age	First choice	Second choice	Notes
Eastern cottontail rabbit	Adult (250 g-1.5 kg)	0.2 ml (100 mg/ml) ketamine; 0.4 ml (5 mg/ml) midazolam; 0.2 ml (10 mg/ml) butorphanol → mix→ give IM → wait 15 min. If still alive, put into isoflurane box[b] until deceased or mask at 5% isoflurane until anesthetized then IC sodium pentobarbital (390 mg/ml) 0.5 ml	2 ml (100 mg/ml) alfaxalone → give IM → wait 15 min If still alive, put into isoflurane box[b] until deceased or mask at 5% isoflurane until anesthetized then IC sodium pentobarbital (390 mg/ml) 0.5 ml	Prolonged struggling and breath holding when placed in isoflurane box without pre-medications and even more prolonged in the isoflurane face mask/induction box
Other mammal species	Neonate, just weaned	Isoflurane box[b] until deceased. They dislike the strong smell and will wipe their faces but are anesthetized within 10 seconds		Young mammals' brains are very tolerant to hypoxia and they hold their breath and struggle in an isoflurane mask at 5% for several minutes
Breath-holding avian species (i.e., diving species, waterfowl, waterbirds)	Adults	3 mg/kg midazolam, 5 mg/kg butorphanol, 5 mg/kg ketamine IM → wait 15 min then isoflurane box[b] until deceased	Debilitated individuals are often too weak to hold their breath and do well in the isoflurane box[b]	
Avian species	Any (except breath-holding species)	Isoflurane box[b] until deceased		Birds have efficient gas exchange in their lungs; they become anesthetized in ~2 breaths
Raccoons/ similar sized canids	Adults	0.3 ml (0.5 mg/ml) dexdormitor; 0.3 ml (10 mg/ml) butorphanol→ mix→give IM; if seizing, start with 1mg/kg midazolam→ wait 15 min, if still alive, put into isoflurane box[b] until deceased or mask at 5% isoflurane until anesthetized then IC sodium pentobarbital (390 mg/ml) 3 ml	Debilitated individuals are often too weak to hold their breath and do well in the isoflurane box[b]	
Aquatic turtles	Any	15-25 mg/kg alfaxalone IM →wait 15 min → 10 mg/kg Propofol IV wait 15 min→ same volume (as propofol) pentobarbital IV wait 15 min → pith	Other anesthesia medications can be used but all must induce anesthesia and end in pithing	Chelonian brains are so tolerant to hypoxia[26-28] that reanimation can happen if using anesthetics/ pentobarbital alone

[a]The Wildlife Rehabilitation Center of Minnesota (WRCMN) admits > 19,000 sick and injured wild animals each year. Asterisks (*) are used to indicate WRCMN data.
[b]See isoflurane box euthanasia technique below.

Basic Wildlife Stabilization

During all of these steps, decreasing stress is the most important thing we can do for wildlife. Stress will kill wildlife and often a hands-off protocol is preferable over an invasive one.

First. Do visual triage. Is there an obvious injury that would render the animal non-releasable? If so, euthanize immediately. Immediately treat any breathing difficulties (place in oxygen chamber) or active hemorrhage (apply bandage to stop bleeding).

Second. Treat thermoregulatory deficits. If animal is hypothermic, warm it up. If animal is hyperthermic, cool it down.

Third. Leave the animal alone! Leave the animal in a quiet, dark room so it can de-stress and relax. This also gives the animal time to adjust internal body temperature.

Fourth. Give fluid therapy. Assume at least 5% dehydration in all animals admitted. Subcutaneous fluids are fast to administer and a great choice in almost all cases.

Fifth. Give pain medications. If there is any injury, appropriate pain management is paramount.

References

1. Code of Federal Regulations. 2018. Available at: https://www.ecfr.gov/cgi-bin/text-idx?SID=0e9fd773e2a078e4c54d159ce4a9be80&node=50:9.0.1.1.4.2.1.2&rgn=div8. Accessed June 1, 2022.

2. Schott R. Extra-label drug use in wildlife rehabilitation medicine. *Wildl Rehabil Bull.* 2017;35:(2)33-36.

3. Duerr R, Whittington JK. Legal responsibilities and restrictions on veterinarians working with wildlife. *Wildl Rehabil Bull.* 2017;35(1):25-37.

4. Kelly A, Bland M. Admissions, diagnoses, and outcomes for Eurasian sparrowhawks (*Accipiter nisus*) brought to a wildlife rehabilitation center in England. *J Raptor Res.* 2006;40(3):231-235. doi:10.3356/0892-1016(2006)40[231:ADAOFE]2.0.CO;2.

5. Molina-López RA, Casal J, Darwich L. Final disposition and quality auditing of the rehabilitation process in wild raptors admitted to a wildlife rehabilitation centre in Catalonia, Spain, during a twelve year period (1995-2007). *PLoS ONE.* 2013;8(4):1-8. doi:10.1371/journal.pone.0060242.

6. Barrows Z, Fix S. Raptors rehabilitated study in Iowa. *Wildl Res.* 1990;26(1):18-21.

7. Orós J, Montesdeoca N, Camacho M, et al. Causes of stranding and mortality, and final disposition of loggerhead sea turtles (*Caretta caretta*) admitted to a wildlife rehabilitation center in Gran Canaria Island, Spain (1998-2014): a long-term retrospective study. *PLoS ONE* 2016;11(2):1-14. doi:10.1371/journal.pone.0149398.

8. Duerr R, Purdin G, eds. *Topics in Wildlife Medicine: Orthopedics*. St Cloud, MN: National Wildlife Rehabilitators Association; 2017.

9. Johnson AL. Fundamentals of orthopedic surgery and fracture management. In: Fossum TW, Hedlund CS, Johnson AL, et al., eds. *Small Animal Surgery.* 3rd Ed. St. Louis, MO: Elsevier; 2007:930-982.

10. Denny HR, Butterworth S. Osteoarthritis. In: Denny JR, Butterworth S, eds. *A Guide to Canine and Feline Orthopaedic Surgery.* 4th Ed. Hoboken, NJ: Wiley-Blackwell; 2000:52-63.

11. Ackermann J, Redig PT. Surgical repair of elbow luxation in raptors. *J Avian Med Surg.* 1997;11(4): 247-254.

12. Chambers JN, Hanley CS, Hernandez-Divers S. Hip luxation in a woodchuck (*Marmota monax*): successful treatment by closed reduction and a modified Ehmer sling. *J Zoo Wildl Med.* 2004;35(4):569-571. doi:10.1638/04-012.

13. Bennett KR, Desmarchelier MR, Bailey TR. Surgical correction of bilateral patellar luxation in an american black bear cub (*Ursus americanus*). *J Zoo Wildl Med.* 2015;46(2):359-362. doi:10.1638/2013-0288R1.1.

14. Newton CD. Etiology, classification, and diagnosis of fractures. University of Pennsylvania. 2018. Available at: http://cal.vet.upenn.edu/projects/saortho/chapter_11/11mast.htm#b. Accessed November 4, 2021.

15. Bull EL, Beckwith RC. Diet and foraging behavior of Vaux's swifts in northeastern Oregon. *Condor.* 1993;95(4):1016-1123. doi:10.2307/1369437.

16. Granger N, Carwardine D. Acute spinal cord injury: tetraplegia and paraplegia in small animals. *Vet Clin North Am Small Anim Pract.* 2014;44(6)1131-1156. doi:10.1016/j.cvsm.2014.07.013.

17. Citti C, Blanchard A. Mycoplasmas and their host: emerging and re-emerging minimal pathogens. *Trends Microbiol.* 2013;21(4):196-203. doi:10.1016/j.tim.2013.01.003.

18. Dhondt AA, DeCoste JC, Ley DH, Hochachka WM. Diverse wild bird host range of *Mycoplasma gallisepticum* in eastern North America." *PLoS ONE.* 1014;9(7). doi:10.1371/journal.pone.0103553.

19. Wellehan JFX, Zens MS, Calsamiglia M, et al. Diagnosis and treatment of conjunctivitis in house finches associated with mycoplasmosis in Minnesota. *J Wildl Dis.* 2001;37(2):245-251. doi:10.7589/0090-3558-37.2.245.

20. Williams ES, Yuill T, Artois M, et al. Emerging infectious diseases in wildlife. *Rev Sci Tech.* 2002;21(1): 139-157. doi:10.20506/rst.21.1.1327.

21. Luttrell MP, Stallknecht DE, Kleven SH, et al. *Mycoplasma gallisepticum* in house finches (*Carpodacus mexicanus*) and other wild birds associated with poultry production facilities. *Avian Dis.* 2001;45(2):321. doi:10.2307/1592971.

22. Stallknecht DE, Luttrell MP, Fischer JR, Kleven SH. Potential for transmission of the finch strain of *Mycoplasma gallisepticum* between house finches and chickens. *Avian Diseases* 1988;45:352-358.

23. Mashima TY, Ley DH, Stoskopf MK, et al. Evaluation of treatment of conjunctivitis associated with *Mycoplasma gallisepticum* in house finches (*Carpodacus mexicanus*). *J Avian Med Surg.* 1997;11:20-24.

24. Newman TJ, Baker PJ, Harris S. Nutritional condition and survival of red foxes with sarcoptic mange. *Can J Zool.* 2002;80(1):154-161. doi:10.1139/z01-216.

25. American Veterinary Medication Association. *AVMA Guidelines for the Euthanasia of Animals: 2013 Edition*. Available at: https://www.purdue.edu/research/oevprp/regulatory-affairs/animal-research/docs/2013_AVMA_Guidelines.pdf. Accessed June 1, 2022.

26. Jackson DC, Ultsch GR. Physiology of hibernation under the ice by turtles and frogs. *J Exp Zool A: Ecol Genet Physiol.* 2010;313A(6):311-327. doi:10.1002/jez.603.

27. Larson J, Drew KL, Folkow LP, et al. No oxygen? No problem! Intrinsic brain tolerance to hypoxia in vertebrates. *J Exp Biol.* 2014;217:1024-1039. doi:10.1242/jeb.085381.

28. Ultsch GR. The ecology of overwintering among turtles: where turtles overwinter and its consequences. *Biol Rev Camb Philos Soc.* 2006;81:339-367. doi:10.1017/S1464793106007032.

Working with Exotic Pet Rescues and Shelters

Cathy A. Johnson-Delaney, DVM

Session #2678

Affiliation: From Washington Ferret Rescue & Shelter, 12522 128th Lane NE, Kirkland, WA 98034, USA.

Abstract: Bird and exotic animal rescues and shelters have need of veterinarians with experience not only with single pet birds and exotics but with animal colonies as well as general shelter expertise. Resources including facility, finances and personnel are often extremely limited. Preventive health program development as well as training of personnel, assistance with facility planning, development of standard operating procedures, regulatory issues interpretation and guidance, establishment of an animal database and records systems, and development of the organization's culture are part of what the exotic animal veterinarian can provide beyond just medical and surgical care. The veterinarian needs to establish clear guidelines for the amount of involvement that can be provided as well as fees for any services, pharmaceuticals, imaging and other equipment usage.

Introduction

Most exotic pet veterinarians will be asked to help with an exotic pet rescue and shelter, as more of these organizations are being formed. And unlike government or major animal charity organization that have a full or part-time veterinarian on staff, many of these are small and run by individuals, and cannot afford full or part-time staffing but turn to a veterinarian on an as needed basis. This master class is primarily discussing the small non-dog, non-cat exotic animal rescues and shelters that do not regularly employ a shelter veterinarian, but rather try to work with a local exotics veterinarian to provide for their facility. The veterinarian fills a unique role in assuring that these species receive appropriate diet, husbandry and medical care. How involved in the actual organization the veterinarian is becomes a labor of love, as most time spent is gratis, although supplies, facility/equipment usage, technician time, diagnostics, and pharmaceuticals used may be compensated often at a deep discount or as negotiated fees. Ferrets, rabbits, small rodents, hedgehogs, sugar gliders, reptiles and birds are increasingly being relinquished to shelters as owners can no longer care for them. Because these species are usually not accepted at municipal and major humane facilities designed for dogs and cats, many exotic pet shelters are run by individuals. Without governmental funding, most of these private shelters depend on donations, adoption fees, and volunteer participation to care for these animals. The veterinarian can be instrumental in educating and training the volunteers about the proper care of these species.

Shelter medicine its own field of veterinary medicine, although primarily it is geared to dog/cat or equine facilities, with veterinarians usually paid for services.[1-4] These are usually linked to municipal or government animal control agencies or private humane facilities operated by or associated with national or international humane organizations. Veterinarians working with shelters have to change from primarily working with individual animals and owners to a more herd health mentality as far as realistic management from a space, person-power, budget, and time consideration.[1-4] Working with shelters that cater to the non-dog, non-cat abandoned animal face even more obstacles as husbandry, dietary, medical and even zoonotic disease issues of the different species are extremely variable. They also rarely receive support from community or governmental sources. Guidelines are beginning to be included in some of the major shelter medicine texts, as well as there now exists the Association of Shelter Veterinarians as well as an American Board of Veterinary Practitioners (ABVP) Certification in

Shelter Medicine.[5-9] There also is a program for Shelter Medicine at UC Davis.[10] For purposes of this discussion, sanctuaries are facilities that take in unwanted or neglected animals to care for them for the life of the animal. Shelters operate with the goal of re-homing the animals.

An Emerging Necessity: Exotic Pet Shelter Medicine

"The good of the many outweighs the good of the few, or the one."[11] While many volunteers and animal lovers may find this quote contrary to their beliefs and reasons for establishing or assisting a shelter, the reality is that there are always far more animals being abandoned and needing care than shelters can maintain for the life of the animal. Most avian and exotic pet shelters would consider themselves no-kill shelters (i.e., that the animal will be maintained until it is adopted or fostered). No-kill shelters must still have in place a policy to humanely euthanatize an animal presented in grave condition or one that is very ill or in pain that cannot be readily treated given the funding, staffing, and resources of the shelter. Shelters must establish a policy of how much of their budget can go to treating any particular animal. This is one of the aspects of shelter medicine that is the most difficult for volunteer caretakers: the decision of when to euthanatize an animal, who makes that decision, and the accepted humane method as well as the policy for disposal of the body. This policy needs to include not only the animals surrendered in grave condition, but those housed at the shelter as well as those in foster care homes. All policies must be in writing, and training sessions should be held on a regular basis. Shelters should have legal documentation of surrender of an animal to their care. Table 1 is a sample surrender form in current usage. Forms utilized by shelters for surrender, fostering and adoption should be reviewed by an attorney.

Shelters frequently begin in someone's home who takes in abandoned pets. While their initial reason to do so is compassion and care, without planning for increased numbers of pets within their home, they are easily and quickly overwhelmed. Most people work outside the home, and additional exotic animals require extra time as well as financial commitment. Individuals soon find they need help maintaining the animals, hence they recruit friends or other owners of the same exotic animal species to help. Individuals also soon find out they are tied down with so many animals, as word usually gets out that they will take that species, and more unwanted animals are surrendered. Those individuals who enlist the help of others to care for these animals usually reach the point where they realize they need to form some sort of organization to continue, particularly if they realize they cannot keep all the pets themselves. They need to develop a method to adopt them to appropriate homes. Individuals who just continue to collect animals in the guise of a shelter but do not enlist the help of others and begin forming an organization and network are hoarders. Hoarders or collectors are usually well-meaning but rarely can house, feed, and provide the needed medical care for these animals. An individual identified as a hoarder frequently comes to the attention of animal control agencies, public health agencies and even mental health agencies. Animals confiscated then need to be rehabilitated and placed.

Sanctuaries and shelters that tend to be longer lived are those that are located in a separate facility from any one volunteer or founder's home. Location of a facility must comply with zoning that will allow for an animal shelter. Most residential neighborhoods particularly within city limits have restrictions on the number of animals that can be maintained, although many do not specify caged pets. Shelters that organize and become 501(c) non-profit organizations also tend to be more stable, and less likely to get into the "hoarding" mode, although policies need to be in place to inspect foster animal placement homes, and to keep foster animals medically checked. Sanctuaries and shelters need to set a limit for the number of animals they can afford, have space to house, feed, and clean, in addition to being able to provide for veterinary care. Volunteers frequently take home more than they can really care for too, so setting up mandatory training programs for fosterers as well as a schedule of medical evaluations of the animals should be incorporated into the policies of the shelter. All policies and training programs should be reviewed annually, and training sessions held regularly.

Table 1. Washington Ferret Rescue & Shelter Surrender Agreement Form.

I/we____have made the decision to surrender the following named ferret(s) and have chosen to enlist the adoption services of the Washington Ferret Rescue & Shelter, hereafter designated as WFRS, that will find new homes as appropriate.

Name of ferret:_______________________________________ Color_______________

Male/Female Age________Neutered: Y/N CD Vaccination Date: ____________________

Name of ferret:_____________________________________ Color_______________

Male/Female Age________Neutered: Y/N CD Vaccination Date: ____________________

Name of ferret:_______________________________________ Color_______________

Male/Female Age________Neutered: Y/N CD Vaccination Date: ____________________

Name of ferret:_______________________________________ Color_______________

Male/Female Age________Neutered: Y/N CD Vaccination Date: ____________________

It is understood that the ferret(s) are being surrendered for their continued welfare and that the WFRS accepts said ferret(s) in an AS IS condition. It is further understood, therefore, that the WFRS assumes full legal ownership, thereby full responsibility, for any current or future illness while in the custody of the WFRS. It is also understood that items surrendered with the ferret(s) become the legal property of the WFRS, and as such will not be refunded nor returned.

Total confidentiality will be observed unless the surrendering party wishes to communicate with the *adopting* party, in which case, *this request must be put in writing* for the adopting party's acknowledgement.

The WFRS reserves the right of refusal in placing any ferret so surrendered if in their judgment certain prerequisites of ownership are not met. No ferret will be placed in any city, county, or state that restricts or bans ownership of ferrets specifically or by inclusion, nor shall any ferret be knowingly placed in a rental/condominium that excludes pet ownership, nor shall any ferret be knowingly placed in a home or institution for the purpose of ritual, rite, or medical experimentation.

The WFRS guarantees to the surrendering party that their pets will receive the utmost care and consideration in the placement with a new owner and that future contact with the same will be done in the interest of continued health and welfare concerns.

The WFRS operates this Shelter as a service to ferrets and as such, is not a business. It is a non-profit 501(c)(3) corporation.

Signed________________ Date____________________
Name (Printed)____
Address____
Phone________________ Email_________________________________

I would like to make a donation of $__________, to help cover the food and medical costs of the ferret(s) I am surrendering to the WFRS. Donations may be tax-deductible where applicable.

Received by WFRS staff member: _______________ Date _______________

The veterinarian's role

The shelters that organize will elect a board, and establish guidelines to govern acceptance of animals, adoption and foster policies, and a budget. Most are staffed by volunteers. It is usually after individuals within the organization have to provide medical care for the animals that the veterinarian with a special interest in that species is approached. The individual veterinarian needs to decide if her/his role will be one of volunteer pro bono, per-diem surgeon, as-needed or regular consultant, and/or board member. In any role, it usually is a labor of love; working with a shelter takes time and commitment.

The first step is usually to negotiate reduced fees for animals owned by the shelter, although most soon realize that a knowledgeable veterinarian can assist not only with the medical care of the animals, but help the board with many aspects including liability and insurance needs (largely due to injuries volunteers or visitors may sustain while handling the animals), dietary and husbandry requirements, preventive health care including parasite control, vaccinations, dental care, among other things. If appointed or elected to the board of a shelter, the veterinarian can also serve as a medical director and set up programs to prevent disease outbreaks as well as training of volunteers to recognize illness. The veterinarian may be crucial to assisting a shelter establish record keeping, particularly of intake, adoption, foster, and medical needs. Vaccination records also must be kept. Many shelters rely on volunteer's memories rather than establish an electronic database and records system.[12] Numbers of animals on-hand at the start and end of the year, numbers of adoptions, fostered, transferred, died, etc. are statistics that are needed not only to create a budget plan for the shelter, but are needed to apply for many types of grants.

In determining compensation, most shelters have extremely limited funding. Providing at-cost medications, supplies, diagnostic testing or at a reduced rate, with minimal fees for actual labor are usually what is negotiated. Actual fees, costs, and discounts need to be established in writing, with a clause to review the fees at least on an annual basis. Establish which shelters, sanctuaries you will work with, and set a limit. It is not unusual for word to get out that you do a service for one shelter, and others try to enlist your services as well. In urban areas, a practitioner could spend all clinical time just working with shelter animals, but economically this is not realistic. Set special appointment or visitation times for shelter animal examinations, procedures and surgeries so that you are not taking time away from regular clinic duties. The veterinarian's duties and services to be provided to the shelter in any capacity should be formalized in writing. This is extremely helpful as volunteer organizations have frequent turnover of people involved.

The veterinarian also may assist in the development of contracts with municipal animal control facilities and humane animal facilities that do not accept the avian/exotic species. This may include facilities inspections from the municipal animal control agency before they will transfer an animal to the private shelter. The veterinarian also is responsible for making sure that animals adopted out from the shelter conform to local animal regulations. Spay or neuter, vaccinations and microchipping are frequent requirements, with documentation provided to the new owner. The veterinarian also is responsible for issuing rabies vaccination certificates and other health certificates for transport as required locally.

When working with any animal rescue organization, there will be a large turnover of volunteers and even board members. Many caregivers will experience compassion burnout, particularly if they have fostered many terminally ill and geriatric animals. The veterinarian can provide assistance and training to shelter volunteers to deal with animal death. Grief counseling can be made available.

General guidelines

The veterinarian needs to make sure the animal care is appropriate for the species being sheltered including psychological needs. Animals arriving at shelters are stressed, many are in sub-optimal nutritional and health status, and may experience separation anxiety from their former owners. Volunteers accepting surrendered animals can follow an intake checklist (Table 2) that ensures an animal in medical need is seen by the veterinarian. This checklist (Table 2) is an intake form in use by the Washington Ferret Rescue & Shelter (WFRS) in Kirkland, Washington State and can be adapted for other shelters and species. In areas with endemic heartworm, heartworm diagnostic testing and preventive medications should be included in an intake evaluation. Some ferret facilities also require Aleutian Disease titers and/or Canine Distemper titers as part of admission process. The veterinarian is instrumental in establishing the needed evaluations for a specific shelter. The veterinarian also trains volunteers in accomplishing the checklist. Isolation and quarantine procedures and facilities should be in place for animals newly surrendered. Each animal admitted to the shelter should have a file that includes the surrender papers, history information, intake record, and medical records. In addition to the stored paper records, computerized record-keeping should be encouraged. When an animal is adopted, fostered, or dies, this information is also kept in that animal's record. Storage requirements for these records should be determined by the Board and may comply with medical records retention requirements in that locality.

The veterinarian assesses the animals and can make the recommendation if the animal is adoptable or should be fostered by a volunteer of the shelter. The shelter needs to set the parameters and training programs that qualify a volunteer to be a fosterer. Volunteering to be a "fosterer" should not be a way to get a free exotic animal. It is recommended that a contract between the shelter and the trained fosterer be signed to ensure the fostered animal returns for regular medical care and veterinary attention. Some shelters that do not have permanent facilities put all animals into foster care defined as housing until a permanent home and owners can be found. Other shelters that have facilities to house animals deemed adoptable (i.e., good temperament, at most middle-aged, and healthy). Animals defined that need to be fostered include behavioral problems, geriatric condition, or continuing medical needs. With this definition, foster care is essentially "pawspice care" and necessitates the shelter maintain frequent contact with the foster caregivers to ensure the health care and well-being of those animals. It is recommended that a contract between the shelter and the trained fosterer be signed to ensure the fostered animal returns for regular medical care and veterinary attention or is still available for adoption where there is no shelter facility. Fostered animals or any being held for adoption all legally belong to the shelter and therefore it is necessary for the shelter to track them, maintain records, and ensure health and well-being. Samples of adoption and fostering agreements are presented in Tables 3 and 4.

The veterinarian can also make recommendations on housing, disinfectants and cleaning methods used, protocols for sanitation, appropriate disposal of waste, appropriate storage of food and other products needed for the care of the animals. The facility and procedures should emphasize stress and noise reduction, no crowding, and no exposure to potential predators.

Table 2. Sample shelter intake record.

WASHINGTON FERRET RESCUE & SHELTER
SHELTER INTAKE RECORD

Ferret Name: ______________________________ **Sex** (M/F): ________ **Date of Arrival**: _______________
Age: ______________________ **Ear Tattooed**? Yes No *(description)*: ________________________________
Color: *sable dk /lt sable hvy/lt silver champagne chocolate albino DEW black (solid) cinnamon*
Markings: *mitts bib kneepads panda hood blaze roan*
Nose: *pink speckled spots black outlined* **T** *on nose*

Intact (Y / N): ________ **Scent Glands** (Y / N): ____**Microchip** (Y / N): ______________________________
Weight on Arrival in grams: __________________**Underweight**: (Y / N) @ 1 week: @ 2 weeks:
Type of food ferret is eating if known:
____

Intake Care (check upon completion):
☐ Nails Clipped
☐ Ears Cleaned *(note anomolies below)*
☐ Ivermectin Injection *(0.2cc)*
☐ Bathed *(if necessary)*
☐ Advantage Flea Control - *2 drops between shoulder blades or Revolution*
☐ Teeth (note condition) ____
☐ Distemper (date) ______**0.5cc Children's Benedryl 20 mins prior* 2nd Distemper (date):*__________
☐ Imrab-3 Rabies (date) **0.5cc Children's Benedryl 20 mins prior**
☐ Fecal Float Result ________________ Date _____________

Physical Inspection:
☐ No Apparent Physical Problems
☐ Inflamed Prepuce ___________________
☐ Swollen Vulva _____________________
☐ Hair Loss ____
☐ Scars/Marks/Scabs
☐ Other ____

Temperament:
☐ Good temperament
☐ Aggressive
☐ Seems to dislike other ferrets
☐ Lap Ferret
☐ Biter

Condition on Arrival:
☐ Excellent
☐ Good
☐ Fair
☐ Poor

Surrendered by:
☐ Owner
☐ Animal Control
☐ Humane Society
☐ Other ____

Shelter volunteer(s) doing intake:

____
Request for Vet Visit (Y/N): ___________________ Date of Vet Visit: ____________________________
Adoptable? Yes No **Foster?** Yes No Temp **Entered in Computer?** Yes No

Table 3. Example contract for adoption.

WASHINGTON FERRET RESCUE & SHELTER
CONTRACT FOR ADOPTION

I. The ferret(s) will be handled in a loving, gentle manner by all members of my household and by any guests that visit my home. The ferret(s) will be fed quality dry ferret food, have access to fresh water, will be kept indoors and will be housed in a cage or ferret-proof room when no one is at home to supervise the activities of the ferret(s). The ferret will be seen regularly by a veterinarian knowledgeable about ferrets.

II. If adopters are unable to spend quality time with the ferret(s), unable to keep them as pets, neglect, abuse, or mistreat the ferret(s), or must give them up for any other reason, the ferret(s) will be returned to, or reclaimed by, the Washington Ferret Rescue & Shelter (WFRS). A full refund of the adoption fee is guaranteed up to 7 days from the adoption date.

III. If the ferret's health is ever in question, the adopter will seek veterinary care. It would be appreciated if you would contact the WFRS if any health problems arise or when the ferret passes on so we may better our understanding of breeding, health, and life spans of ferrets. Ferrets will be vaccinated yearly against Canine Distemper or as antibody titer drops and it is required by State Law to have the ferret(s) vaccinated against Rabies once a year as required by regulations in the location of the adopter's residence. The next Canine Distemper shot due:______ _____The next Rabies shot is due:______________________

IV. Adopter understands that the WFRS cannot and does not make any presentations or guarantees, either expressed or implied as to the temperament, habits, or background of the named ferret(s). Therefore, in consideration of receiving the ferret(s), I hereby release the WFRS from any and all claims of liability for the actions of the ferret(s) being adopted.

This agreement applies to the following ferret(s):

____________________ ____________________ ____________________ ____________________

Signature:____ Date:___________

Volunteer Signature:_____________________________________ Date:___________

Questions? Call or email:____

A copy of this contract will be kept on file at the WFRS.

Table 4. Example contract for fostering.

WASHINGTON FERRET RESCUE & SHELTER
CONTRACT FOR FOSTERING

I. The ferret(s) will be handled in a loving, gentle manner by all members of my household and by any guests that visit my home. The ferret(s) will be fed quality dry ferret food, have access to fresh water, will be kept indoors and will be housed in a cage or ferret-proof room when no one is at home to supervise the activities of the ferret(s).

II. If adopters are unable to spend quality time with the ferret(s), unable to keep them as pets, neglect, abuse, or mistreat the ferret(s), or must give them up for any other reason, the ferret(s) will be returned to or reclaimed by the Washington Ferret Rescue & Shelter (WFRS).

III. If the ferret's health is ever in question, the fosterer will contact WFRS. The WFRS will direct the Fosterer to a WFRS-approved veterinarian or emergency facility. The WFRS may or may not authorize funds if the fostered ferret is taken for medical care without WFRS being notified for approval of costs otherwise the Fosterer must pay for veterinary care. The Fosterer understands that the WFRS cannot and does not make any presentation or guarantees, either expressed or implied, as to the temperament, habits, or background of the named ferret(s). The Fosterer must bring the fostered ferret(s) in to the WFRS at least on an annual basis for medical examinations, regular treatments and vaccinations. The WFRS expects the Fosterer to attend training sessions as they are offered at the WFRS. The Fosterer understands that the ferret(s) legally belong to the WFRS which is responsible for their health and well-being.

Therefore, in consideration of fostering the ferret(s), I hereby release the WFRS from any and all claims of liability for the actions of the ferret(s) being fostered.

This agreement applies to the following ferret(s):

________________ ____________________ __________________ ________________

Signature:____Date:_____________

A duplicate copy of this contract will be kept on file at the WFRS

Specific Species Shelters: Special Considerations for Health Management[5-9]

Depending on endemic disease problems in a geographic area, and whether animals are predominantly coming from indoor or outdoor homes, backyard breeders or pet stores, the following lists of potential disease problems within a shelter will vary. The following are guidelines.

Ferret shelters

Anorexia with/without gastrointestinal ulceration and subsequent diarrhea or hemorrhage, just due to the stress of being surrendered along with changes in the food or habitat is often termed “shelter shock”. The ferret suffers severe weight loss during the first week of arrival, despite all other health parameters being normal. These ferrets can be placed on famotidine at 2.5 mg/kg PO q24h (Pepcid Complete, Consumer Pharmaceuticals Co., Fort Washington, PA), and weighed daily. Hand-feeding soup (Carnivore Care, Oxbow Pet Products, Murdoch, NE) is used to nurse them through the first few days. In some cases, the ferret will be taken home by one of the volunteers to nurse through this period and help it become adjusted to its new life. Newly arrived ferrets are kept separate from the main colony until the medical issues listed below have been addressed and the ferret seems to be adjusted to shelter life. Attempts at this time are made to find the ferret suitable cage mate(s) or play groups for socialization.

Communicable disease and parasitic problems that are of particular importance in a ferret shelter include fleas, earmites, heartworm, coccidia, *Cryptosporidium*, influenza, SARS-CoV-2, canine distemper, and possibly enteric bacteria such as *Salmonella* or *Campylobacter* sp. Occasionally fungal skin infections may be found. Rabies in some locales may be considered if the ferret has been outdoors.

Non-communicable disease health concerns: anorexia, gastrointestinal ulceration, diarrhea, weight loss: stress, diet, and husbandry changes, geriatric conditions including heart disease, neoplasia, dental disease, trauma, anal/rectal complications following anal sacculectomy. If unneutered, surgical neutering must be done prior to adoption or fostering. Anal sacculectomy can be considered an elective surgery, but one that may increase the adoptability of the ferret. Most ferrets over the age of 4 years will need attention for adrenal disease treatment.[9]

Rabbit shelters

Many rabbits are brought to shelters that are feral (i.e. were once pets but were turned loose frequently in urban parks). These rabbits may be extremely anxious and frightened in addition to facing diet changes, caging of some type, and human handling. Use of anti-anxiety medications such as midazolam (Midazolam, Baxter Healthcare Corp., Deerfield, IL) when these rabbits are first brought to the shelter may help the rabbit adjust. Many of these rabbits are intact, may be pregnant or lactating and reproductive status must be included in the initial physical examination. Rabbits surrendered should be kept separate from the main shelter until disease status is determined and the rabbit is eating well with normal stools. The first few weeks a rabbit is admitted to the shelter it should be closely observed for urination, defecation, weight changes, and signs of stress. Upon arrival, the rabbit should be weighed and an individual intake record should be started. Full evaluation of each rabbit may be delayed for a few hours after transportation to allow the rabbit to become less frightened. Feral rabbits captured from outdoors may need to be kept in cooler rooms than rabbits accustomed to indoor home temperatures. Rabbit shelter volunteers should strive to prevent abrupt and loud noises as these may severely frighten the already tense rabbits.

Husbandry concerns include caging areas with solid footing rather than flooring entirely of bars or mesh. Substrates such as recycled newspaper pellets may make sanitation easier, although all wet and soiled bedding must be removed from the caging on a daily basis. Various commercial rabbit pellets, fresh hay, herbs, and greens must be stored to prevent spoilage and access by insects or rodents. Ventilation of any housing area should allow fresh air and circulation adequate to prevent ammonia and dusts from accumulating. Rabbits noted with upper respiratory signs should be housed separately from the resident shelter animals. Respiratory disease problems will be exacerbated by stress and introduced rabbits may potentially carry bacterial strains different from the resident rabbits. Shelters may require *Pasteurella* serology and/or culture and treat individual rabbits and determine appropriate adoption status. Surgical neutering should be delayed until respiratory signs are cleared, and the rabbit has adjusted to its new housing and diet. Adoption criteria include temperament, age, and existing medical conditions.

Communicable disease and parasitic problems that are of particular importance in a rabbit shelter: fleas, cuterebra, ear mites, fur mites, coccidia, *Pasteurella* sp., *Treponema*, *E. cuniculi*, dermatophytosis, abscesses.

Non-communicable disease health concerns include stress, anorexia, enteropathy, dental disease like malocclusions and abscesses, urinary calculi, sore hocks, dermatitis secondary to urine scalding, fur matting, other illnesses or injury such as trauma, wounds, and fractures. A rabbit presented with head tilt will need to be evaluated for otitis interna or neurologic problems due to multiple etiologies.

Other exotic mammals

Small mammal shelters are usually individuals or clubs organized by fanciers of specific species. Among the more common surrendered small mammals are guinea pigs, rats and sugar gliders, although a smaller number of hamsters, gerbils, fancy mice, and hedgehogs may also be abandoned at municipal animal control facilities, pet stores, or just turned loose to fend for themselves. All of these animals have unique diet, husbandry and medical needs. In the author's experience, most of these rescued animals are cared for in individual homes rather than designated shelter facilities. With all the rodent species, rescuers need to be aware of the potential to develop severe allergies to rodent antigens and should be encouraged to minimize dusts, exposure to urine/feces, and consider ventilation and filtration systems for rooms housing more than two or three rodents. Rodents should be housed in species-specific rooms and not mixed. The veterinarian may consider multiple-animal fee rates and care plans based on previous working relationships with these individuals. Table 5 lists medical conditions that need to be considered with these rescued species. All stressed rodents may develop enteropathies and have fatalities due to *Clostridium piliformes* (Tyzzer's disease) or *Clostridium perfringens*. Rodents may also carry the zoonotic tapeworm, *Rodentolepsis* (*Hymenolepis*) *nana*. With rats, and to a lesser extent mice, *Streptobacillus moniliformes* is usually carried, and can cause rat bite fever in humans, which can be extremely serious.

Table 5. Species-specific rescued animal concerns.

Animal	Concern
Guinea pigs	Anorexia (stress, change in diet, environment), enteropathy, hypovitaminosis C, dental disorders, dermatophytosis, pododermatitis including overgrown toenails, lice, respiratory disease (*Streptococcus* sp., *Bordetella bronchiseptica*, *Chlamydophila caviae*, etc.); urolithiasis (cystitis). Guinea pigs do not tolerate changes readily, and many rescued cavies require hand-feeding and nursing care to adapt them to a new home.
Rats	*Mycoplasma pulmonis* (other pathogens within the chronic murine respiratory complex, including Sendai virus, Sialodacryoadenitis virus; CAR bacilli, *Corynebacterium kutscheri, Streptococcus pneumoniae, Bordetella bronchiseptica*); dermatophytosis, mange mites, enteropathies (frequently dietary changes), pinworms, neoplasia.
Hamsters	Trauma including wounds, fractures, demodectic mange mites (immunosuppression, stress), neoplasia, enteropathy especially with *Clostridium perfringens*. Cheek pouches should be checked for impaction, infection, neoplasia. Hamsters that have been neglected may be malnourished, torporous, hypothermic, and dehydrated. Hamsters should be housed individually. Dwarf hamsters have different dietary requirements than Syrian (golden) hamsters, although other types of environmental requirements are similar.
Gerbils	Malnutrition, pneumonia and skin conditions such as dermatophytosis and nasal dermatitis related to high humidity (urine, feces, dripping water bottles and other sanitation issues), degloving injuries to tail from improper handling, enteropathy, trauma including fractures.
Hedgehogs	Mange mites, enteropathy, pododermatitis (poor sanitation and accumulation of feces), torporous due to cool ambient temperature, dehydration, malnutrition, neoplasia, pneumonia, ataxia associated with condition currently called wobbly hedgehog syndrome. Many hedgehogs initially may have been obese and when they are abandoned or face a major dietary change may become anorexic with secondary hepatic lipidosis.
Sugar gliders	Malnutrition, osteomalacia/fibrous osteodystrophy, torporous due to cool ambient temperature, dehydration, trauma (wounds, fractures), enteropathy particularly due to *Giardia* sp., pouch infections

Herpetologic shelters

Reptile rescues may range from individuals housing all the animals within their homes to those who can build a free-standing reptile sanctuary complete with a rainforest room. Most also utilize individual foster families as housing for adoptable or hospice reptiles. Foster families apply to the program and undergo training on recognizing medical problems as well as proper husbandry for the species selected to foster. Often they are provided with caging and lights appropriate to the selected species from the rescue organization. Once an animal is fostered there are quarterly examinations made by the adoption committee until the animal is adopted or deceased. Half-way homes also serve as quarantine facilities since newly acquired reptiles are segregated from other rescued reptiles. Unfortunately, a foster home system restricts the number of animals that can be accommodated. Shelters frequently stop accepting mature iguanas due to the large number of iguanas needing to be placed and the small number of adoptions of mature iguanas.

Reptiles surrendered to shelter and rescue organizations frequently fall into two categories: those that are too large/too aggressive to keep and those that have medical conditions created by poor husbandry. Often owners of reptiles are ill-equipped, particularly financially, to properly care for these animals. Medical and behavioral evaluations are important in determining immediately adoptable, adoptable after medical treatment/socialization, permanent fosters/hospice animals and those that should be humanely euthanatized. Legal issues surround venomous snakes and lizards, crocodilians and snakes over 12 feet in length in many jurisdictions. In many states/municipalities these animals are subject to confiscation by state/municipal animal control personnel. While some jurisdictions will require euthanatization of these animals, others place them in sanctuaries with people knowledgeable on how to properly care for these species.

All newly acquired reptiles should be given a physical examination by trained personnel. Any medical conditions are noted and the foster family is authorized to seek veterinary care and treatment from local veterinarians experienced with reptiles, that the organization/rescue has pre-approved. Most medical conditions seen by the author involve inadequate uVB light exposure, imbalanced nutrition, hypothermia and internal parasitism. Once improper husbandry and diet practices are corrected and parasitism addressed, these reptiles are usually adoptable. An exception may be reptiles infected with *Nannizziopsis* sp., boid inclusion disease, or *Cryptosporidium*, as these may pose a risk to other reptiles within a home or facility.

Those animals that are deemed too aggressive can be fostered with specially trained volunteers for socialization. Sometimes, retraining is effective but often these animals are deemed too dangerous and euthanized per shelter/sanctuary policy. In an effort to ameliorate hormonal aggression and subsequent abandonment, the author knows of one rescue organization that sponsors a spay/neuter program for iguanas. This program has decreased the number of iguanas seeking placement in that facility.

Reptile specific veterinary concerns include internal/external parasites (particularly in wild-caught animals), anorexia (stress, new food items, changes in environment), hypocalcemia/hyperphosphatemia (renal disease, metabolic bone disease), hepatic lipidosis (carnivorous lizards), bacterial abscesses, sepsis, respiratory disease in chelonians (e.g., *Herpesvirus*, *Mycoplasma* sp., *Bordetella bronchiseptica*), dystocia, neoplasia. Chelonians should be isolated from snakes because of possible *Entamoeba* cross-infection.

Avian Rescues and Sanctuaries

There are many well-known avian sanctuaries and rescue organizations across the United States. Veterinary concerns for avian rescues include quarantine procedures for new acquisitions, general medical screenings, geriatric screenings, disease outbreak protocols, nutritional requirements of the species housed as well as individual medical attention. Even minimally funded shelters need to have quarantine procedures, excellent sanitation protocols, predator resistant caging and adequate nutrition.

Frequently parrots are given up because of behavioral issues (e.g., screaming, aggression, feather picking). Behavioral assessment is very important, especially in determining adoptable versus sanctuary (unadoptable) candidate. Most birds surrendered come from single bird households, and many have underlying medical problems as well.

Specific veterinary concerns include internal parasites, viral infections/exposure (circovirus, bornavirus- PDD, polyoma, herpesviruses), respiratory disease (*Aspergillus*, *Chlamydophila*), hypovitaminosis A and dietary insufficiencies, dystocia, hepatic lipidosis, cardiac disease, neoplasia, and psychological disorders.

A recent tragedy occurred at a local avian rescue. It was a large facility housing 400 cockatoos, macaws, African greys, and Amazons. It was located at the founder's house, and the website as well as on tours and with contacts, and house-call visits by avian veterinarians over the years, it was promoted as being a 501(c)(3) facility. The public was not allowed on the premises; surrenders would be left at the gate. They used pheasants in the larger flight cages to pick up food detritus and act as rodent control. All birds were housed outdoors even though some needed winter shelter. The founder apparently was turning away offers of help with animal care recently, and when contact was lost, she was found deceased in her home, and all but 1 of 400 birds were dead as well.

Acknowledgments: The following organizations are acknowledged: Washington Ferret Rescue & Shelter, Kirkland, WA; Best Friend Rodent Rescue, Everett, WA; Oregon Tiger Sanctuary, Eagle Point, OR; Pacific Primate Sanctuary, Haiku, Maui, HI.

References

1. Kreisler RE, Spindel ME, Rishniw M. Surveys of salary, benefits, and job responsibilities for veterinarian employed in the field of shelter medicine in the United States conducted in 2011 and 2018. Top Companion Anim Med. 2020;39:100430:1-11.

2. Powell L, Reinhard CL, Serpell J, Watson B. A survey of veterinary student and veterinarian perceptions of shelter medicine employment. J Vet Med Educ. 2021;e20210112. doi:110.3138/jvme-2021-0112.

3. Powell L, Reinhard CL, Serpell J, Watson B. Workplace relations and opportunities for career development impact the retention of veterinarians in shelter medicine. Front Vet Sci. 2021;8:732105. doi:10.3389/fvets.2021.732105. eCollection 2021.

4. Laderman-Jones BE, Hurley KF, Kass PH. Survey of animal shelter managers regarding shelter veterinary medical services. Vet J. 2016;210:68-76.

5. Association of Shelter Veterinarians. Available at: https://www.sheltervet.org. Accessed March 31, 2022.

6. Hanley, CS. The care of small mammals in the animal shelter. In: Miller L, Zawistowski S, eds. Shelter Medicine for Veterinarians and Staff. 2nd Ed. Ames, IA: Wiley-Blackwell; 2013:185-200.

7. Nicholson DA. Avian care in the animal shelter. In: Miller L, Zawistowski S, eds. Shelter Medicine for Veterinarians and Staff. 2nd Ed. Ames, IA: Wiley-Blackwell; 2013:225-246.

8. Isaza N, Isaza R. Reptile and amphibian care in the animal shelter. In: Miller L, Zawistowski S, eds. Shelter Medicine for Veterinarians and Staff. 2nd Ed. Ames, IA: Wiley-Blackwell; 2013:201-224.

9. Johnson-Delaney CA. Rescue centre medicine. In: Johnson-Delaney CA, ed. Ferret Medicine and Surgery. Boca Raton, FL: CRC Press; 2017:437-448.

10. University of California Davis Koret Shelter Medicine Program. Shelter Medicine 101 VME 424-01. Available at: http://www.sheltermedicine.com. Accessed April 3, 2022.

11. Kirk JT. Spock quote. Wrath of Khan. Star Trek II [Film]. Hollywood, CA: Paramount Pictures; 1982.

12. Vinic T, Dowling-Guyer S, Lindenmayer J, et al. Survey of Massachusetts animal shelter record-keeping practices in 2015. J Appl Anim Welf Sci. 2020;23(4):385-401.

Section 37

Technical Topics

Christina Miller, RVT, BSc;
Elizabeth Vetrano, RVT
Session Chairs

Food, Function and Fecals: Gastrointestinal Anatomy, Physiology and Nutrition of Rabbits, Guinea Pigs and Chinchillas

Micah Kohles, DVM, MPA

Session #2655

Affiliation: From University of Nebraska-Lincoln, Lincoln, NE, Oxbow Animal Health, Murdock, NE 68407, USA, Woodland Animal Hospital, Hickman, NE 68372, USA.

Introduction

The gastrointestinal (GI) function of rabbits, guinea pigs, chinchillas, and other exotic companion mammal (ECM) species is complex and dependent on proper nutrition. Understanding the GI anatomy and physiology of ECM species is key to properly educating owners, in addition to treating and diagnosing many common diseases as most we see in clinical practice are directly or indirectly related to nutrition. In this lecture we focus on the most common ECM species and discuss their GI tracts and what diets are on the market including the good, bad and ugly.

Rabbits

Domestic rabbits (*Oryctolagus cuniculus*) are strict herbivores, classified as hindgut (cecum and colon) fermenters, concentrate selectors, and are adapted to ingest large amounts of high fiber food. As prey species, maintaining a small body size is advantageous. Their unique GI system, with 12- to 16-hour GI transit time and ability to differentiate particulates of fiber, allows them to remain small and active, while surviving on high-fiber, low-energy density plant materials. GI disease in this species is a very common presenting issue in clinical practice. It is often associated with dysbiosis and can present as both primary and secondary disease.

They are mostly crepuscular and nocturnal feeders. They are anatomically and physiologically adapted to handle significant amounts of low-energy density fibrous food and to effectively utilize the nutrients found in a high-fiber diet; however, due to their small body size, they are unable to store large amounts of food material and therefore utilize a unique process (wash back colonic separation mechanism) to separate fiber based on size. The easily fermentable component of the diet is retained, while the slowly fermentable components of the diet (preeminently cellulose based plant fiber) are rapidly eliminated. This separation process allows for the production of two types of stool: a dry, hard, high-fiber fecal pellet and an enlarged, soft fecal pellet that is covered in mucous and referred to as a cecotroph (cecal, night stool, etc.) which is immediately ingested. It should be noted that when rabbits ingest cecotrophs, normal mastication does not occur and the cecotrophs are swallowed intact. Rabbits, like all herbivores, lack mammalian enzymes to break down the cellulose components of their plant-based diets and thus have a symbiotic relationship with a diversity of gut microbes and flora. Some of the most common issues with domestic rabbits, both in companion and commercial animals, are related to the GI system. The symbiotic relationship rabbits have with the microflora in the cecum allows them to effectively ferment structural plant carbohydrates which serves as their main energy source.

Rabbits have a large cornea and laterally placed eyes, which give them a panoramic field of view when grazing and foraging. Food is selected based on smell and tactile perception of the lips, as their visual field does not include the area just under their nose. As concentrate selectors, rabbits will choose the most tender, succulent plant parts which are nutrient dense and lowest in fiber to meet their nutritional needs, owing to a high metabolic rate. Rabbits have hypsodont or open-rooted (constantly growing) incisors and cheek teeth (premolars and molars).

Rabbits' incisors are designed to apprehend and cut through vegetation. The large upper incisors have two smaller incisors behind them (peg teeth). The occlusion of the upper and lower incisors allows the teeth to be constantly sharpened. The premolars and molar teeth, also known as cheek teeth, are designed to process hard, fibrous foods. The fibrous food is ground between the cheek teeth with jaw movements up to 120 per minute. This repetitive, rapid chewing movement helps to maintain proper dental health. If low-fiber mixed cereal or seed-based foods are fed with minimal or no hay, the rapid chewing activity is no longer required, which can result in dental disease, such as malocclusion and root elongation, as well as GI issues. As the teeth elongate secondary to inadequate wear, the jaws are forced apart and the radiographic appearance of the diastema increases. Over time, the jaws begin to no longer occlude properly, and mandibular movement is reduced. The condition deteriorates as the teeth become mobile and maloccluded, resulting in the development of spurs that may lacerate and ulcerate the tongue and buccal cavity. A suitable diet of high-fiber hay and high-fiber, uniform pellets will help control the growth of the teeth and prevent overgrowth and malocclusion. As concentrate selectors, rabbits, like deer, have a predisposition to "select" the lower fiber, more tender portions of their diets which, if they are fed cereal based or mixed pellets, can lead to selective feeding and malnutrition.

The stomach is J-shaped and large compared to most monogastrics. It comprises roughly 15% of the GI tract and has both a well-developed cardiac sphincter that prevents vomiting and a muscular pyloric area. The pH of the stomach is low and ranges from one to six. The pH is affected by a variety of factors, including the site of determination, the presence or absence of cecotrophs, the duration of time from feed intake to measurement, and the age of the rabbit. The pH of weanling rabbits is higher (5-6.5), which permits the passage of bacteria through the stomach to the hindgut to colonize the cecum.

The small intestine is made up of the duodenum, jejunum (longest), and ileum, is short compared to other species at ~3 meters in length and comprises ~12% of the total GI tract. Transit time through the jejunum and ileum is 10-20 and 30-60 minutes, respectively. The duodenum is the largest of three segments but shortest in length (~3-5 inches in rabbits). The distal end of the ileum is thickened and referred to as the sacculus rotundus, which is unique to the rabbit and marks the intersection of the ileum, colon, and cecum. This location signals the beginning of the large intestine and links to the ampulla coli, which forms an intersection of the distal ileum, cecum, and proximal colon. Located at the junction of the ileum and the large intestine is the ileocecal valve, which regulates the movement of material from the small intestine into the large intestine and prevents the retrograde movement of intestinal contents.

The small intestine is the site where the greater part of digestion and absorption takes place by passive and active transport throughout the mucosa. Like monogastric animals, rabbit's stomach aids in initiation of digestion through secretion of hydrochloric acid and pepsin. Pancreatic enzyme production, such as amylase, is minimal, with most coming from saliva or bacterial production within the cecotroph. A key function of the small intestine is the absorption of nutrients contained in cecotrophs. Cecotrophs contain essential nutrients, including volatile fatty acids, hydration, protein, vitamins and amino acids which help to meet nutrient requirements. They are coated in a protective mucus which is applied by the goblet cells in the descending colon that protects them from the acidity of the stomach prior to absorption in the small intestine. The transit time of the small intestine is rapid compared to other herbivorous species and fiber is quickly moved to the cecum and large intestine.

The rabbit cecum is very large and has a capacity roughly ten times that of the stomach, making up around 40% of the total GI tract. It is a thin-walled organ that folds on itself multiple times and has an internal surface made up of a long spiral fold (sometime referred to as spiral valve) that is continued into the beginning of the colon, an area known as the *ampulla coli*. The distal tip of the cecum is known as the vermiform appendix. This tube-like area is narrow and thickened compared to the cecum and contains significant amounts of lymphoid aggregates which secrete bicarbonate ions into the cecum. The bicarbonate ions act as a buffering agent for the volatile fatty acids formed during cecal fermentation.

The colon is divided into three sections, the ascending, transverse, and descending colon. Because the *fusus coli* (located anatomically at what is referred to as the transverse colon) forms such a natural division between the morphologically and functionally distinct ascending and descending sections of the rabbit colon, many physiology texts have abandoned the three-section description and simply use the terms proximal and distal colon. The colon is a very functional component of the hindgut and is characterized by sacculations (haustra) and bands (taeniae). The ascending colon contains small protrusions, approximately 0.5 mm in diameter, that are commonly referred to as warzens. These wart-like protrusions are believed to be unique to rabbits. They potentially represent an increase in the surface area of the colon that would favor absorption and also may assist in mechanical separation of gut contents. The *fusus coli* has a mucosa that is four to five times thicker than the descending colon and contains ganglion aggregates. The *fusus coli* is commonly referred to as the pacemaker of the hind gut and controls the retrograde and normograde peristaltic activities that occur during the formation of soft and hard feces. The descending colon's mucosa is smooth, and while it contains numerous goblet cells, does not contain obvious specialization and usually contains hard fecal pellets.

There is an almost continuous flux of material between the cecum and the proximal colon. The mixture of fluid and nutrients is moved through vigorous peristalsis in both normo- and retrograde directions. These contractions result in large indigestible fibrous particles accumulating in the center of the colon where they are rapidly transported along the colon to the rectum for defecation, usually within four hours of ingestion. Conversely, smaller, more fermentable fibrous particles accumulate at the periphery of the colon where through retrograde peristaltic contractions of the haustra, they are returned to the cecum for further fermentation.

Periodically there is a dramatic change in the peristaltic activity of the cecum and colon as the retrograde movement of smaller fiber particles ceases and a large quantity of the cecal content is expelled into the colon. These contents are excreted, usually once to twice daily, as soft, mucous covered cecotrophs. This process is controlled by the *fusus coli*. Because rabbits do not completely ferment fiber, the ability of the colon to rapidly eliminate large indigestible fiber particles and retain smaller, more digestible fiber for further fermentation, makes the rabbit an extremely efficient herbivore, capable of surviving on very low-quality forage.

The cecotroph fermentation process depends heavily on an appropriate diet and the action of resident bacteria (especially *Bacteroides* spp. and *Clostridium* spp.) and protozoa which are vital to the GI health of the rabbit. The diversity of bacterial flora and other organisms within the cecum is substantial and includes many bacterial species dependent on location within the cecum lumen or wall, unidentified anaerobic species, ciliated and flagellate protozoa, and a rabbit specific yeast. The effects of bacterial fermentation include the production of volatile fatty acids (VFA) including acetic, formic, propionic, and butyric acids. These are a major energy source for rabbits. A portion of these VFAs are contained in cecotrophs and utilized upon re-ingestion, while another portion is absorbed directly across the cecal mucosa.

There is still much to learn and understand about the microbial populations of the rabbit hindgut. Fermentation produces volatile fatty acids, vitamins B, C, and K, and amino acids which are eventually digested in the small intestine after cecotrophy, which occurs once or twice a day, usually at night. Over one hundred additional strains of anaerobic bacteria have been isolated from the cecum mucosa of rabbits, and of those, very few have been identified.

In healthy and physically able rabbits, the large, soft mucous-covered cecotrophs are ingested directly from the rectum (midnight snack). Multiple factors, including stimulation of rectal mechanoreceptors, perception of the cecotroph odor, and the blood concentrations of various metabolites and hormones, all stimulate the rabbit to directly ingest the cecotrophs. Higher-fiber diets increase cecophagy, whereas high-protein and low-fiber diets decrease cecophagy. Once eaten, the acidity of the stomach breaks down the protective mucous covering of the cecotroph. Continued digestion occurs in the small intestine where cecotrophs represent an important component of the diet providing protein, water and vitamins.

Guinea Pigs

Guinea pigs (*Cavia porcellus*) are obligate herbivores and have a large tongue and narrow oral cavity. Dental disease is a common malady, similar to rabbits and chinchillas, and proper diet (specifically, ample fiber intake) is key to decreasing the likelihood of dental and GI issues. Guinea pigs are hindgut fermenters and are coprophagic. While the belief is that coprophagy is an important function in guinea pigs, its contribution to overall nutritional needs has not been fully characterized. Coprophagy in rabbits is a key source of B vitamins and a means of optimizing the availability of proteins; however, there is uncertainty if guinea pigs are truly cecotrophic, as unlike rabbits, they are not reliant on the key amount of B vitamins found in cecotrophs. Small herbivores utilize one of two primary mechanisms for the colonic separation mechanism necessary for coprophagy. This capability permits rapid passage of less digestible food particles while retaining micro-organisms and more digestible food particles in the cecum, to allow adequate opportunity for fermentation. Rabbits utilize a multi-directional peristaltic motion to separate components of fiber based on size and mix these with other ingesta, bacteria, and water to concentrate more fermentable particles and eliminate larger particles of fiber. Guinea pigs and chinchillas, on the other hand, utilize a process sometimes referred to as "mucus trap," in which bacteria from the cecum are trapped in the mucus of the colon with few to no food particles and returned to the colon through antiperistalsis. The mucous trap process is less efficient than the mechanisms utilized by rabbits. Guinea pigs and chinchillas have a colonic furrow in the ascending colon, in which the concentration of bacteria and nitrogen is twice as high as in the lumen.

The stomach of guinea pigs contains the same four regions, the cardia, fundus, body, and pylorus. The small intestine of guinea pigs is made up of three sections (duodenum, jejunum, and ileum), as found in most species. They are visually non-distinguishable, with the jejunum being the longest section (~90-100 cm). The cecum is the largest component of the guinea pigs' GI tract and contains upwards of 65% of total contents. The flora is similar to the rabbit, being primarily gram positive with anaerobic lactobacilli being the primary organisms in the large intestine. As in the rabbit, a large number of additional organisms including yeast, protozoa, and other bacterial species are prevalent. Gastric emptying time is ~2 hours and total GI transit time is ~20 hours (8-30 hours), and considerably longer (up to 60 hours) if coprophagy is considered.

Chinchillas

Chinchillas (*Chinchilla lanigera*) are obligate herbivores, hindgut fermenters, and cannot vomit, similar to rabbits and guinea pigs. They are similar to the guinea pig than the rabbit in terms of GI anatomy and physiology. Their dentition is also similar to guinea pigs than rabbits, as they have a larger grinding surface owing to a naturally higher fiber and coarser diet than rabbits. As compared to the guinea pig, the jejunum and descending colon are longer. The colon is highly sacculated and the cecum large, yet the cecum holds less contents of the overall large intestine in chinchillas compared to rabbits and guinea pigs.

Nutrition

The discussion of the GI system of herbivorous exotic companion mammals above makes it clear how integral proper diet is to the health of all small herbivores. The keys to a proper small herbivore diet are the following:

1. High-quality high-fiber grass hay (~70-75%).
   - A variety of grass hays should be fed free choice, available at all times. Any high-quality grass hay such as timothy, orchard, meadow or oat is appropriate.
   - Diversity in hay type offering is strongly recommended as this mimics the natural diversity of plants these species would forage for in wild environments.
   - These species will also use hay for bedding and nesting material. It is important to differentiate these uses and ensure the animals always have access to clean fresh hay for ingestion.

2. High fiber uniform pellet (~20-25%).
   - Timothy or alfalfa-based, depending on life stage. Always feed only the recommended amount.
   - It is important to feed only uniform products and avoid products that contain added seeds, fruits, or vegetables, or that utilize additional ingredients (typically starch or carbohydrate based) required for certain manufacturing techniques.

3. Fresh or dried greens (~10-15%).
   - Only appropriate species should be fed, and all should be carefully cleaned. (Organic products are preferred.)

4. Treats (~0-5%).
   - Low-sugar, high-fiber treats.

5. Unlimited fresh, clean water.
   - Multiple sources, such as water bottles and crock, should be available at all times.

High-quality, high-fiber grass hay should be available at all times. Fiber is essential to maintain proper gut health, and to stimulate gut motility (insoluble fiber only). Although there are multiple varieties available on the market, there is minimal nutritional difference between them, and therefore, different species, such as timothy, orchard, oat, and mixed are appropriate. Hay is a product of nature, and therefore, there will be variances in color, texture, feel, and taste. Because of this, it is important to transition between hay bags and offer different varieties of hay. Hay not only provides a key component of the nutritional needs of rabbits and other small herbivores, but it also encourages natural behaviors such as foraging and grazing. Stimulating these natural behaviors may diminish boredom-based behaviors, increase activity, and also provide a sense of security to the animal.

There are many diets available to pet owners, the two most common being seed/cereal-based muesli mixes and hay based, high-fiber uniform pellets. Rabbits, guinea pigs and chinchillas are all concentrate selectors, and if offered a mixed pellet, will selectively feed on certain items from the mix, which will lead to improper and unbalanced nutrition. These mixes are often high in carbohydrates and simple sugars, which can lead to obesity and GI issues, as well as low in fiber, which slows GI motility and decreases normal teeth wear. Uniform hay-based fortified pellets, high in fiber and low in protein, are recommended, as they eliminate selective feeding, thereby guaranteeing a proper vitamin and mineral intake.

Fresh and dried greens are an important and enjoyable component of a rabbit's diet and should be considered, not an option, but an essential part of a healthy animal's diet. They also provide an excellent source of water and enrichment. Varying types and textures is a great enrichment tool. It is recommended to completely wash and clean all greens, and organic products are preferable.

Although treats are not a required component of these animals' nutrition, limited appropriate treats can be beneficial for strengthening the pet/owner bond, training, and enrichment. It is important to limit the number of treats and ensure they are low in protein, fat, calcium, and sugar. Some examples of inappropriate treats are nuts, seeds, popcorn, bread, crackers, and yogurt drops. High-fiber grass and alfalfa-based treats are appropriate in limited amounts, as are small amounts of dried fruits and vegetables. There are far more inappropriate treats on the market, but good products do exist.

Unlimited fresh, clean water is vital, and the water requirements of rabbits have been estimated at 50-100 ml/kg/day depending on their diet. Fresh, clean water should be always available to small herbivores. It is a good idea to offer herbivores water in more than one manner, such as a sipper bottle and bowl/crock, to stimulate increased water intake. Water should be changed daily, as bottles/bowls/crocks will become dirty with normal drinking. Actual water consumption can be significantly affected by diet and other environmental factors.

Recommended Reading

1. Alworth L, Harvey S. Chinchilla anatomy, physiology and behavior. In: Suckow M, Stevens K, Wilson R, eds. *The Laboratory Rabbit, Guinea Pig, Hamster and Other Rodents.* Waltham, MA: Academic Press; 2012:955-966.

2. Bjornhad G, Snipes R. Colonic separation mechanism in lagomorph and rodent species-a comparison. *Zoosystem Evol.* 1999;75:275-281.

3. Brewer N, Cruise L. Physiology. In: Manning PJ, Ringler DH, Newcomer CE, eds. *Biology of the Laboratory Rabbit.* 2nd Ed. Waltham, MA: Academic Press; 1994:63-70.

4. Carabano R, Piquer J, Menoyo D, Badiola I. The digestive system of the rabbit. In: de Blas C, Wiseman J, eds. *Nutrition of the Rabbit.* 2nd Ed. Cambridge, MA, CABI Publications; 2010.

5. Cheeke P. *Rabbit Feeding and Nutrition.* Orlando, FL: Academic Press; 1987.

6. Davies R, Davies J. Rabbit gastrointestinal physiology. *Vet Clin North Am Exot Anim Pract.* 2003;6: 139-153.

7. De Blas C, Wiseman J. *Nutrition of the Rabbit.* 3rd Ed. Boston, MA: CABI; 2020.

8. Franz R, Kreuzer M, Hummel J, et al. Intake, selection, digesta retention, digestion and gut fill of two coprophageous species, rabbits (*Orytolagus cuniculus*) and guinea pigs (*Cavia porcellus*), on a hay only diet. *J Anim Physiol Anim Nutr.* 2010;95(5):564-570.

9. Griffiths M, Davies D. The role of the soft pellets in the production of lactic acid in the rabbit stomach. *J Nutr.* 1963;80:171-80.

10. Hargaden M, Singer L. Guinea pig anatomy, physiology and behavior. In: Suckow M, Stevens K, Wilson R, eds. *The Laboratory Rabbit, Guinea Pig, Hamster and Other Rodents.* Waltham, MA: Academic Press; 2012:575-602.

11. Harkness J, Murray K, Wagner J. 2002. Biology and diseases of guinea pigs. In: Fox J, Anderson L, Loew, Quimby F, eds. *Laboratory Animal Medicine.* 2nd Ed. San Diego, CA: Academic Press; 2002:203-246.

12. Irlbeck N. How to feed the rabbit (*Oryctolagus cuninulus*) gastrointestinal tract. *J Anim Sci.* 2001;79: 343-436.

13. Jenkins J. Feeding recommendations for the house rabbit. *Vet Clin North Am Exot Anim Pract.* 1999;2(1): 143-151.

14. Jilge B. The gastrointestinal transit time in the guinea pig. *Z Verdsuchstierk.* 1980;22:204-210.

15. Quesenberry K, Orcutt C, Mans C, Carpenter J. *Ferrets, Rabbits & Rodents: Clinical Medicine and Surgery.* 4th Ed. St. Louis, MO: Elsevier; 2020.

16. Snipes T, Clauss W, Weber A, Hornicke H. Structural and functional differences in various divisions of the rabbit colon. *Cell Tissue Res.* 1982;225:331-346.

17. Stevens C, Hume I. Contributions of microbes in vertebrate gastrointestinal tract to production and conservation of nutrients. *Physiol Rev.* 1998;78:393-427.

18. Sohn J, Cuoto M. 2012. Rabbit anatomy, physiology and behavior. In: Suckow M, Stevens K, Wilson R, eds. *The Laboratory Rabbit, Guinea Pig, Hamster and Other Rodents.* Waltham, MA: Academic Press; 2012:195-215.

19. Wolf P, Schroder A, Wenger A, Kamphues J. Nutrition of the chinchilla as companion animal-data, influences and independences. *J Anim Physiol.* 2003;87:129-133.

20. Wingerd B. Rabbit anatomy and dissection guide. Eden Prairie, MN: Blue Door Publishing; 2007:57-67.

Food, Function and Fecals: Gastrointestinal Anatomy, Physiology and Nutrition of Ferrets and Rodents

Micah Kohles, DVM, MPA

Session #2901

Affiliation: From University of Nebraska-Lincoln, Lincoln, NE, Oxbow Animal Health, Murdock, NE 68407, USA, Woodland Animal Hospital, Hickman, NE 68372, USA.

Introduction

The gastrointestinal (GI) function of ferrets and other exotic companion mammal (ECM) species is complex and dependent on proper nutrition. Understanding the GI anatomy and physiology of ECM species is key to properly educating owners, in addition to treating/diagnosing many common diseases as most we see in clinical practice are directly or indirectly related to nutrition. In this lecture we focus on ferrets, rodents, hedgehogs and other and discuss their GI tracts and what diets are in the market including the good, bad and ugly.

Ferrets

Domestic ferrets (*Mustela putorius furo*) are members of the order Carnivora, suborder Caniformia, family Mustelidae, subfamily Mustelinae. The scientific name *Mustela putorius furo* roughly translates to mouse eating smelly little thief. There are three species of ferrets (also known as polecats in Europe and Asia): the European polecat (*Mustela putorius*), the Steppe or Siberian polecat (*Mustela eversmanni*), and the black-footed ferret (*Mustela nigrans*). They are related to martens, fishers, otters, wolverines, long tailed weasels, least weasels, mink, badger and skunks. Ferrets are obligate carnivores, meaning they are designed to kill and eat whole prey items, which include all parts of the killed animal. They are specialized, with elongated bodies and short limbs, for the pursuit of warm-blooded prey in holes and tight spaces. Ferrets tend to eat small meals throughout the day and will hide cache excess food in their dens or other hiding places.

Ferrets are obligate carnivores with a very short intestinal tract and are sometimes referred to as super carnivores, because of anatomic, physiologic and behavior traits adapted to a strict carnivore lifestyle. The GI system of a ferret is essentially a long tube from the stomach to the rectum which allows food to pass too quickly for efficient absorption. The entire GI tract is under vagal and sacral innervation. The short length of the overall GI system of ferrets results in a rapid transit time of approximately 3-4 hours.

The gut flora is simple, and therefore GI upset or diarrhea secondary to antibiotic usage is uncommon. Like humans, ferrets are spontaneous secretors of hydrochloric acid which differs from many other predators such as dogs and cats. In addition, they spontaneously produce proteolytic enzymes and the GI tract remains active even under anesthesia. Ferrets do have the ability to vomit, but often do not even in the presence of obstruction or other pathology. Because of their rapid transit time and need for small frequent meals, a prolonged fasting, greater than 3-6 hours, is neither required nor recommended. It should be noted that ferrets with insulinomas may become hypoglycemic much quicker, even less than 3 hours, and therefore require special pre-operative care.

Ferrets have binocular vision, a very broad auditory range, and extremely sensitive olfactory specialization which make them excellent predators. They also have a hinge like temporomandibular joint, large mastication musculature and a large bite force to body mass ratio which allows them to predate larger sized prey.

Ferrets have typical carnivore dentition and oral anatomy with elongation of the carnassial blade, reduction or loss of post carnassial molars, and shortening of the rostrum in comparison to skull length. The upper incisors are slightly longer than the lower incisors when the mouth is closed. The canines are very pronounced, as in other carnivores. The lower premolars as well as first and second upper premolars all have two roots whereas the fourth upper premolars, the carnassial teeth, have three roots. The lower second molar is very small single rooted tooth. Dental formula: I (3/3) C (1/1) PM (3/3) M (1/2) = 34

Ferrets' teeth are precisely interdigitated, which is inline with the bladed shape of their teeth. Their teeth are designed to shear and tear animal tissue and are not designed for grinding. Previous research has pointed to commonly fed extruded diets predisposing ferrets to dental disease or exacerbating what could be consider normal dental wear. It is possible that the shape of their teeth and design for shearing are not in line with the chiseling and grinding process utilized to effectively masticate dry extruded kibble.

The stomach of ferrets is simple and is shaped similar to the dog. The stomach fits in the curve of the liver and has an identifiable pylorus. The two limbs of the pancreas are connected at the midline near the pylorus. It can expand to accommodate large amounts of food even though ferrets typically do not gorge themselves like some other carnivore species. The antrum of the stomach provides the mixing and propulsion of contents for gastric emptying.

The small intestine is quite short (~182-198 cm in length) and consists of three sections all of which contain goblet and villi cells. This short length is one factor that results in such a decreased GI transit time in ferrets. The major duodenal papilla is found ~3 cm from the pylorus and serves as the common opening for the bile and pancreatic ducts. The small intestine is made up of the duodenum, jejunum and ileum. Atropine does not block jejunal motility, as this movement is mediated by hormonal activity. Gastric acid secretion is only reduced by 30% in the presence of atropine. The ileum and jejunum are indistinguishable on gross examination and may be referred to as the jejunoileum which ends at the ascending colon.

The entire large intestine of a ferret, which includes the colon and rectum, is approximately 10 cm in length and lacks both a cecum and ileocolic valve. This junction is inferred by the presence of the anastomoses of the jejunal artery with the ileocolic artery. The ascending, transverse, and descending colon make up the entire colon with the descending portion being the largest.

Nutrition

It is important to look at the evolutionary diet of ferrets to try and better understand their nutritional needs. Although there is no documented proof, it can be assumed that a ferret's diet would reflect that of the polecat. This would be composed of mostly small and medium sized mammals, but could also include a smaller portion of birds, carrion, reptiles, amphibians, eggs, crustaceans, mollusks, insects, worms and fish. Although some evidence reveals ferrets will feed on a small amount of seasonal plant-based foods, their diet is obviously that of animal prey. They are, as we have discussed, strict obligate carnivores.

In view of the previous ferret GI discussion, it is abundantly clear how unique and inefficient their GI tract is and how important feeding an appropriate diet is to their health. To compensate for their inefficient GI system ferrets, require a very concentrated diet which is high in protein and fat as well as low in fiber and carbohydrates. The only carbohydrate sources in a ferret's natural diet are from the gut contents of their prey. Fat, when metabolized, releases twice as much energy as carbohydrate or protein metabolism and therefore should be the main source of calories in a ferret's diet.

While there has been some advancement in ferret diets, the majority of animals are still fed a dry extruded kibble. Most of these diets contain higher than recommended levels of plant based starches which are required for kibble shape and stability as part of the extrusion process. This process consists of finely grinding ingredients and then mixing them into dough which is extruded through a tube under significant heat and pressure. The heating and processing of the animal proteins can potentially make them less digestible. Dry extruded kibbles that contain grains and other plant based starch sources could be related to: 1) Urolithiasis: Due to diets high in plant proteins 2) Insulinomas: Relating to the potential impact of carbohydrates on predilection for the formation this disease 3) Excessive dental wear: Dry extruded kibble has been shown to cause an excessive amount of dental wear in relation to semi moist or natural diets.

Unfortunately, there are limited feeding options currently available to ferret owners. Although feeding a whole prey or raw diet is not owner-friendly, this is the best possible alternative for providing a complete natural diet for ferrets. It is important to educate owners about the potential risks of raw diets, such as bacterial (Salmonella, et.) and parasitic contamination which can be a zoonotic issue. There are multiple semi moist diets on the market as well which, if they fit the nutritional profile, might be applicable but could predispose the ferret to periodontal disease if proper dental care is not maintained.

Unlimited fresh, clean water is vital as ferrets require approximately three times as much water as dry matter. It is a good idea to offer ferrets water in more than one manner, such as a sipper bottle and bowl, to stimulate increased water intake. Some ferrets enjoy playing in their water, so a heavy crock/bowl is recommended. Actual water consumption can be significantly affected by the diet provided (dry versus moist) as well as other environmental factors. Supplements or additives should not be added to ferrets' water.

Rats

The Norway or brown rat (*Rattus norvegicus*) belongs to suborder Myomorpha (mouse-like rodents). Originating in temperate regions of central Asia, rats are nocturnal, burrowing, and colonial. Like most other myomorph rodents, rats are naturally omnivorous, eating grasses, seeds, grain, and occasionally invertebrates and can practice coprophagy based on diet. Rats are quiet, clean, and easy to maintain. They make excellent pets, especially if socialized from an early age.

Rats feed primarily at night and sleep during the day but can adapt to environmental conditions. They tend to avoid new foods therefore diet changes must be made gradually. Diet should be based on a commercial pelleted feed (rodent block) supplemented with a large variety of fresh and dried fruits and vegetables and smaller infrequent amounts of animal and insect-based proteins. provided in moderation. Seed-based diets should be avoided to prevent selective feeding, malnutrition, and obesity. Limited amounts of seeds, nuts, raisins, etc. should be provided only as treats and to encourage foraging behavior. Protein recommendations vary from 16-20%. Rats practice coprophagy, which provides nutrients such as B-vitamins and bacterial protein. Young rats are reared communally, and nursing responsibilities are shared. Weaning occurs at 21 days of age.

Like other rodents, rats are monophydont (have only one set of teeth) and only their incisors grow throughout their life. Their dental formula is 2(I 1/1, C0/0, P 0/0, M3/3). The ratio of upper incisor to lower incisor length is approximately 1:3, which inexperienced practitioners often mistake for overgrowth of the lower incisors.

Food is grasped in the forepaws, chipped by the incisors, and ground down by the molars. Only one side of the mouth is in apposition at a time during mastication. Cheek tissue can be drawn into the diastema, closing off the back of the mouth to allow gnawing without debris passing into the oral cavity.

The esophagus enters the stomach in the middle of the lesser curvature. They are monogastrics with the stomach divided into two parts by a limiting ridge. Situated at this junction, the limiting ridge prevents a rat from vomiting. Myomorphs such as the rat are unable to vomit also, in part, due to the crural sling (muscle) and the innervation of the diaphragm; therefore, pre-operative fasting is not required. The non-glandular forestomach is lined by squamous epithelium, while the distal glandular part is lined by fundic glands containing chief and parietal cells. Like other myomorphs, rats have a relatively large cecum and long colon. The cecum is not septate as in other species. GI transit time for the rat is reported to be between 8-16 hours depending on diet factors.

Mice

The domesticated mouse (*Mus musculus*) originated in temperate Asia. Mice belong to the suborder Myomorpha (mouse-like rodents) and are primarily nocturnal. They are social and prefer to live in small groups dominated by a single adult male. Mice are omnivorous in the wild, eating grasses, seed, grain and occasionally invertebrates. Mice naturally hoard food items, thus food consumption is hard to measure. Diet should be based on a commercial pelleted feed (rodent block) supplemented with a large diversity of fresh/dried fruits and vegetables and smaller infrequent amounts of animal and insect-based proteins. Seed-based diets should be avoided to prevent selective feeding, malnutrition, and obesity. Limited amounts of seeds, nuts, raisins, etc. should be provided only as treats and to encourage foraging behavior. The diet should contain 4-5% fat and 16% or more protein for maintenance; 7-11% fat and 20-24% protein for breeding. Fresh water should be available at all times (monitor water bottles for malfunction). Like all rodents, mice practice coprophagy, which provides nutrients produced via hindgut fermentation.

Mice have elodont incisors and anelodont check teeth. The mouse dental formula is 2(I 1/1, C0/0, P 0/0, M3/3), upper/lower incisor length ratio is 1:3 and incisors grow throughout their life. Mice are monogastric, but there is a division (the "limiting ridge") between the forestomach and the glandular stomach. The forestomach is lined by squamous epithelium and the glandular stomach by cardiac, fundic, and pyloric glands. Like other myomorphs, mice have a relatively large cecum and long colon. Because of the lower esophageal sphincter, the limiting ridge, crural sling, and innervation of the diaphragm, mice cannot vomit. The liver of the mouse is composed of four lobes: median (divided into left and right portions), left and right lobes, and caudate lobe. The life span of mice averages 1.5-2 years; up to 3 years is reported.

Gerbils

The Mongolian gerbil (*Meriones unguiculatus*) is a Myomorph rodent native to the desert regions of Mongolia and northeastern China. It is a social animal that is active during the day and night. Gerbils are adapted to a desert environment, require little water and produce only small amount of concentrated urine. Their feces are small, dry, and hard. For this reason, gerbils are nearly odorless and require less cleaning than do other rodents.

Their dental formula is 2(I 1/1, C0/0, P 0/0, M3/3). Wild gerbils have a variable, seasonal diet of leaves, seeds, and insects. If captive gerbils are often fed a seed-based diet, which is nutritionally incomplete. A pelleted rodent chow with 15-20% protein is recommended instead. For reproduction, dietary protein may need to exceed 20%.

Although wild gerbils can obtain almost all the water they need from available fruit and vegetable matter, commercial pellets do not contain adequate moisture to meet these needs. The captive diet should be supplemented with small amounts of fresh/dried fruit and/or vegetables such as apples and carrots, occasional seeds and insect-based nutrients as well as fresh, clean water available at all times. Fat in the diet should be 4% or less, as gerbils are prone to high cholesterol levels. Gerbils appear to have a higher need for magnesium than rats may develop alopecia and increased susceptibility to seizures when fed a Mg-deficient diet. Gerbils are not typically coprophagic when fed a nutritionally complete diet.

Hamsters

Hamsters are Myomorph rodents of Middle Eastern and Asian origin. A number of different species of hamster are kept as pets: the Syrian or golden hamster (*Mesocricetus auratus*); the closely-related dwarf hamsters—the Russian dwarf, Siberian or Djungarian hamster (*Phodopus sungorus*), the Campbell's Russian dwarf hamster (*Phodopus campbelli*), and the Roborovskii (Robo) or desert hamster (*Phodopus roborovskii*); and the Chinese or striped hamster (*Cricetulus griseus*). Hamsters are generally solitary and nocturnal, although dwarf hamsters may live in family groups.

Hamsters are nocturnal, burrowing animals that prefer a deep layer of bedding. The word hamster comes from the German word hamstern, which means to hoard. Hamsters naturally hoard food therefore the cage must be cleaned frequently to remove hoarded food and prevent ingestion of moldy food items. Hamsters are desert adapted however they are not heat tolerant; they burrow to escape heat. They will hibernate at low environmental temperatures, which may be mistaken for extreme debilitation or death. Hamsters live 1.5-3 years depending on species.

Hamsters are omnivorous and coprophagic. The feeding pattern of adults is typically a 5-minute feeding period, followed by a 2-hour fast. Diet in the wild includes grains, seeds, plants, and some invertebrates. Dietary protein requirements are lower for hamsters than for mice/rats. Pet hamsters are usually fed a commercial hamster mix containing 16-20% protein, 60-65% carbohydrate, and 5-7% fat, which can be supplemented with fruit, vegetables, nuts, and animal protein (e.g., egg, cheese, insects). Seed-based diets are common but lend towards selective feeding. Commercial pellets are less likely to result in nutritional deficiency, obesity, and diabetes. Hay is offered to increase fiber consumption, and sugary treats are kept to a minimum. Hamsters should have wood, branches, etc. to gnaw on for enrichment, and fresh water must be available at all times (water bottle malfunction is relatively common).

The hamster dental formula is 2(I1/1, C0/0, P0/0, M2-3/2-3). The four prominent incisors are elodont while the cheek teeth are anelodont. The crown of the mandibular incisors is longer than the crown of the maxillary incisors—commonly mistaken to be overgrown. The yellow color of the incisors is due to iron pigments. The cusps of cheek teeth retain food and thus hamsters are prone to dental caries formation (males > females).

Hamsters have large cheek pouches, invaginations of the oral mucosa used for transporting food, bedding material, and young. When swimming, hamsters can inflate their cheek pouches in order to increase buoyancy. Cheek pouches extend to the shoulders/scapulae and are retracted by the buccinator muscle, which evolved from the trapezius muscle. Hamsters empty their cheek pouches by pushing cheek contents forwards with the paws. When weighing a hamster, the weight of food in the cheek pouches must be taken into consideration.

Hamsters have a distinct non-glandular forestomach (*pars cardiaca*) that resembles the rumen. Like a rumen, it has high pH and contains microorganisms for fermentation. The glandular stomach (*pars pylorica*), is like that of a monogastric animal. The differential in pH between the two compartments indicates that food does not mix between them. The presence of VFAs (mainly acetic acid) indicates that some fermentation does go on here, although food only stays in the stomach for about 1 hour. The forestomach is lined by squamous epithelium while the glandular stomach region is lined by glandular mucus membranes. Stomach bacteria are mainly gram-positive with some gram-negative coliforms. The two chambers are separated by a muscular sphincter which may regulate passage of food between the two sections. The distal esophagus has a muscular sphincter and enters the stomach just cranial to this constriction. The lesser curvature of the stomach is greatly reduced such that the cardia and the pylorus are very close; therefore, vomiting is impossible for hamsters.

The cecum is sacculated, containing bacterial microflora including *Bacteroides* spp. and *Lactobacillus* spp. plus a few coliforms and clostridia. The pH in the cecum is higher than in the non-glandular stomach, suggesting more active fermentation. The liver has four lobes: right and left medial, and right and left lateral. Hamsters are permissive hibernators; low environmental temps stimulate food gathering, and at 41°F and below they enter deep sleep.

Degus

The degu (*Octodon degus*) is native to the western foothills of the Andes in northern and central Chile. It is the most populous mammal in Chile, where it is considered an agricultural pest. Degus are rodents belonging to the suborder Hystricognathi (porcupine-like rodents), based upon jaw musculature and skull structure. They belong to the family Octodontidae, and genus *Octodon*, which refers to the figure-eight shape of their cheek teeth. Its other names include brush-tailed rat and trumpet-tailed rat. Degus are semi-fossorial, diurnal, and do not hibernate.

Degus are adapted anatomically and behaviorally to a fibrous, low-carbohydrate diet. The wild diet includes a wide variety of plants, grasses, roots, seeds, fruit, and livestock droppings. Captive diet should consist of high-quality hay, limited high-fiber herbivore rodent pellets (e.g., chinchilla pellets), a small amount (20% of daily intake) of greens such as kale, collard, romaine lettuce and dandelion greens, broccoli, green beans, and roots such as beets, carrots, parsnips, sweet potatoes, and turnips. Degus produce the enzyme L-gulonolactone oxidase and therefore do not require dietary supplementation with vitamin C.

Improper diet is a major cause of disease in captive degus. To reduce the risk of diabetes and cataracts, foods containing sugar (e.g., apples, raisins, other fruits, breakfast cereal, honey treat sticks) should be avoided. Note that most pelleted feeds (including guinea pig chow) contain molasses, which is used as a binder. Starchy foods may also pose a risk.

Degus are similar anatomically and to both guinea pigs and chinchillas. They possess large, powerful incisors which are yellow to orange from iron pigmentation. The dental formula is 2(I 1/1, C 0/0, P 1/1, M 3/3). Both the incisors and the cheek teeth are elodont (aradicular).

The degu stomach is simple and glandular, and the intestine is the primary site of nutrient absorption. Degus possess a large, functioning, haustrated cecum where microbial fermentation of ingesta occurs. The colon lacks haustra, and colonic absorption of water from digesta results in very dry feces. GITT is very fast at only 5.1 hours. Degus are cecotrophic and coprophagic: one study showed that 38% of feces produced in 24 hr period were reingested; 87% of this activity occurred at night. Cecotrophy and coprophagy begin as early as 3 days of age.

Prairie Dogs

The black-tailed prairie dog (*Cynomys ludovicianus*) is a stout, short-legged ground squirrel native to the Great Plains of the United States. Prairie dogs belong to suborder Sciuromorpha (squirrel-like) rodents. They are herbivorous, having a functioning cecum. In their natural habitat, prairie dogs eat primarily grasses, prairie herbs and weeds, and occasionally roots and insects. Prairie dogs are diurnal and spend much of the day foraging for food; they do not store food in their burrows.

Captive prairie dogs should receive unlimited grass hay and a measured quantity of pelleted food; prairie dog pellets are preferred, but rodent block and rabbit pellets are also suitable. Alfalfa, fresh greens and vegetables can be offered as treats. Calorie consumption for the prairie dog varies throughout the year, increasing in the fall; caloric intake should be regulated to prevent obesity.

Fresh water should always be available, provided daily in a bowl or sipper bottle. Fluid consumption varies depending on diet. Wild prairie dogs can get most of the water they need from the plants they eat; however, dry pelleted diets necessitate more water intake than fresh foods do.

Prairie dogs are more difficult to care for than other pet rodents. As a native species, they are also illegal to keep in some states. Prairie dogs are relatively hard to handle and may bite. To syringe feed or medicate, handlers may need to use a burrito wrap or heavy leather gloves. Gavage feeding is complicated by the presence of a palatial ostium; with a 5-Fr nasogastric tube it is possible but unlikely to be very effective.

Prairie dogs are in the squirrel family. Their incisors are elodont (aradicular) whereas their cheek teeth are anelodont. The dental formula for the prairie dog is 2(I 1/1, C 0/0, PM 2/1, M 3/3). Incisor enamel is normally orange-yellow in color. Prairie dogs are hindgut fermenters that require adequate dietary roughage. They are also coprophagic. Normal droppings resemble those of guinea pigs, dry and oval in shape. Prairie dogs have a triad of scent glands surrounding the anus. Their white papillae resemble fly larvae when they are winked. These are not removed as an elective procedure in prairie dogs.

True hibernation does not occur in prairie dogs, but dormancy may occur for brief periods during cold weather. Dormancy can also be stimulated in captivity by sudden drops in temperature. Prairie dog pups appear above ground at approximately 5-6 weeks and are weaned at about 5-7 weeks. The majority of prairie dog pups in the pet trade are harvested each spring from the wild. Prairie dogs do not breed well in captivity.

Hedgehogs

The African pygmy hedgehog (*Atelerix albiventris*) is native to West and Central Africa, one of many hedgehog species found in Africa, Europe, Asia, and the Middle East. Hedgehogs are monogastric omnivores, consuming a variety of invertebrates, frogs, lizards, snakes, mice, eggs, fruit, and fungi. They are generally solitary and nocturnal. They inhabit a wide range of habitats including grassland, scrub, savannah, and suburban gardens. They live in brush piles, rock crevices, and burrows. They are able to climb, dig and swim. The African pygmy hedgehog can hibernate during periods of cool, dry weather, remaining torpid for periods of up to six weeks but emerging during warm intervals.

Hedgehogs exhibit a unique behavior known as self-anointing or anting, which can be triggered by any new or strange object or substance. The new object is licked until thick, frothy saliva accumulates. The saliva is then carefully placed onto the spines of both sides of the body with the tongue. This elaborate process may continue for up to 20 minutes. The behavior may serve as a sexual signal (confined to the breeding season in the wild) or protection against predators. Substances that cause self-anointing behavior are typically novel or irritating (e.g., hedgehogs have been observed to rub toads against their spines). This behavior has been observed in hedgehogs as young as 15 days of age.

Hedgehogs have 36 teeth. The dental formula is: 3/2 incisors, 1/1 canines, 3/2 premolars, and 3/3 molars. Hedgehogs have a simple stomach and no cecum. The GI transit time averages 12-16 hours.

African pygmy hedgehogs appear not to have a physiologic need to hibernate and will not do so if kept warm and well fed. They may enter hibernation if the environmental falls below 18 °C (65 °F). Metabolism, heart rate and respiratory rate will decrease, while synthesis of endogenous heparin increases, preventing thrombus formation. This decrease in metabolism renders hedgehogs more susceptible to infection, therefore, pet hedgehogs should be prevented from hibernating.

The natural diet of hedgehogs includes insects, worms, snails, slugs and occasionally small vertebrates and fruit. The nutritional requirements of hedgehogs are not known. They can be successfully maintained on dry low-fat cat food and commercial hedgehog diets are also available for use as the staple diet but quality of ingredients should always be evaluated. Supplement the diet with earthworms, mealworms, crickets, and a small amount of chopped fruits and vegetables. Hard-boiled or scrambled eggs, evaporated milk, cottage cheese, wax worms, pinky mice, and vegetable or meat baby foods provide additional variety to the diet. An insect or meat-only diet should be avoided in order to prevent metabolic bone disease.

Hedgehogs can learn to use a sipper-type watering system or water can be provided in a shallow pan. Provide fresh water daily. Offer food in the evening. Limit the amount to what an adult will consume by morning. Use heavy crocks and food bowls that cannot be tipped. Ad lib feeding, over-consumption, and inactivity commonly lead to obesity in African hedgehogs. The problem can become so severe that the patient can no longer fully roll up. Obesity can lead to other metabolic derangements and health problems, including hepatic lipidosis. Treat obesity by gradually reducing the quantity of food until the amount offered is consumed overnight. Limit foods high in fat: meal worms, wax worms, etc. Offer live insects in the cage to promote normal hunting and burrowing activity.

Critical Care Feeding of Exotic Companion Mammals

Micah Kohles, DVM, MPA

Session #2902

Affiliation: From University of Nebraska-Lincoln, Lincoln, NE, Oxbow Animal Health, Murdock, NE 68407, USA, Woodland Animal Hospital, Hickman, NE 68372, USA.

Introduction

Exotic companion mammals require diets specifically formulated for their nutritional needs. During times of anorexia or poor nutritional status, assisted feeding may be required and their requirement for specifically formulated diets does not change.[1-6] For many of these species, anorexia of even 12-24 hours in duration can have serious health implications and even lead to mortality, so providing vital nutrients when voluntary intake ceases is crucial. Powdered recovery diets mixed with water also provide essential hydration as well as routes to administer medication, in addition to nutritional support. Case-specific calculations of feeding amounts using animal body weight, appropriate metabolic factors, and activity and growth multipliers may help avoid over- or under-feeding. While duration of assisted feeding can vary, goals of recovery feeding should always be to return the animal to voluntarily consuming its standard diet.

What is Critical Care or Recovery Feeding?

Recovery or assisted feeding products should be formulated and designed with the intent of hand feeding animals unwilling or unable to eat on their own. Calorically dense, nutritionally complete powdered products reduce volume intake while still providing key nutrients for recovery and health. Because these products are meant to be mixed with water, they also supply vital hydration to animals in sub-optimal nutritional and dehydration status. When selecting a recovery or assisted feeding formula, it is vital to use a formula that aligns with the animal's nutrient needs. When targeting exotic companion mammals (ECM), products designed for dogs and cats may not be appropriate. It is also important to select products formulated with ingredients that appropriately mimic the normal diet to encourage voluntary feed intake. Biologically relevant ingredients are easily digested by the animal and enhance palatability to stimulate appetite appropriately during recovery.

General ingredient recommendations specific to ECM nutritional needs are as follows:

- Herbivores: grass-hay-based formulas with added plant-based sources of protein, fat as well as omega fatty acids, prebiotics, and amino acids for recovery; natural flavors for appetite stimulation.
- Omnivores: balance of plant, animal, and insect ingredients that are biologically relevant and appropriate for these species.
- Carnivores: animal-protein-based with high caloric density and balanced amino and omega fatty acids profiles; plant-based protein isolates should be avoided.

Artificial colors, preservatives and flavors have no nutritional benefit and should be avoided.

When to Assist Feed

Assisted feeding in ECM is appropriate, or even vital, in a wide variety of situations and should occur anytime an animal becomes anorexic. Anorexia is a non-specific clinical sign and investigation into root causes should begin immediately. For certain ECM species, especially hind gut fermenters such as rabbits, guinea pigs and chinchillas, anorexia bouts lasting even 18-24 hours constitute a potential emergency. Assisted feeding can and should occur any time an animal is anorexic or has potential to go off feed including, but not limited to, use as or during times of the following:

- Appetite stimulant
- Malnutrition
- Stress
- Support during specific disease states
  - Gastrointestinal stasis
  - Dental disease
  - Anemia
  - Pancreatitis
  - Insulinoma
  - Metabolic disease
  - Cancer support
- Severe weight loss
- Geriatric patients
- Medication carrier
- Pre-surgery and post-surgery
- Receiving protocols

While assisted feeding in the clinical setting may be best, it is not always practical for long-term patients. Teaching owners of pets assisted feed or having handouts and videos to send home can be a great tool to ensure the animals best chance of recovery. This can also be made easier by having assisted feeding products for sale in the clinic or on hand to send with patients. Stress alone can trigger anorexia in many ECM species, so pets entering a new home, changing diet, and otherwise entering a period of stress may also benefit from 1-2 days of assisted feeding.

How Much to Feed?

Most companies that make critical care feeding products will have charts or websites with instructions on how to calculate feeding amounts. Charts often are oversimplified for convenience sake because they are typically used as a quick-reference guide in clinical situations. The most accurate way to calculate feeding doses is via species- and situation-specific calculations. One example of an available reference is Oxbow's Vet Connect website (https://www.oxbowanimalhealth.com/vet-connect), which utilizes specific formulas to calculate individual feeding recommendations. They can be used with even greater detail specific to the animal in treatment.

The first step in calculating an animal's energy requirement is determining metabolic body weight (MBW; $BW_{kg}^{0.75}$). Using Kleiber's law, this equation states, for the vast majority of animals, the metabolic rate scales to the 3⁄4 power of the animal's mass (body weight in kilograms). Using MBW, the resting energy requirement (RER) can be calculated by multiplying by a factor specific to species or groups of species: MBW × factor. These factors are well defined for dogs and cats in the literature, but for ECM and other exotic species, field metabolic rates have much more recently been investigated and validated.[7]

Metabolic energy requirements (MER) are calculated by multiplying RER by an appropriate activity or growth multipliers: RER × multiplier. Research has been conducted by Omaha's Henry Doorly Zoo, that determined safe multipliers for exotic mammals and reptiles at 1-2 and 1-3 for exotic birds. Other multipliers may be used for various physiologic states or metabolic goals and include the following:

- Adults at maintenance = 1.0
- Young/growing = 1.5
- Weight gain = 1.2-1.8
  - Use *ideal* MBW
- Weight loss = 0.8-1.0
  - Use *ideal* MBW
- Seniors = adjust for weigh loss or gain as necessary

Once you have calculated the MER (kcal), product feeding amounts can be calculated. Grams of product to feed are calculated by dividing MER by kilocalories per gram of product: MER/kcal per gram.

- Kcal/gram (Oxbow Critical Care)
  - Herbivore: 2.67
  - Omnivore: 3.6
  - Carnivore: 6

If you prefer to calculate feeding recommendations on a volumetric basis, dividing feeding grams by grams per tablespoon of dry product can be calculated: MER ÷ kcal per gram ÷ grams per tablespoon.

- Grams/tablespoon (Oxbow Critical Care)
  - Herbivore: 9
  - Omnivore: 8
  - Carnivore: 4

Example

A 0.5-kg, 3-month-old guinea pig eating only Oxbow's Critical Care

MBW: $0.5^{0.75} = 0.59$

RER: $0.59 \times 140 = 83.24$

MER: $83.24 \times 1.5 = 124.87$ kcal

Grams Critical Care feed: 124.87 kcal ÷ 2.67 kcal per gram = 46.78 grams Critical Care per day

Tablespoons Critical Care feed: 46.78 grams Critical Care ÷ 9 grams per tablespoon = 5.20 tablespoons Critical Care per day

How Long to Assist Feed?

Duration of assisted feeding is extremely case-specific. The most generic answer would be to continue assisted feeding until the animal has returned to voluntary dietary intake. During this time, investigation into root causes should be carried out to determine if further medical diagnostics or treatment is necessary. If after 3-5 days patient status has not improved, underlying issues are likely preventing full recovery. While most Critical Care and assisted feeding formulas are designed to be short-term solutions, there are cases in which a recovery formula can be fed on a longer-term basis under close supervision. However, goals of recovery feeding should always be to return the animal to voluntarily consuming its standard diet.

What About Hydration?

Chronic and acute dehydration are constant threats for ECM species. Understanding the role hydration plays in animal well-being and recovery is essential to properly support ill and diseased animals. Independent of the route of hydration support, monitoring patient hydration status and ensuring appropriate levels of hydration support are key.

Rabbits

- Typical water intake is about 50-150 ml/kg (22-46 ml/lb) daily
- 2.5-kg rabbit on a dry diet will take approximately 160 ml of water, or 65 ml/kg, daily
  - This amount is approximately twice the intake of an average human adult derived from both food and drink
- Summary: 50-100 ml water/kg body weight
  - Range: 50-150

Guinea pigs

- 21.7 ml/100 g BW/day in 6-week-old male guinea pigs (312 ± 13 g) that were individually housed on sawdust[8]

- 7.5 ml/100 g BW/day in male guinea pigs (698 ± 19 g) fed a non-purified diet containing 20 percent crude protein[9]

- Summary: 75-125 ml water/kg body weight
  - Range: 75-215

Hamsters and gerbils

- Male and female golden hamsters consume, on average, 8.5 ml water/100 g BW/day
  - Males consumed 5 ml water/100 g BW/day, while females consumed 14 ml water/100 g BW/day[10]

- Chinese hamsters water intake 11.4 ml/100 g BW/day for males and 12.9 ml/100 g BW/day for females

- Water intake for golden hamsters was found to be 4.5 ml/100 g BW/day in males and 13.6 ml/100 g BW/day for females

- Summary: 75-125 ml water/kg body weight
  - Range: 45-140

Acknowledgments: I thank Cayla J. Iske, PhD for her contributions to this manuscript.

References

1. Lichtenberger M, Lennox AM. Exotic companion mammal emergency medicine and critical care. *Proc Annu Conf Assoc Exot Mamm Vet.* 2009.

2. Paul-Murphy J. Critical care of the rabbit. *Vet Clin North Am Exot Anim Pract.* 2007;10(2):437-461.

3. Fisher PG. Standards of care in the 21st century: the rabbit. *J Exot Pet Med.* 2010;19(1):22-35.

4. Graham JE, Doss G, Beaufrère H, eds. *Exotic Animal Emergency and Critical Care Medicine.* John Wiley & Sons; 2021.

5. Hawkins MG, Graham JE. Emergency and critical care of rodents. *Vet Clin North Am Exot Anim Pract.* 2007;10(2):501-531.

6. de Matos R, Morrisey JK. Emergency and critical care of small psittacines and passerines. *Sem Avian Exot Pet Med.* 2005;14(2):90-105.

7. Nagy KA, Girard IA, Brown TK. Energetics of free-ranging mammals, reptiles, and birds. *Annu Rev Nutr.* 1999;19(1):247-277.

8. Liu CT. Energy balance and growth rate of outbred and inbred male guinea pigs. *Am J Vet Res* 1988;49: 1752-1756.

9. Tsao CS, Young M. Effect of dietary ascorbic acid on levels of serum mineral nutrients in guinea pigs. *Int J Vitam Nutr Res.* 1989;59:72-76.

10. Fitts DA, St. Dennis C. Ethanol and dextrose preferences in hamsters. *J Stud Alcohol.* 1981;42(11):901-907.

How to Create Mutual Respect and Mutual Trust in the Examination Room

Jan Hooimeijer, DVM, RPBC

Session #2573

Affiliation: From Consultancy Practice for Birds, Schoolstraat 7 A 7941 CA Meppel, The Netherlands.

Abstract: Addressing welfare and wellbeing of parrots is our responsibility as avian veterinarians. It all starts in the examination room by acting as a positive role model making a difference, creating mutual respect and trust with the bird. It is an exciting learning experience for the parrot and for the caretaker. Showing love can be considered as the opposite of showing respect.

Showing respect is acknowledging our body language, the cognitive abilities and the intelligence of parrots as prey animals. Ignoring the negative effects of human body language and human attitude is not showing respect. Acknowledging the need for privacy is another way to make the parrot experience that we show respect.

Within the Consultancy Practice for Birds, the 5-step protocol is developed as a tool to create mutual respect and mutual trust, reducing the stress of novel experiences including handling. Understanding parrot behavior and behavioral problems is crucial to be able to show respect for the birds.

Definitions are crucial to describe and to understand parrot behavior. Within the Consultancy Practice for Birds, definitions are formulated concerning normal behavior, undesired behavior, desired behavior and enforced behavior. The definition of displacement behavior is formulated by Niko Tinbergen. Using displacement behavior within the examination room is described as a method to work with parrots showing behavior problems based on insecurity.

Parrots are, in general, kept under extreme unnatural circumstances in captivity. Wing clipping can make it possible to offer parrots more freedom as companion birds and to become part of the family flock inside and especially benefit outdoors from the social interaction, exercise, sun and fresh air. The disadvantages of having a parrot on the shoulder are discussed.

Introduction

The major dogma within veterinary science is to do no harm. Working with animals as veterinarians, it is essential to know and understand the basics of handling and manipulating the animal in a professional manner. A professional manner is inclusive of avoiding unnecessary stress as well as avoiding trauma to the animal, the handlers, and caretakers. Reducing stress is an important responsibility for anyone caring for animals.

In general, there is still a lack of awareness that parrots are non-domesticated prey animals being kept in captivity under unnatural and often bird unfriendly circumstances.[1] In general, parrots are still not acknowledged as highly intelligent animals with remarkable cognitive abilities. This is the reason for a wide variety of behavioral problems depending upon the individual parrot, species, and circumstances surrounding the bird.

In the context of the veterinary clinic we have to realize that owners may be nervous and uncertain about what to expect during the consultation, examination and treatment. We need to consider that the attitude and insecurity of the owner can have a negative influence on the behavior and the welfare of the parrot. In general, owners do not realize that their presence can create insecure behavior, and not the presence of a knowledgeable avian veterinarian. In my examination room I always compare the behavior of a child in the presence of the nursery school teacher with the behavior of the same child at home in the presence of the parents.

We often have to discuss basic behavioral issues with the caretaker before approaching and dealing with the parrot. It is important that owners demonstrate happy and positive behavior themselves to prevent insecure behavior in the parrot.

The caregiver is asked not to look at the parrot under any circumstance, especially when the parrot is showing insecure behavior. The owners are to look at the whole exam room, and to me, instead.

When the parrot shows fear, every time the caregiver is giving attention (including simply looking at the bird), we have to make the caretaker aware of the rewarding effect of any response that is confirming the fear and uncertainty. For that reason, caretakers are not allowed to reassure the parrot because it is in general not productive, especially when owners are showing that they are not comfortable, themselves, with the situation. This reassuring is rewarding the parrot's fear and creates a stressful situation.

The first impression of the veterinarian towards the parrot is very important. It is the first opportunity to make the parrot experience whether the veterinarian is showing respect for the bird or not.

It is crucial to start with showing to be happy and positive and not paying any attention to the parrot entering the examination room and during the conversation with the caregiver. It means that the bird is not put on the table in the middle of the attention but put aside in its cage or on a perch or branch. It is providing the parrot the opportunity to observe and experience that the veterinarian is the opposite of intimidating and feels comfortable in the presence of the parrot as if the bird is not there. To me this is the first step to make the bird experience that I show respect for the bird.

Behavior

Behavior in nature is based on innate behavior and founded on learning, experience, intelligence and cognitive abilities. Behavior is driven by the urge to survive as an individual and as a species.

Working with parrots in captivity we need to understand that the behavior of parrots is determined by different factors including hereditary factors, as well as being determined by living in captivity under extremely unnatural circumstances.[1] To understand behavior we have to learn about the principles of ethology developed by the founders of ethology and Nobel prize winners Niko Tinbergen, Konrad Lorenz, and Karl von Frish. A well-known ambassador of ethology is Jane Goodall who has shown that showing respect for chimpanzees in nature is rewarded by a relationship based on mutual respect and mutual trust.

Niko Tinbergen has formulated the four questions below as a tool to understand and interpret behavior.

1. What is the function of the behavior?
2. What is causing the behavior?
3. How to explain the behavior within the context of ontogeny?
4. What is the evolutionary origin of the behavior?

Looking for differences between showing love and showing respect, after answering the four questions, it is obvious that showing love is behavior that is driven by the need for reproduction to ensure the future of the species. Hormones do play a crucial role within a mutual love relationship meant to produce offspring. Relationships based on mutual respect and trust are not based on hormones and not driven by the need to create offspring. Creating a relationship with a parrot based on love can be considered as unnatural and must be very confusing for the bird. Creating a relationship based on mutual respect and mutual trust results in a long term relationship, ensuring the welfare of the bird. Relationships based on mutual respect and mutual trust are crossing the borders between species as Jane Goodall has shown. The relationships between humans and non-human domesticated animals are crossing the lines between species all the time. The same for many examples of relationships between individuals of different species.

Parrots are built and behave as prey animals

Parrots are prey animals in nature. The fear of being killed as a prey animal determines a major part of normal behavior in nature. Fear of getting killed as prey animals can also be expected as normal behavior within captivity.

A typical anatomic feature of parrots as prey animals is the positioning of their eyes. The eyes of parrots are positioned at the side of their head enabling parrots to observe the whole environment. The eyes of predators are positioned in a way enabling the animals to look ahead, using binocular vision. The eyes of humans are positioned as in predators like dogs, cats, owls and birds of prey having binocular vision. The perception of a parrot looking at humans that are talking and smiling to the parrot, showing their teeth, can be considered as being intimidated by a predator. Looking through the eyes of parrots we have to consider that our body language and attitude determines in different ways the body language and behavior of parrots. Working with parrots without understanding the consequences of the specific characteristics of parrots as prey animals is not showing respect for the birds and is a reason for parrots to develop insecure or defensive behavior. This insecurity is cause for a wide range of behavior problems.

Parrots need privacy

In nature, parrots are hiding in the foliage of the forest canopy. Other bird species exhibit the same behavior to avoid being noticed by predators. This explains the captive parrot's natural need to search for places to hide and their need for privacy. At the same time we know that the need for privacy is not restricted to prey animals. Lack of privacy can be considered as affecting the welfare of animals including humans.[2] Approaching a parrot within the context of the examination room we need to acknowledge that invading their territory, especially while sitting in their cage, can be a reason to create insecure behavior and therefore behavior problems.

Parrots are intelligent birds with amazing cognitive abilities

Learning more about the intelligence and cognitive abilities of parrots we can conclude that parrots do understand our posture and attitude. Parrots learn by observing other birds and other animals including humans as part of their environment. Parrots draw their conclusions and behave accordingly. Dealing with parrots is like dealing with children that are eager to show and demonstrate what they have learned. The model/rival technique, described and used by Pepperberg is acknowledging the intelligence and observational learning skills of parrots.[2]

Allowing myself as an author some anthropomorphism, from the viewpoint of parrots, it must be frustrating to experience not being appreciated for their intelligence and talents but to be considered as just beautiful, funny, and cute.

Understanding behavior is crucial to be able to show respect. It is important to educate caretakers before working with the bird to allow them to observe, in an educated way, the proper approach, and to learn from the interaction between the bird and the knowledgeable veterinary technician and avian veterinarian.

Definitions of Behavior

Normal behavior

Normal behavior is species and individual-specific behavior that is not meant to manipulate the caretaker. Examples include: Eating, drinking, grooming, playing, sitting in the cage, sitting on the hand. It is the experience of the author that the outcome of any response is rewarding normal behavior and the bird will become insecure and develop behavior to manipulate the caretaker to get attention. The behavior that started as normal behavior can easily become an undesired behavior meant to provoke a response.

Desired behavior

Desired behavior is created or manipulated by the caretaker in a way that the animal does not experience that the caretaker wants the animal to perform that specific behavior. It can be compared with manipulating the behavior of children with autism, knowing that those children are very sensitive and will respond in a negative way when they experience a command and even a friendly request. Rewarding the desired behavior, including rewarding their intelligence, is showing respect for the bird and will create self-esteem, mutual respect and prevent undesired behavior.[3]

Undesired behavior

Undesired behavior is meant to manipulate the behavior of the caretaker. Undesired behavior is meant to elicit a positive or negative response. In general, undesired behavior is meant to create a problem. This may include screaming, biting, feather picking, talking, and whistling to get attention. The outcome of rewarding undesired behavior is obvious. Any response like showing fear, frustration, anger, concerns and laughter can be considered as a reward and reason to develop even more undesired behavior.

Enforced behavior

Enforced behavior is the result of a command, any kind of force, or behavior, out of fear for punishment. Even a request can be considered as a command. In general, behavior that is motivated to get a reward can be considered as enforced behavior, making it the opposite of desired behavior.

Displacement behavior

Tinbergen[4,5] as one of the founders of ethology has described displacement behavior in gulls showing territorial behavior. Instead of a fight or flight response, the gulls started to preen themselves, started to pull at grass as if they were collecting materials to improve the nest and looked away from other gulls. In a conflict situation in nature (as well as in captivity) parrots have the choice between a fight response or a flight response. Displacement behavior can be considered as the ultimate alternative behavior when a fight or flight response are not the best choices, or are not possible. Displacement behavior seems to be completely out-of-order but is having the effect that the other parrot or the human caretaker becomes confused and overwhelmed, creating a new situation. Typical examples of displacement behavior are nail biting, scratching, grooming behavior, flapping the wings, shaking the tail, shaking the head, or looking at the sky with one eye.

Grooming and preening have been associated with normal behavior that birds perform when they are comfortable. Within the context of a conflict situation the same behavior can be shown as displacement behavior.[6] When we are responding to this behavior in a way that rewards displacement behavior, we can expect that this normal behavior in a conflict situation becomes undesired behavior to shape a response.

Working with parrots we can expect to see displacement behavior in different conflict situations. The most common conflict situation is when parrots as prey animals are approached by humans in an intimidating manner. We can also observe displacement behavior of parrots that are inside or on top of their cage demonstrating territorial behavior. Under those circumstances it is important to know how an avian species analyzes and combines its perceptual experiences to understand what is happening in the surrounding world.[7]

There are reasons to hypothesize that parrots utilize extensive information processing, that is at least to some degree, they do not simply react mindlessly to environmental stimuli. They actually process the stimuli and choose to react in a certain way.[8]

When displacement behavior is not recognized or not understood and caretakers are responding to displacement behavior in a way that is considered as rewarding, displacement behavior can easily result in abnormal repetitive behavior without the primary trigger that was causing the conflict situation to begin with.[9] It can be one of the explanations when we are dealing with complex behavioral problems like feather destructive behavior. It can be assumed that the type of displacement behavior has a genetic component and that the type of displacement may differ between species accordingly. It can also help explain why some species are more prone to feather destructive behavior.[10] It can be postulated that the way displacement behavior is expressed also will be influenced by environmental factors and observational learning ability of the parrots. Because of that, it has been argued that hand feeding baby parrots is contraindicated because birds need to learn by observing normal behavior of their parents, including normal displacement behavior in conflict situations.[11]

How to deal with displacement behavior

First, it is important to recognize and acknowledge displacement behavior and to consider that the parrot is demonstrating this behavior as they perceive a conflict. Any direct response to the bird at that moment can be considered a reward and can create maladaptive behavior. By showing displacement behavior as the caretaker or veterinarian, we are interrupting the conflict situation.

5-Step Dutch Behavioral Protocol Working With and Manipulating a Parrot

The most important aspect of dealing with parrots is the attitude of the person or persons caring for the parrot. A basic protocol for veterinarians and caregivers of parrots, creating mutual respect and mutual trust, can be summarized in five steps. This protocol can also be used with any parrot exhibiting a behavior problem.

Step 1

Make the bird observe without being the center of attention. Acknowledge the fact that parrots are prey animals by presenting a non-intimidating body language and posture. This way the parrot will observe that you are not a threat. It is important to show that you are comfortable in the presence of the parrot. This improves your perception as a positive role model. Show it all without any interaction with the parrot; don't even look at the bird. Ensure that you will have the bird's complete attention when you are doing your stuff and be silly, playing with a toy, throwing a little ball against the ceiling or whatever. Parrots feel comfortable with people who are comfortable with parrots and with themselves. Your behavior around the parrot will tell them all about you. This means that people often have to act and play a role around the parrot. Your behavior is demonstrating that you respect the parrot, resulting in mutual respect towards the veterinarian. What we create by doing so, at step 1, can be considered as desired behavior.

Step 2

The outcome of step 1 is creating a reason to tell, at step 2, the parrots that it's the most beautiful creature on this planet and to show how precious the bird is to you. Of course always use a non-dominating posture. The reward can be considered as reinforcing the behavior that has been created and manipulated by our attitude at step 1.

Step 3

Establish yourself as the best teacher the parrot can wish for. Show respect for his intelligence by telling yourself, in detail, what there is to see around you (i.e., birds in the trees, toys on the table, colors, shapes, sizes, objects). You also can describe the color of the tail and the beak. You can also describe your own body parts or your shoes. Parrots truly enjoy observing and learning. Acknowledging the intelligence is rewarding the bird for his intelligence, again in a non-intimidating way.

Step 4

After acknowledging/rewarding the intelligence, allow the bird to touch your pen, toy, paper or towel and praise the bird enthusiastically. Only let them touch items when you tell the bird, "now you may touch." Then they are in general very gentle when touching objects with their tongue. Ask the bird to participate in the learning process by offering him new, small, non threatening objects. Tell the bird that he/she may touch and feel the objects and allow it to investigate. Praise each positive move. At step 4 you also can present your hand and tell the bird that he/she may step up and praise the bird for doing so. This is the opposite of giving a command or making the bird step up motivated by a treat.

Step 5

Show the parrot that accepting new situations is okay because there are reasons to trust you under the circumstances that you have created. When the parrot shows fear, do not reassure the bird. Reassurance can easily create more fear and insecurity. It is the experience of the author that after steps 1 through 4 that mutual respect and mutual trust have been established. When the parrot fears a place or object like the towel, be understanding and start over again with step 1 of the protocol and follow through. Soon she/he will experience that there is no reason for fear. Then tell the parrot how smart and wonderful he/she is. This will help to reinforce the self-esteem that all parrots need to be able to develop normal behaviour. The outcome is to create a situation in which a parrot can accept novel situations. It can be a physical examination, taking a blood sample, grooming, wing clipping, gently towelling, or other situations.[3,12]

Dealing with Behavior Problems of Parrots

Dealing with behavior problems is looking for a diagnosis.[11] It is important to consider the circumstances and to understand the reason for the behavior and its consequences. In general, undesired behavior of parrots in captivity can be considered as behavior that started as a normal behavior, but under unnatural circumstances. Dealing with undesired behavior we have to consider that undesired behavior will become a greater problem when the consequence of that behavior is experienced by the parrot as a reward. Any response, (whether positive, negative, emotional, and even subconscious) to the undesired behavior can be considered a reward for the undesired behavior. Even ignoring the undesired behavior may be perceived as a reward.

Preventing undesired behavior

Prevention of undesired behavior requires understanding the reason for that behavior, including examining the circumstances and consequences of said behavior, as well as the natural behavior and body language of parrots as a social, intelligent prey animal. Prevention of undesired behavior involves ensuring their health and welfare by providing balanced and species-appropriate certified organic nutrition, sunlight (or appropriate artificial lighting), fresh air and exercise. Prevention of undesired behavior is encouraged by creating an enriched environment in which the parrot is allowed to express its intelligence and skills; providing toys and providing food in a way that it is stimulating parrots to express normal foraging behavior. Preventing undesired behavior is enabled with social interactions with other birds, other animals and humans. Taking a bird outside for a walk, a bike ride, a picnic or a family visit will help to prevent undesired behavior. Preventing undesired behavior also includes the understanding that the cage of a bird can be considered by the birds as their nesting site. In nature, birds spend their days and nights at the same location when they are breeding as part of their reproductive cycle. When at the nesting site, the bird may be defensive of intruders and express the need to demonstrate territorial behavior. Parrots benefit from a more natural day-night routine in which the bird gets 10-12 hours of sleep in a special small sleeping cage that is located in another room of the house, away from the location the bird spends the daylight hours. The understanding is that birds in nature do not sleep at night where they spend their daytime hours, foraging or drinking. When interacting with the bird, sitting on the shoulder is not encouraged as this may reinforce situational insecurity.

Redirecting undesired behavior by demonstrating displacement behavior

We have to consider that any response to undesired behavior can be perceived as a reward. Instead of responding to or ignoring the behavior, the advice is to act in a way that there is no problem with the demonstration of displacement behavior. Part of the displacement is "looking away" and demonstrating behavior showing no problem with the undesired behavior.

The result of demonstrating displacement behavior is that insecurity and aggression disappears. Over the years, the author has observed this behavior both as a birdwatcher in nature and as an avian veterinarian working with parrots in captivity. In a situation of undesired behavior, step 1 of the 5-step behaviour protocol, can be considered as showing displacement behavior, creating desired behavior that can be rewarded as step 2. Implementing the protocol in this order, undesired behavior is not rewarded. Having the bird watch us when we are not looking at the bird, acknowledging and rewarding their intelligence and rewarding the behavior with touching, feeling and even biting a specific object can all be done within 30 seconds. Using this protocol, undesired behavior is redirected into desired behavior and the bird is rewarded for that and immediately rewarded for their intelligence.

Wing Clipping

In the wild, flying is essential for the survival of parrots. Flying enables parrots to find food, water, and a safe place to sleep. Flying is a way to escape from predators; flying in flocks also offers protection from predators. The well known exception is the kakapo (*Strigops habroptilus*), which lives in a habitat that rendered flight expendable. For parrots in captivity that are kept as companion birds, flying is no longer essential for their survival. Within the Consultation Practice for Birds, wing clipping can be recommended for safety reasons but is mainly recommended because it makes it possible to provide the parrot more freedom and social interactions both inside the house and during the daily outside activities with the caretaker. Wing clipping may give the bird more freedom to become more a part of the family (flock). In the experience of the author, wing clipping done with a parrot-friendly attitude helps to prevent behavior problems and even prevents parrots from ending up in parrot rescue facilities.

Hand-Rearing Baby Parrots

The need to hand rear baby parrots in the past was the result of management failures concerning housing, nutrition and care complemented with a lack of knowledge about the natural breeding behavior of parrots. Over the years, hand-rearing baby parrots became a commercial activity because it encouraged females to produce more eggs.

The Dutch Parrot Foundation, supported by the Consultancy Practice for Birds has been successfully campaigning in the Netherlands against the hand-rearing of baby parrots and the trade in unweaned baby parrots. Since July 1, 2014, it is no longer allowed to separate a baby parrot from the parents until the baby parrot is able to eat completely on its own. This legislation has included parrots on the existing list with dogs, cats, rabbits, chimpanzees and other species where separating juveniles from their parents has been regulated for many years. This new regulation is based on data that has become available during the past 15 years demonstrating the negative consequences of separating baby parrots from their parents. A range of behavioral problems and welfare issues have been recognized as consequences of this practice. As described in other species, separating juveniles before or during the first imprinting phase can lead to irreversible behavior problems.

Some of these behavioral problems can occur after months or even years, especially when the birds become sexually mature. Insecure behavior, phobic behavior, biting behavior, feather picking and self-mutilation can become serious consequences of separating a baby parrot from the parents.[13-15]

Conclusions

We have to learn from parrots what they have known for tens of millions of years, by looking at the world through their eyes. Parrots are among the most charismatic, intelligent, and social animals on the planet with a considerable life expectancy when we are doing a good job of caring for them. Consider and acknowledge parrots as prey animals. By adjusting our body language instead of showing an intimidating posture, creating insecure and defensive behavior, we show respect for parrots. Showing respect for their social skills, intelligence and cognitive abilities is another crucial part of making the birds experience our respect. Their natural learning process is by observing the behavior and skills of their parents, other members of the flock and humans in captivity. Never demonstrate insecurity or fear to parrots, as this will make the bird insecure and create behavior problems. Parrots are very perceptive of our insecurities. Using the 5-step behavior protocol creates a relationship based on mutual trust and respect.

It is vital to reward desired behavior. It is vital to reward the intelligence and acknowledge the cognitive abilities in parrots. It is vital not to reward normal behavior. It is vital not to create and not to reward undesired behavior. It is important to recognize displacement behavior and to use displacement behavior when we are in a conflict situation.

Taking care of the health, welfare, well-being and conservation of parrots in the world is our primary responsibility as avian veterinarians. The main focus should be on preventing health/behavior problems and to ensure their welfare in captivity. Parrots are honest and show whether you are doing a good job, just as children show whether their teacher is doing a good job. In that way parrots have become my best teachers because they correct me when I am wrong. I am grateful for the parrots in my life because they still teach me how to become a better avian veterinarian.

If Not Us, Who? If Not Now, When?

Acknowledgments: I thank Harrison's Bird Foods, The Netherlands for sponsoring the parrot events in the Netherlands. I thank the staff of the Consultancy Practice for Birds for their support and understanding. Last but not least I thank the owner of Roxy and Charlie for the support during the past years. Without this support, nothing could have been accomplished over the years.

References

1. Van Sant F. Seeing the rain forest through the trees. *Proc Annu Conf Assoc Avian Vet.* 1993;228-232.

2. Pepperberg IM, Hooimeijer J. Understanding and appreciating avian cognition to address behavioral issues. *Proc Annu Conf Assoc Avian Vet.* 2019;239-247.

3. Hooimeijer J. Dealing with behavior problems of parrots. Proc DVG. 2008.

4. Tinbergen N. *Spieden en Speuren in de Vrije Natuur.* Amsterdam, The Netherlands: Uitgeverij Ploegsma; 1959.

5. Tinbergen N. *Curious Naturalists.* New York, NY: Doubleday; 1958.

6. Sprijt BM, van Hooff JARAM, Gispen WH. Ethology and neurobiology of grooming behavior. *Physiol Tev.* 1992;72:825.

7. Zucca P. Mind of the avian patient: cognition and welfare. *Proc 8th Conf Europ Assoc Avian Vet.* 2007; 357-365.

8. Pepperberg IM. The African grey: how cognitive processing might affect allospecific vocal learning. In: Balda RP, Pepperberg IM, Kamil AC, eds. *Animal Cognition in Nature.* San Diego, CA: Academic Press; 1998.

9. Garner JP. Perseveration and stereotype-systems-level insights from clinical psychology. In: Rushen J, Mason G, eds. *Stereotypic Animal Behavior: Fundamentals and Applications to Welfare.* Wallingford, England, UK: CABI; 2006:121-152.

10. Seibert LM. Feather-picking disorder in pet birds. In: Luescher AU, ed. *Manual of Parrot Behavior.* Oxford, England: Blackwell Publishing; 2006:255-265.

11. Orosz S. Diagnostic workup of suspected behavioral problems. In: Luescher AU, ed. *Manual of Parrot Behavior.* Oxford, England: Blackwell Publishing; 2006:233-245.

12. Hooimeijer J. A Practical behavioural protocol for dealing with parrots. *Proc Annu Conf Assoc Avian Vet.* 2003;177-181.

13. Engebretson M. The welfare and suitability of parrots as companion animals: a review. *Anim Welf.* 2006;15:263-276.

14. Schmidt R. The influence of the breeding method on the behaviour of adult African grey parrots. Inaugural Dissertation zur Erlangung der Doktorwurde der Vetsuisse-Falkultat der Universitat Bern. 2004.

15. Hooimeijer J. Becoming a role model in the examination room handling parrots. *Proc Annu Conf Assoc Avian Vet.* 2005:97-109.

Exotic Companion Small Mammal Enrichment

Stefanie Kotschwar, RVT, BScH (Zoology, Nutrition), MSc (Zoo Conservation), AAS (Vet Tech), Diploma (Web Design and Equine Science)

Session #2610

Affiliation: From Lynwood Animal Hospital, 107-30 Stafford Rd, Ottawa, ON, K2H 1W8, Canada.

Abstract: There has been increased discussion in the animal world of providing enrichment to all animal species regardless of whether they are privately owned or are part of a zoological institution. Enrichment is important to our pet exotic small mammals to enhance their quality of care. The end goal is to provide an environment where species can express a whole range of species-appropriate behaviors and avoid any mental frustration or boredom. There are some minor pitfalls to providing enrichment such as time, money, accidents; however, these are far outweighed by all the positive benefits provided such as decreased anxiety, increased learning ability, and health monitoring. Enrichment can be divided into two general types: social and non-social enrichment. Social enrichment includes interaction with conspecifics (if they are a species found naturally in social groups) or with the owner. Non-social enrichment includes toys or materials made available to animals, as well as the actual enclosure and its location. While many general enrichment concepts are reviewed, there are some species-specific ideas that also are presented. While there are many exotic small mammal species, the focus of this presentation is on species frequently seen in veterinary clinics, such as rabbits (*Oryctolagus cuniculus*), guinea pigs (*Cavia porcellus*) and other small rodents.

Introduction

We know that providing enrichment in our exotic small mammal's life will positively affect their welfare, but have you really thought about what it means or how you can implement it? Do you know how to guide your client's into providing environmental enrichment for their own pets? There are many factors in deciding what kind of enrichment to provide to different species and how to assess whether it is providing a benefit or if adjustments need to be made. While it is easier to provide environmental enrichment at home, we can certainly provide it to our exotic small mammal patients during their short- or long-term visit to the clinic.

Not only does environmental enrichment provide a multitude of benefits to our exotic small mammals, but we as human caretakers also experience some benefits from these enrichment programs in that there is an improvement in the human animal bond. In addition, engaging in enrichment for our pets provides a source of distraction and entertainment for us as caretakers.[1] You cannot have a healthy animal if you are ignoring their mental health and their need to express normal behaviors. Enrichment is one easy way to ensure you are doing your best to meet your patient's mental needs, in addition to providing veterinary care.

What is Enrichment?

There are many definitions of enrichment depending on your source, but in simple terms enrichment can be defined as: to enhance the quality or value of something. In the case of our exotic companion animals, it means to enhance the quality of care by providing animate, inanimate, and nutritional environmental modifications for their well-being.[2] The focus of enrichment is to enhance an animal's mental health, which in the long-term will have an impact on their physical health as well. The goal is to provide a beneficial environment that will encourage our pets to express species-appropriate behaviors and mental activities, as well as to reduce abnormal behaviors and increase their ability to deal with changes or stressful situations.[1,2] It is important to express to an owner that enrichment is not an optional item, but rather, a necessary aspect of captivity.[3]

Benefits of Enrichment

Research has shown that there are multiple benefits of providing enrichment to not just our exotic companion animals, but all animals. Some of those benefits include the following:

- Increased behavioral diversity
- Decreased abnormal behaviors
- Decreased anxiety
- Improved learning abilities
- Reduced depression
- Mental stimulation
- Increased exercise
- Increased bonding to caretakers

Even small changes such as scattering or hiding food can have a positive impact on the behavioral expression of our exotic animal companions. Not only does enrichment enhance an animal's quality of life, but it is also a tool an owner can use to monitor the health of their animal. An animal that has historically enjoyed engaging in enrichment opportunities that is suddenly not interested in interacting with any of the provided stimuli might be indicating that it is not feeling well or is in pain.[1-5] Providing enrichment in a hospital setting can therefore make an animal more comfortable during its short visit and even extended stay by being able to express normal behaviors. This can in turn help veterinary staff assess the physical and mental well-being of the animal.

Potential Shortfalls of Enrichment

While there are many positive impacts to our exotic animal companions, we must acknowledge that there can be some potential issues of implementing enrichment activities. It might take some trial and error to find the best type of enrichment that will suit a particular species and stimulate normal behaviors. It can also take some time to build up to a proper enrichment protocol, since not only are you trying out which items work best for a particular animal, but it may also take time and money to create and obtain the different items.

Cost can be another concern as commercially made items can come with a hefty price tag. While obtaining items from pet stores or online are options, you can always recycle items you find at home or the veterinary clinic and create your own unique items that are tailored to the specific pet's preferences and even save some money in the process.[2]

In a clinic setting you will be limited by the time and material available, as well as the health status of your patient. It is always beneficial to have a conversation with the owners ahead of time to find out what kind of environmental enrichment their pet prefers.

Despite exercising caution, accidents can happen with new items, and animals can get injured with any of the newly offered enrichment items. It is important to always monitor an animal when you are offering a new item in their enclosure to ensure they are not going to get injured. Always make sure that items do not have small holes where a pet can get stuck and that the materials used are not toxic.[2]

One last important thing to remember is that while enrichment is aiming at reducing stress in the long term, it is not possible to eliminate all fear; it is, however, essential to reduce conditions of fear.[6] You need to be able to recognize the signs of a stressed animal as in the beginning stages some animals can be stressed with changes to their environment and might require a slower approach.

Types of Enrichment

Types of enrichment can be divided into two broad categories: social and non-social enrichment.[1] Social enrichment refers to either conspecific interactions or interactions with owners or other animals. Prior to introducing a new animal, it is important to consider the species' social preferences in the wild. Many different species such as rabbits, rats and chinchillas are very social and enjoy living with conspecifics, but individual preferences should be observed. On the other hand, species such as hamsters are not social and problems can arise when these animals are kept in groups. If keeping animals together is not an option, it can be beneficial to have animals close together but in separate cages. Animals can then safely interact at a distance. This means that any animals that are bonded will benefit from being at the veterinary clinic together if feasible.

Interactions with the owner or other caretakers (or the veterinary staff) need to be gentle and can include time spent petting, holding, or playing with the pet. These types of interactions are based on trust and need to be built slowly. Social interaction with the owner is important to reduce an animal's anxiety when around humans.[1,2] Providing positive reinforcement during those interactions will have a greater impact, even if you only have the ability to spend a couple of minutes interacting with the pet.

Non-social enrichment refers to the environment and all items a pet is exposed to within its enclosures or general living space; it includes things such as toys, beds, substrates, noise, and light.

General Enrichment Concepts

Any enrichment provided should take into consideration the natural history of the species and the range of normal behaviors that they exhibit. For most of our small exotic mammals we have a general idea of their wild counterparts and can extrapolate into care at home. While these guidelines can be implemented in a hospital setting, space, time, and medical conditions can limit the extent of environment enrichment.

Figure 1 showcases some enrichment options.

Figure1. Examples of environmental enrichment.

Enclosure

While there are resources outlining the minimum size of cage an owner should provide to their pet, they should consider obtaining the largest sized enclosure they are able to afford and have room for in their home. This allows for greater movement and even the ability to assign specific areas for particular functions.[3] Consider adding vertical space in the enclosure when possible and if appropriate for your particular species. This can be done by including ropes, hammocks, and extra floors.

The substrate you use will depend on the species you have. Many species like to dig so providing a deep enough substrate to allow for this behavior is recommended. Research has shown that animals prefer hard surfaces over wire surfaces, plus wire flooring is not typically recommended as it can cause pain and wounds on the pet's feet.[2,3] In the hospital we are going to be limited by the size of the already existing cages, but as it is for a short stay it will not be a problem.

Bedding

The amount of bedding provided can be enriching for animals such as rodents by providing a deep enough layer to allow for digging. One research study showed a preference for about 10 cm (4 inches) of soft bedding. Note that having a deeper layer does seem to have a greater impact since it can come into contact with the water bottle and get wet, potentially leading to hypothermia.

The ideal bedding material is dust free, non-palatable, absorbent, and free of toxic components. Different species and different medical conditions might require you to change the type of bedding that you use.[2] Again, for any animal in hospital we want to provide bedding that we can clean up easily and avoid potential cross contamination. Offering towels or soft fleece blankets are a great option.

Toys, huts, boxes, tubes

Providing different toys with different functions and different materials can encourage play in the enclosure. Ensure no metal or glass toys are provided as they could injure the animal.[3] Many rodents like to gnaw on things so providing safe sticks or wooden blocks to encourage this behavior is recommended. Rabbits like to toss toys around so providing stacking cups or other plastic toys can keep them entertained.

Huts, boxes and tubes of different sizes and materials are also enjoyed by many species to hide in during the day, when they are resting or if they are stressed. You also want to make sure you can easily clean the huts provided.[2,3] Ferrets particularly enjoy running through tubes so they should always have access to this enrichment.[3]

Exercise

While having species-appropriate items in the cage can encourage exercise, placing animals in a separate pen for exercise can further stimulate this activity by encouraging exploration. The exercise pen can be kept indoors or even outside, weather and safety permitting.

Rodents also enjoy running on an exercise wheel, so providing one can be a form of enrichment for them. Ensure that they are cleaned often and are of an appropriate size and material.[3]

Certain species such as ferrets, and rabbits can be trained to wear a harness and be taken out into the backyard for extra exercise while still being kept safe. Always monitor your pet while outside to ensure it does not get injured, overheat, or manage to escape.[3]

Larger exotic mammals that are hospitalized should be allowed to spend time out of their small cages as long as medically allowed and you can provide a safe area where they cannot get injured or escape from.

Light

Some exotic animal companions such as rodents and rabbits are usually more active during dusk and dawn, so they benefit from decreased light levels to encourage activity.[2]

Puzzle feeders and foraging toys

When we look at the natural history of the wild counterparts of our exotic animal companions, we find that they spend most of their waking time searching and foraging for food. We can encourage this behavior easily by spreading the food out around the enclosure instead of placing it all in a bowl. This encourages animals to move around and explore to find food items.[7] You can further challenge your pet by making more complicated toys. Simple strategies for this are using paper bags or egg cartons to hide food in or even using an empty toilet paper roll and filling it with hay and treats is a great alternative. There is also the option of acquiring commercially made puzzle feeders and adapting them to your pet. The ball treat dispensers are great for our larger exotic pet animal companions to distribute their pellets.

Training

Training is another great strategy to provide enrichment by stimulating your pet and increases the owner-pet bond. Find a high reward treat for your pet and start slowly by conducting short training sessions and slowly creating the behavior you want. This can be a fun experience and you can also train your pet to be comfortable with procedures such as regular nail trims and accepting medication via syringe.[8] Training sessions do not have to be long; they can be as short as 5-10 minutes and can easily be done during the pet's visit to the hospital.

Species-Specific Enrichment Suggestions

Each species will require different tactics to provide enrichment in their lives and Table 1 shows-examples of easy items to incorporate but it is not meant to be an all-inclusive list.[1-9] Please remember to always supervise your pet when you provide a new enrichment item. You want to ensure it does not stress them out and that all parts are safe.

Table 1. Enrichment examples for a variety of species.

Species	Enrichment examples
Ferret	Multi-level cage connected with ladders and ramps
	Appropriately sized tubes to run in
	Boxes with shredded paper
Rabbit	Supervised yard time for exercise
	Plastic toddler toys
	Box filled with shredded paper and hay for digging
	Chew sticks
Guinea pig	Hiding food in bedding
	Multiple homes to hide in
	Brushing
Chinchilla	Use of rope perches in the cage
	Dust baths
	Compressed hay cubes
Rodent	Tube maze within cage
	Objects to chew

Conclusions

As this paper discusses, environmental enrichment is a necessity for all our exotic animal companions to ensure they can express their normal behaviors and are able to cope with any changes in their lives without too much added stress. Many different options are available, and it is important to work with your selected species to be able to encourage them to express their normal behaviors.

References

1. Banks RE, Sharp JM, Doss SD, et al. *Exotic Small Mammal Care and Husbandry.* Ames, IA: Wiley-Blackwell; 2010:11-19.

2. Brandao J, Mayer J. Behaviour of rodents with an emphasis on enrichment. *J Exot Pet Med.* 2011;20(4): 256-269.

3. Pilny AA. Small exotic companion mammal wellness management and environmental enrichment. *Vet Clin North Am Exot Anim Pract.* 2015;18:245-254.

4. Lidfors L. Behavioural effects of environmental enrichment for individually caged rabbits. *Appl Anim Behav Sci.* 1997;52:157-169.

5. Oxbow Animal Health. All About Enrichment. Available at: https://www.oxbowanimalhealth.com/blog/all-about-enrichment-handout. Accessed March 15, 2021.

6. Morris CL, Grandin T, Irlbeck NA. Companion Animal Symposium: environmental enrichment for companion, exotic, and laboratory animals. *J Anim Sci.* 2011;89:4227-4238.

7. Rommers JM, Bracke MBM, Reuvekamp B, et al. Cage enrichment: rabbit does prefer straw or a compressed wood block. *World Rabbit Sci.* 2014;22:301-309.

8. Harris LM. Ferret wellness management and environmental enrichment. *Vet Clin North Am Exot Anim Pract.* 2015;18:233-244.

9. Belz EE, Jamilyn S, Kennell R, et al. Environmental enrichment lowers stress-responsive hormones in singly housed male and female rats. *Pharmacol Biochem Behav.* 2003;76:481-486.

Index

D

E

F

G

Addendum

ExoticsCon Virtual AAV Sessions

Previously published on the Veterinary Information Network (VIN) November 11-13, 2022

Avian Biochemistry

Bob Doneley, BVSc, FANZCVS (Avian Medicine)

Session #3000

Affiliation: From School of Veterinary Science, The University of Queensland, Gatton, Queensland, Australia.

Abstract: Birds often are presented to veterinarians only when they are near-terminally ill. In these cases, a quick answer is needed. In addition, birds are limited in their range of expression of clinical signs. In avian medicine it has become common to conduct exhaustive diagnostic tests on patients, often with minimal attention to a complete history and physical examination. Selection of diagnostic tests should be based on a solid understanding of the species in question, the results of a thorough history taking and physical examination, and a shortened list of probable differential diagnoses based on this workup. While avian biochemistry analysis can provide insight into internal organ function and physiology, it does not provide a precise diagnosis. Plasma biochemistry analysis is just one tool to facilitate reaching a diagnosis. Understanding what causes changes in biochemistry profile patterns is the key to using this tool more effectively.

Introduction

Veterinarians who treat birds are faced with challenges that are often not encountered by their counterparts who treat dogs and cats. Many birds are presented to veterinarians only when near-terminally ill, and a quick answer is often the difference between life and death. Birds are often limited in their range of expression of clinical signs and many clinicians, through no fault of their own, lack the experience to conduct a thorough physical examination. The combination of these factors has led to an increasing tendency in avian medicine to conduct exhaustive diagnostic tests on patients, often with scant attention paid to a complete history and a careful physical examination, and with little attempt to refine or focus the diagnostic efforts. The selection of diagnostic tests should be based on a solid understanding of the species in question, the results of a thorough history taking and physical examination, and a shortened list of probable differential diagnoses based on this workup.

Before proceeding with diagnostic tests, the clinician should first ask the following questions:

- Are the test(s) appropriate to the patient (species, age, sex, etc.) and its clinical signs?
- Has the test been validated to ensure that the result obtained is likely to be both accurate and meaningful?
- Are the physical risks to the patient and cost of the test(s) to the client justified by the likely clinical value of the results?

If the answer to these questions is yes, then diagnostic testing should proceed.

One of the most used diagnostic tests in avian medicine is the analysis of serum biochemistry analytes. This involves the measurement of specific groups of chemicals within the blood and the interpretation of the results obtained. These chemicals include the following:

- Metabolites, which are those chemicals that are produced as the end-products of various metabolic processes within the body
- Tissue enzymes, which catalyze chemical reactions within the body without being altered themselves
- Electrolytes, including sodium, potassium, and chloride
- Minerals such as calcium, phosphorus, and magnesium
- Bile acids, which are produced in the liver from cholesterol and used in the emulsification of dietary fats

The levels of these chemicals in the blood can be influenced by either physiologic or pathologic processes. Physiologic variations can be due to age, sex, body fat to muscle ratio, nutritional status, reproductive status, and species; however, pathologic processes, including cellular damage or abnormal function of an organ system (or systems), often produce significant changes in blood levels.

Causes of Plasma Biochemistry Abnormalities

There are three major causes of abnormal clinical biochemistries: (1) normal variation between species and individuals, (2) artifacts, and (3) pathology.[1] These causes are discussed below.

Normal variation between species and individuals

There are over 9,000 species of birds, with major differences in anatomy, physiology, form, and function. Some are carnivorous, some are nectivorous, some are granivorous and some are omnivorous. It is unrealistic to expect that they would all conform to a relatively narrow spectrum of biochemical values.

Other variations arise between individuals of the same species. These variations occur because of differences in age, sex, diet, husbandry, physiologic events (such as reproduction), etc.

Post-prandial elevation in uric acid is normal in carnivorous birds. It is not seen in granivorous birds.

For this reason, some clinicians recommend establishing a set of normal values for individual birds during annual health examinations, and then using these values as a comparison should the bird become ill.

Artifacts

When interpreting a biochemistry analysis, care must be taken to distinguish between abnormal results due to disease, and abnormal results due to other factors. These other factors, referred to as artifacts, can occur for a variety of reasons, including the following:

- Physiologic changes,
- Previous therapy,
- Clinical condition of the patient,
- Collection method, and
- Storage and transport of the sample.

Physiologic changes

Stress due to transport and handling of the patient can lead to a release of endogenous corticosteroids, resulting in changes in the hemogram and in blood glucose.

Lipemia, while occasionally seen in diseases of the liver and reproductive system, can also occur naturally in the reproductively active female. Regardless of the cause, lipemia can cause false elevations in bile acids, protein, calcium, phosphorus, and uric acid. It may also falsely decrease amylase. Postprandial lipemia is uncommon in pet birds, so fasting will not help; the clinician needs to check with the laboratory if the sample submitted was lipemic before interpreting these biochemistries.

Previous therapy

Before interpreting biochemistries, the clinician should consider if any treatment given prior to the sample collection could have influenced the results. Therapy given by another veterinarian or to stabilize a crashing patient can have marked effects. Parenteral fluids can dilute biochemistries; exogenous corticosteroids can markedly elevate aspartate aminotransferase (AST), creatine kinase (CK) and lactate dehydrogenase (LDH); intramuscular injections, particularly of irritant drugs, can do the same.

Clinical condition of the patient

Trauma, starvation and dehydration can all have marked effects on biochemistries and need to be considered when interpreting results. Trauma can cause elevations in AST and CK and possibly glucose; starvation can lower glucose and elevate AST and CK if protein catabolism has begun; dehydration can elevate uric acid and urea.

Collection method

Ideally, sample collection should be performed in such a manner that it has minimal impact on the patient while providing an artefact-free sample suitable for analysis. This usually requires venipuncture to be performed on a minimally stressed patient. Inexperienced clinicians may need to consider gaseous anesthesia to collect a good sample without the bird struggling.

Venipuncture should be performed on a large vein (e.g., the jugular) using a needle that is large enough to minimize hemolysis while being small enough to minimize trauma to the blood vessel wall. Hemolysis can cause elevations in bile acids, LDH, CK, alkaline phosphatase (ALP), potassium and phosphorus. Glucose and albumin may be decreased. Calcium may be elevated or decreased, according to the methodology used.

Toenail clipping should be discouraged. Not only is it unduly painful but crush artifacts and contamination with uric acid and bacteria from droppings on the perch can cause elevations in uric acid and decreased glucose if testing is delayed long enough for bacterial growth to occur in the sample.

Storage and transport of the sample

Blood collected for biochemistry analysis should be placed immediately into a lithium heparin tube. Ideally, miniature tubes as used in medical pediatrics should be used. The sample should be gently rolled or rocked; clotting must be avoided, but hemolysis must be as well. If the analysis is to be performed in-house, it should be processed immediately. If a delay is likely, or if the sample is to be shipped to an outside laboratory, the sample should be centrifuged, and the plasma harvested. Sending whole blood to an outside laboratory can result in decreased glucose (as cell metabolism continues) and hemolysis.

EDTA tubes are unsuitable for biochemistry analysis, but can be used for hematology, lead analysis and fibrinogen determination.

Some pearls to help you get the best possible results

Interpretation of avian and biochemistry results is only part of the art of using hematology and biochemistry to assess your patients. To get the most out of a submitted sample it is important to provide your laboratory (external or in-house) with the best possible sample in the best possible condition. The following are some hints for doing that.

a. Submit the largest size sample you can without compromising your patient's wellbeing. Generally, we can collect 10% of the patient's blood volume (equal to 1% of its bodyweight), but the amount we collect in pediatric and anemic patients is less. It is always worth speaking to your lab to find out what size sample they need but negotiate with them if they require volumes larger than 0.5ml of whole blood.

b. Avoid hemolysis by using an appropriate size needle (e.g., 23- to 27-gauge) and using gentle negative pressure on the plunger to prevent turbulence during collection. Remove the needle before placing the sample into the collection bottles.

c. Make several blood smears before placing the sample into anti-coagulant. Rather than using another microscope slide to push the blood along the slide, use a cover slip to drag the sample along the slide.

d. Use BD Microtainers® for your sample, but don't under or over fill the tubes. Under-filling the tube mixes excessive anticoagulant with the blood; overfilling may result in a clotted tube.

e. Use lithium heparin tubes (green top) for biochemistry and Na EDTA tubes (purple top) for hematology.

f. Ensure the blood and anti-coagulant are mixed immediately by rolling the tube along the palm of your hand or gently inverting the tube 10 times. Do not shake the sample, as this may result in hemolysis.

g. If sample processing is likely to be delayed more than a few hours, centrifuge the lithium sample, and separate the plasma from the red cells. Decant the plasma and place in another lithium heparin bottle. This prevents artifacts associated by prolonged contact time with the erythrocytes.

h. Avian plasma samples frequently are yellow due to carotenoid pigments, not bilirubin. Pink or red plasma is usually indicative of hemolysis.

i. Avoid over-interpretation of results. Look for significant elevations or decreases, not changes that could be within the range of error or normal variations for the machine or patient. Be aware of artefactual changes affecting some parameters (e.g., hyperkalemia due to hemolysis).

j. Always treat your patient, not your test results!

Pathology

Liver enzymes

Detection of liver disease through biochemistry is complicated by the fact that there are no specific liver enzymes that can be evaluated conclusively in every case. Typically, decreases in hepatic enzyme activities do not have clinical relevance except in the cases of end-stage liver disease, leaving elevated levels as the screening tool for liver disease.

Liver disease can be broadly classified into three conditions: hepatocellular rupture, decreased hepatic function, and cholestasis. These conditions can occur either separately or concurrently.

Hepatocellular rupture

A hepatocellular rupture releases intracellular enzymes, which then reach elevated levels in the blood. These so-called leakage enzymes include:

- Aspartate aminotransferase (AST). This cytosolic enzyme is distributed in high concentrations throughout hepatic parenchyma, skeletal muscle, heart, brain and renal parenchyma, but the highest concentrations are found in skeletal muscle and liver. Significant elevations therefore usually represent either muscular or hepatocellular damage. AST, therefore, must be interpreted alongside Creatine kinase (CK, also known as creatine phosphokinase, CPK), released from damaged muscle, to distinguish between the two. CK is specific and sensitive for muscle damage in birds; in general, an elevated AST with a normal CK indicates hepatocellular rupture; however, CK has a much shorter half-life than AST; a single-point muscle injury (e.g., an injection) 4-7 hours before sample collection could duplicate this biochemistry pattern. Although AST has long been considered to be the most useful liver enzyme, it cannot be considered in isolation as an indicator of liver disease.

  A recent study[2] showed there was poor correlation between AST/CK and histopathology in diagnosing liver disease. Since there are many confounding influences on the AST levels, this commonly used screening test may not be as useful as previously thought.

- Glutamate dehydrogenase (GLDH). A mitochondrially-bound enzyme, GLDH is both specific and sensitive for the detection of liver disease that involves hepatocellular rupture. GLDH is found in many tissues in the body, including hepatocytes, kidney, intestine, muscle, and salivary gland; however, most of serum GLDH originates from hepatocytes. As GLDH is a large molecule bound to hepatocyte mitochondria, practically none is liberated in generalized inflammatory diseases of the liver such as infectious diseases. Liver diseases in which necrosis of hepatocytes is the predominant event, such as toxic liver damage, are characterized by high serum GLDH levels. It therefore has a high specificity for liver disease, but a low sensitivity (as damage must be severe before elevation occurs). This enzyme has an extremely short biologic half-life (< 1 hour) in racing pigeons, up to 72 hours in raptors.[3]

- Sorbitol dehydrogenase (SDH). SDH is found in the highest concentration in the liver. It is a cytoplasmic enzyme that catalyzes the conversion of fructose to sorbitol. It is a very specific indicator of liver disease in all species, although increases can occur with primary or secondary liver disease. SDH activity has a short half-life (< 12 hours) and an increase suggests acute hepatocellular injury.[4]

- Alanine aminotransferase (ALT) and alkaline phosphatase. These are not considered useful in detecting liver disease in birds.[1] ALT in birds is very nonspecific for the liver, and normal levels have been shown in cases with severe liver damage. ALP elevations are more commonly associated with osteoclastic conditions in birds (e.g., growth, trauma, repair, osteomyelitis, neoplasia, nutritional secondary hyperparathyroidism, and egg-shell deposition).

Decreased liver function

Decreased liver function can occur with any number of liver diseases, not all of which involve hepatocellular rupture. Chronic cirrhosis, amyloidosis and hepatic lipidosis can all have an adverse effect on liver function without causing any cellular damage. In these cases, a liver function test is necessary to detect the problem.

- Bile acids are produced in the liver and are excreted in bile into the small intestine where they act to emulsify fat. Most of the bile acids are then resorbed in the small intestine, enter the portal system and are taken up by the liver to be recycled. Elevated levels occur when there is significant impairment of the liver's ability to remove bile acids from the portal circulation. Normal bile acids may not reflect a lack of liver disease, as there must be a high enough reduction in functional hepatocytes to observe a significant increase in bile acid concentration A two- to four-fold increase in bile acids indicates such a decrease in liver function.

- Total protein concentration is substantially lower than that of mammalian species. The total protein value determined using a refractometer is frequently inaccurate in birds, due to interference by high concentrations of other refractive compounds in plasma, such as chromogens, lipids, and glucose. Biuret analysis of total protein is therefore recommended in birds.

- The BCG method for albumin determination has not been validated in birds caused, in part, by use of human albumin standards and controls, which have different binding affinity for the dye than does avian albumin. Gel electrophoresis is the recommended method of albumin determination in avian species.[3]

Cholestasis

Cholestasis occurs when the biliary system is partially or totally obstructed. This can be seen with biliary neoplasia, pancreatic disease, or diffuse swelling of the entire liver.

Gamma glutamyl transferase (GGT) is an enzyme found in the epithelial cell membranes of the bile ducts and kidneys. GGT activity is more likely to be increased in cholestatic conditions and biliary epithelial disorders in birds as it is in mammals; GGT is not sensitive to hepatocellular damage alone. The clinical utility of GGT in the diagnosis of biliary conditions in birds has not been adequately evaluated but may be elevated in conditions such as biliary carcinomas.[2]

Bilirubin is not produced in birds (therefore they cannot become jaundiced); they utilize biliverdin instead. There are no commercial assays for biliverdin.

Kidney function

The avian kidney filters a large volume of fluid (approximately 11 times the entire body water each day for a 100 g bird) and then reclaims most filtered water by tubular and cloacal reabsorption. The glomerular filtration rate decreases with dehydration, and with injection of arginine vasotocin, the avian analogue of mammalian antidiuretic hormone. Avian kidneys shunt blood from reptilian to mammalian-type nephrons when GFR is decreased, thus increasing a bird's ability to concentrate urine.[3]

- Uric acid is the end-product of protein metabolism in birds and is an oxidized form of hypoxanthine. It is produced primarily in the liver from purine metabolism, enters the circulation, and is then secreted by renal tubules (> 90%) or filtered in the glomerulus (< 10%). Due to this active renal tubular secretion, blood uric acid levels are not notably affected by dehydration until GFR is decreased to the point that uric acid is not moved through the tubules, which may occur in severe dehydration. Raptors and other carnivores have higher reference intervals for uric acid and marked increases in plasma uric acid concentration may be observed postprandially. Significant loss of renal tubules will therefore result in elevations of uric acid.

At first glance it appears that uric acid offers a sensitive and specific test for renal disease; however, there are several confounding factors. First, species differences: carnivorous birds have higher normal uric acid levels than granivorous birds. Second, age: juvenile birds may have lower levels than adults. Third, although significant elevations usually indicate renal disease, normal levels do not mean the kidneys are normal; mild increases could indicate early renal disease or dehydration (or both). There must be severe renal damage before uric acid levels begin to rise.

Because of this relative insensitivity of uric acid in detecting renal disease, levels are best interpreted alongside a determination of the bird's water intake and loss and a physical examination. To distinguish renal disease from dehydration, the patient's hematocrit, total protein, and blood urea nitrogen (BUN) should be evaluated concurrently. Dehydration can lead to decreased GFR, in turn leading to elevated levels of BUN; this same decrease in GFR can lead to elevations of uric acid without primary renal disease being present. It is therefore prudent, in cases of an elevated uric acid level, to rehydrate the patient over 2-3 days before definitively diagnosing renal disease. Persistent hyperuricemia after fluid therapy, and with hematocrit, total protein and BUN returning to normal, confirms a diagnosis of renal disease.

- Creatinine levels are normally low in birds and may be below the minimum detectable limit of the assays in the laboratory. It is generally accepted as being of little or no value in evaluating renal function in birds.
- Phosphorus elevations are sometimes seen in birds with renal disease and should be compared with calcium (the Ca: P ratio).

Blood glucose

Glucose is an essential energy source for nearly every cell in the body. Blood levels are governed by its intake, absorption, the interactions of hormones controlling carbohydrate metabolism (insulin, glucagon and somatostatin), the body's metabolism, its ability to store glucose and its excretion. As disorders of glucose metabolism involve so many organ systems, it is treated here as a separate entity.

Hyperglycemia may be a normal physiologic process (e.g., in juvenile birds); however, elevated levels are usually related to increased production or release (e.g., stress) or failure of tissues to take it up out of the blood (diabetes mellitus). Iatrogenic hyperglycemia occurs when corticosteroids are administered, or intravenous dextrose is given. Female reproductive disease may also elevate blood glucose, but this may be an indirect result due to inflammation affecting the endocrine pancreas.

Although a positive urine glucose test is an indicator of hyperglycemia, stress hyperglycemia may cause up to 3+ glucosuria on a urine dipstick. Since avian urine/urates are retro pulsed into the colorectum, false-positive glucosuria and proteinuria may occur due to fecal contamination, especially in frugivorous birds. Results of clinical chemistry tests, including repeated blood glucose, glucagon, and insulin concentrations, urinalysis, and clinical signs all should be considered prior to making a diagnosis of diabetes mellitus.

Hypoglycemia may result from poor handling of blood samples (i.e., artefactual rather than pathologic), or with decreased food intake (starvation, anorexia), increased glucose usage (septicemias, neoplasia, and multi-organ failure) or decreased production (liver disease). Carnivorous birds can maintain fasting blood glucose levels much longer than granivorous species. Smaller granivorous species may become hypoglycemic after a 12-hour fast, especially if debilitated

Lipids

Cholesterol and triglycerides are lipids synthesized in the body and used for a variety of metabolic processes. As they are insoluble, they are transported through the body bound to lipoproteins: high-density lipoprotein (HDL), low-density lipoprotein (LDL), and very low-density lipoprotein (VLDL).[5]

Cholesterol, synthesized in the liver, plays a major role in the production of bile acids, steroid hormones, and vitamin D. It also provides stability and flexibility to cellular plasma membranes. Triglycerides, fatty acids attached to glycerol, are synthesized in the liver and intestinal mucosa and stored in fat. They are used as an energy source for metabolic processes.

VLDLs transport cholesterol, triglycerides and protein around the body. Once in circulation triglycerides are released to cells for energy; as the triglycerides are removed the cholesterol component increases until it is higher than the triglyceride component, forming LDL. This is then taken up by cells where the contents are either stored, used for cell membrane structure, or converted into other products such as steroid hormones or bile acids. Excess cholesterol in the cells and blood vessels is taken up by HDL, transporting it back to the liver for subsequent conversion to bile acids.

While cholesterol and triglycerides are necessary for normal body functions, high levels (associated with high fat diets, lack of exercise, obesity, and reproductive activity) may predispose birds to disease.

Elevations in cholesterol concentrations have been reported to be associated with a variety of disease conditions in birds, including hepatic lipidosis, atherosclerosis, hypothyroidism, bile duct obstruction, and diabetes mellitus. Decreased plasma cholesterol concentrations have been associated with some cases of hepatic disease (such as aflatoxicosis), decreased dietary fat, *E. coli* endotoxemia, and spirochetosis.

Elevated triglycerides are frequently associated with reproductive activity in female birds. Very high levels that are persistently elevated have been associated with ovarian pathology.

Elevated VLDL, LDL and HDL concentrations, especially when combined with elevated cholesterol and triglycerides, are considered to be likely markers for atherosclerosis and hepatic lipidosis. More research needs to be done in this field. Most birds transport cholesterol mainly in the form of HDL (unlike humans, where LDL is predominant). Therefore, higher HDL values may not be associated with decreased risks of developing atherosclerosis as it is in humans; in fact, the converse may be true.

Amylase and lipase

Amylase and lipase have been proposed as useful parameters in the detection of pancreatic disease.[1] There is still considerable discussion of the incidence of pancreatic disease and the specificity of these enzymes. Significant elevations of these enzymes, when accompanied by clinical signs of gastrointestinal dysfunction (e.g., vomiting, ileus, diarrhea, coelomic pain) should lead the clinician to consider pancreatic disease as a differential diagnosis; however, normal levels do not preclude a diagnosis of pancreatic disease, nor do abnormal levels confirm such a diagnosis.

Electrolytes

The predominant intracellular and extracellular anions and cations in birds are similar to those in mammals.[3]

Sodium

Sodium must be interpreted with the knowledge of the patient's hydration status. It may be elevated with decreased water intake or dehydration through renal disease, vomiting or diarrhea. Sodium may also be lost through the gastrointestinal tract or the kidneys. Other causes of hyponatremia include over-hydration, end-stage liver disease and congestive heart failure.

Chloride

Chloride is interpreted alongside sodium. It may be elevated with vomiting or regurgitation, although this is uncommon; low levels are usually associated with regurgitation or vomiting, renal disease congestive heart failure and other conditions which cause water retention.

Potassium

Potassium may be decreased with vomiting and diarrhea, and elevated with dehydration, hemolysis, tissue damage, or poor sample handling. Renal disease in birds has been associated with hyperphosphatemia and hyperkalemia; however, hypophosphatemia and hypokalemia also have been reported. The degree of artifactual change in potassium values appears to be species-specific and warrants immediate separation of plasma from RBCs in all avian samples.

Calcium

Avian total calcium values can be much higher under normal physiologic circumstances than would be tolerated by a mammal. Dramatic increases in plasma total calcium concentration are seen in reproductive, oviparous females due to estrogen-induced transport of calcium-bound yolk proteins to the ovary. Reproductive pathology such as egg binding and egg yolk coelomitis also can result in marked total hypercalcemia. Marked hypocalcemia may be caused by malnutrition or reproductive abnormalities such as chronic egg laying.

Conclusions

While avian biochemistry analysis can provide an overview on the function of internal organs and physiology, it cannot provide a precise diagnosis. As such, plasma biochemistry analysis should be regarded as just one tool in the clinician's arsenal in reaching a diagnosis. Understanding what causes the change in pattern in biochemistry profiles is the key to using this tool more effectively.

References

1. Doneley B. *Avian Medicine and Surgery in Practice: Companion and Aviary Birds*. 2nd Ed. Boca Raton, FL: CRC Press; 2016.

2. Hung CS-Y, Sladakovic I, Divers SJ. Diagnostic value of plasma biochemistry, haematology, radiography and endoscopic visualisation for hepatic disease in psittacine birds. *Vet Rec.* 2020;186(17):563. https://doi.org/10.1136/vr.105214

3. Harr KE. Clinical chemistry of companion avian species: a review. *Vet Clin Pathol.* 2002;31(3):140-151. https://doi.org/10.1111/j.1939-165X.2002.tb00295.x

4. Williams SM, Holthaus L, Barron HW, et al. Improved clinicopathologic assessments of acute liver damage due to trauma in Indian ring-necked parakeets (*Psittacula krameri manillensis*). *J Avian Med Surg.* 2012;26(2):67-75. http://www.jstor.org/stable/41682462

5. Beaufrère H, Cray C, Ammersbach M, Tully TN. Association of plasma lipid levels with atherosclerosis prevalence in Psittaciformes. *J Avian Med Surg.* 2014;28(3):225-231. http://www.jstor.org/stable/24624173

Antifungal Treatments in Birds

Gunther Antonissen, DVM, PhD, DipECPVS,
An Martel, DVM, PhD, DipECZM (Wildlife Population Health),
Dominik Fischer, DVM, PhD, DipECZM (Wildlife Population Health)

Session #3001

Affiliation: From the Department of Pathobiology, Pharmacology and Zoological Medicine, Faculty of Veterinary Medicine, Ghent University, 9820 Merelbeke, Belgium (Antonissen, Martel), the Clinic for Birds, Reptiles, Amphibians and Fish, Faculty of Veterinary Medicine, Justus Liebig University Giessen, 35392 Giessen, Germany and Veterinary and Biological Science Department, Zoo Wuppertal, 42117 Wuppertal, Germany (Fischer).

Abstract: The use of antifungal drugs in birds is characterized by major challenges such as a large interspecies and interindividual variability in the pharmacokinetics, which can significantly affect drug safety and efficacy, and limited knowledge on avian antifungal treatment protocols. Therefore, different factors need to be considered in the rational drug selection of antifungal therapy in birds. This masterclass is aimed to provide insight into the interrelation of antifungal drug formulation, administration route, therapeutic-toxic range, antifungal resistance, and treatment outcome in fungal diseases, focusing in particular on aspergillosis in birds. The antifungal activities of the three major antifungal drugs used in avian medicine, azole derivatives, polyenes, and allylamines, are related to the synthesis or direct interaction with ergosterol. Conventional antifungal therapies in avian medicine are frequently associated with a lack of efficacy and high toxicity. Innovative drug formulations can help to reduce the intrinsic toxicity and enhance efficacy. Because the majority of systemic fungal infections require long-term therapy, oral administration of antifungal drugs is preferred, with intravenous administration being reserved for the initial phase of treatment in cases of acute aspergillosis or severely debilitated birds. Finally, alternative treatment strategies including topical administration of antifungals, such as through nebulization, and combining different antifungal drugs in treatment protocols show promising results in birds.

Adapted from: Antonissen G, Martel A, Fischer D. Antifungal drugs. In: *Current Therapy in Avian Medicine and Surgery.* Elsevier; In prep.

Introduction

An antifungal drug is a pharmaceutical fungicide or fungistatic used to treat and prevent mycosis such as aspergillosis, megabacteriosis, and candidiasis. The use of antifungals in birds is characterized by interspecies and inter-individual variability in the pharmacokinetics, affecting drug efficacy and toxicity. In the last decades, newer and less toxic antifungals, including the azoles and echinocandins, have been developed. Aside from the chemical structure, treatment safety is also affected by antifungal drug formulation, route and frequency of administration; however, the knowledge of avian antifungal treatment is limited, therefore treatment protocols are often developed empirically, based on case reports, or extrapolated from humans or other animal species. Rational usage of antifungal drugs in avian species is much more than only selecting an appropriate dosage.[1] The objective of this masterclass is to discuss the mechanism of action and toxicity of different antifungal drugs; the impact of different formulations and drug delivery strategies on antifungal treatment; antifungal resistance; and some practical guidelines on administration and treatment.

Mechanism of Action of Different Antifungals

The mechanism of action of all three major groups of antifungal agents used in avian practice (i.e., polyenes, azole derivatives and allylamines) involve the inhibition of synthesis or direct interaction with ergosterol, the predominant component of the fungal cell membrane. The polyene macrolide antibiotics, such as amphotericin B, nystatin, and pimaricin (or natamycin) act primarily by binding to ergosterol and forming pores in the plasma membrane, increasing membrane permeability, eventually leading to cell death. Polyenes can be fungistatic or fungicidal, depending on the dosage. Polyenes have a broad antifungal spectrum, including a variety of yeasts (e.g., *Macrorhabdus ornithogaster, Candida* spp., *Rhodotorula* spp.*)* and molds (e.g., *Aspergillus* spp.*, Mucor* spp.*, Penicillium* spp.). Azoles inhibit the enzyme cytochrome P450-dependent 14-α-sterol demethylase, a key enzyme in ergosterol biosynthesis, resulting in a disruption of fungal membrane structure and function. Azoles are classified into two groups: the imidazoles with two nitrogens in the azole ring (such as clotrimazole, miconazole, enilconazole, and ketoconazole) or triazoles (such as itraconazole, fluconazole, posaconazole and voriconazole). Most azoles have a broad-spectrum activity against most yeasts and filamentous fungi, and except voriconazole, most azoles used in avian medicine are fungistatic at the doses used in birds and need several days to reach steady-state concentrations. Allylamines, such as terbinafine, inhibit the enzyme squalene epoxidase, resulting in an ergosterol depletion and squalene accumulation. Allylamines have fungicidal activity against dermatophytes, molds and certain dimorphic fungi and fungistatic activity against yeasts.[1,2]

Besides the antimicrobial agents affecting fungal sterols, inhibitors of the glucan synthesis, such as echinocandins (e.g., caspofungin, micafungin, and anidulafungin) also affect the fungal cell wall. Echinocandins exhibit fungicidal activities against *Candida* spp., but are fungistatic for *Aspergillus* spp. Finally, 5-fluorocytosine is a fluorinated pyrimidine with inhibitory activity against many yeasts, including *Candida* spp. and *Cryptococcus neoformans*, and against *Aspergillus* spp.[1,2]

Toxicity

Amphotericin B is characterized by a dose-dependent nephrotoxicity in mammals as a result of renal vasoconstriction, reduced glomerular filtration rate, and direct toxic effect on the membranes of the renal tubule cells. In contrast to mammals, nephrotoxicity is lessened in birds due to their faster elimination rate; however, systemic use of amphotericin B deoxycholate should be limited and reserved for severe cases of aspergillosis. Nystatin is used primarily to treat candida infections of mucosa, skin, and the gastrointestinal tract. Nystatin should not be given if gastrointestinal epithelium is not intact as systemic absorption may occur.[1]

The most common observed side effects associated with azole administration in avian patients are gastrointestinal signs, such as anorexia and vomiting, and liver dysfunction. Toxicity of azoles is associated with the drug specific binding to the fungal cytochrome P450 enzyme instead of the vertebrate analogue. In general, azoles are well tolerated in birds; however, remarkable species differences are observed. The apparent sensitivity to azoles experienced by different bird species might be explained by differences in the drug's pharmacokinetics.[1,2] Of the imidazoles, enilconazole and clotrimazole rarely cause side effects when used topically; however, in humans it has been reported that clotrimazole induces edema, erythema, vesication, desquamation, pruritus, and urticaria after cutaneous application. Prudent usage of ketoconazole is recommended, since adverse liver, renal, gastrointestinal, and circulatory side effects have been reported following parental application.[2]

The use of terbinafine is generally associated with limited side effects. These side effects can include mild gastrointestinal symptoms and hepatobiliary dysfunction.[3]

Drug Formulation

Several antifungal drugs are characterized by their insolubility in water at physiologic pH, poor oral bioavailability, and limited formulation approaches. Different formulations and drug delivery strategies have been developed to improve the antifungal pharmacokinetics with targeted delivery, rapidly followed by sustained release and prolonged retention of high drug concentration localized at the infection site. For example, to reduce the intrinsic toxicity and to enhance efficacy of amphotericin B, different lipid formulations have been developed (i.e., amphotericin B lipid complex, liposomal amphotericin B, and amphotericin B colloidal dispersion);[1,4] however, in-depth information on the pharmacokinetics, pharmacodynamics, and clinical use of amphotericin B lipid formulations in birds is lacking. In addition, nanotechnology holds great promise to overcome the limitations of antifungal drug solubility and toxicity. In falcons, it was demonstrated that itraconazole-loaded nanostructured lipid carriers penetrate deeply in the lungs and air sacs, which is a prerequisite for pulmonary treatment of aspergillosis.[5]

Drug Resistance

Antifungal susceptibility testing may be helpful for monitoring therapy in avian species. Acquired resistance of *A. fumigatus* strains, isolated from companion and wild birds, is reported for amphotericin B, itraconazole, voriconazole and caspofungin. With even some isolates being resistant to more than one antifungal drug.[6] Although historically used, significant resistance of *C. albicans* to itraconazole has been reported. Finally, evidence of nystatin and of amphotericin B resistance in *Macrorhabdus ornithogaster* have been reported.[7]

Combined Antifungal Treatment

Treatment of fungal diseases are often complicated by poor response and high toxicity. Therefore, double-drug combinations might be promising to enhance efficacy. For example, synergistic interaction between terbinafine and itraconazole has been observed in vitro for *A. fumigatus*, resulting in a reduced MIC and MFC;[8] however, since amphotericin B and the azoles target the same site, simultaneous administration might result in an antagonistic interaction, where azoles prevent the synthesis of ergosterol, making the amphotericin B that usually binds the ergosterol in the cell membrane inactive. In avian medicine, the use of certain aerosolized disinfectants such as F10 (a blend of quaternary ammonium and biguanide compound; Health and Hygiene (Pty) Ltd, Johannesburg, South Africa) and lufenuron has been reported as effective addition to the inhalation therapy with azoles (e.g., enilconazole), but scientific studies on the safety and efficacy of these substances are lacking.[9] Finally, different routes of administration can also be combined, such as systemic triazole therapy (itraconazole or posaconazole) combined with inhalation therapy or topical intranasal clotrimazole or enilconazole.[10]

Practical Guidelines on Administration and Treatment

Drug selection and administration route is mainly dependent on the course and the severity of the disease. Depending on the drug formulation, antifungal drugs need to be given regularly, once (q24h) or twice (q12h) daily, to achieve a systemic effect. In most cases, treatment needs to be continued for at least 3-4 weeks, in severe cases 3 months or even longer. Furthermore, a combination of several approaches is usually required. Antifungal drugs may be administered orally, via nebulization, intratracheally, intramuscularly, subcutaneously, intravenously, and topically on fungal lesions.

Oral treatment is the most common drug application for systemic treatment of aspergillosis or microsporidia infections as well as for the treatment of gastrointestinal yeast. For the treatment of aspergillosis, the oral formulations of itraconazole and voriconazole are mostly used, which are reabsorbed from the gastrointestinal tract following oral administration. In contrast, gastrointestinal yeast infections, such as *Candida* spp. and megabacteria (*Macrorhabdus ornithogaster*) infections, are usually treated with formulations which are not absorbed from the gastrointestinal tract but act locally.[2]

Nebulization of antimycotic drugs (e.g., amphotericin B, terbinafine, or the triazole drugs enilconazole and clotrimazole) has been regularly reported for the treatment of aspergillosis and is a valuable and stress-free addition to the therapeutic regime in absence of major side effects. The antifungal medication is mixed with saline solution and nebulized using a commercial inhalation device, providing sufficiently fine and respirable aerosol droplets, able to enter air capillaries.[1,2,11,12]

In severe cases of systemic fungal infections, antifungal drugs may also be administered parenterally via the subcutaneous, intramuscular or intravenous routes. Likewise, parenteral injections are an alternative administration route to oral medication in patients with vomiting or regurgitation, those with acute courses of aspergillosis, or individuals with unusual localizations of fungal infections (e.g., in joints, bones or the spinal cord). Parenteral solutions are available for amphotericin B, fluconazole and voriconazole, but the therapeutic effect of fluconazole against *Aspergillus* spp. has been reported to be low. Voriconazole has been used in recent years successfully in various species, with less side effects and larger clinical utility compared to amphotericin B; however, a faster onset of therapeutic action has been reported for amphotericin B compared to the triazole drugs. This may explain why this agent is often used in the initial phase of medical therapy of aspergillosis but not for a prolonged period.[1]

Topical administration of antifungal drugs by applying injectable solutions directly to the fungal lesion is suggested in addition to nebulization for the treatment of fungal lesions inside the trachea or syrinx. Therefore, parenteral solutions such as amphotericin B or voriconazole may be applied intratracheally, orthograde via the glottis using plastic catheters. In some cases of tracheal aspergillosis, also surgical debridement may be required. Likewise, topical treatment and surgery may be required for the treatment of air sac aspergillomas. Furthermore, a topical antifungal treatment may also be applied to the mucous membranes of the oropharynx in cases of candidiasis, to the outer skin and feather follicles in cases of fungal dermatitis and folliculitis, as well as to eyes and nose in case of ocular fungal infection, rhinitis or sinusitis.[1]

Finally, ocular fungal infections including diseases of conjunctiva, cornea and/or anterior eye chamber are treated using antifungal eye drops, even though they are usually not commercially available, but may be prepared by a pharmacist on request. Similar formulations may be used for the treatment of fungal rhinitis and sinusitis in addition to inhalation therapy and systemic antifungal medication; however, nasal flushes using enilconazole or F10 SC solutions have been reported as beneficial additions to the treatment of fungal rhinitis and sinusitis.[9]

References

1. Antonissen G, Martel A. Antifungal therapy in birds: old drugs in a new jacket. *Vet Clin North Am Exot Anim Pract.* 2018;21(2):355-377.

2. Krautwald-Junghanns ME, Vorbruggen S, Bohme J. Aspergillosis in birds: an overview of treatment options and regimens. *J Exot Pet Med.* 2015;24(3):296-307.

3. Bechert U, Christensen JM, Poppenga R, et al. Pharmacokinetics of terbinafine after single oral dose administration in red-tailed hawks (*Buteo jamaicensis*). *J Avian Med Surg.* 2010;24(2):122-130. doi.org/10.1647/2008-052.

4. Phillips A, Fiorello CV, Baden RM, et al. Amphotericin B concentrations in healthy mallard ducks (*Anas platyrhynchos*) following a single intratracheal dose of liposomal amphotericin B using an atomizer. *Med Mycol.* 2018;56(3):322-331.

5. Pardeike J, Weber S, Zarfl HP, et al. Itraconazole-loaded nanostructured lipid carriers (NLC) for pulmonary treatment of aspergillosis in falcons. *Eur J Pharm Biopharm.* 2016;108:269-276.

6. Spanamberg A, Ravazzolo AP, Denardi LB, et al. Antifungal susceptibility profile of *Aspergillus fumigatus* isolates from avian lungs. *Pesqui Vet Brasil.* 2020;40(2):102-106.

7. Baron HR, Leung KC, Stevenson BC, et al. Evidence of amphotericin B resistance in *Macrorhabdus ornithogaster* in Australian cage-birds. *Med Mycol.* 2019;57(4):421-428.

8. Mosquera J, Sharp A, Moore C, Warn P, Denning D. *In vitro* interaction of terbinafine with itraconazole, fluconazole, amphotericin B and 5-flucytosine against *Aspergillus* spp. *J Antimicrob Chemother.* 2002;50(2):189-194.

9. Bailey TA. Raptors: respiratory problems. In: Chitty J, Lierz M, eds. *BSAVA Manual of Raptors, Pigeons and Passerine Birds.* Gloucester, UK: British Small Animal Veterinary Association; 2008:223-234.

10. Campitelli M, Zeineddine N, Samaha G, Maslak S. Combination antifungal therapy: a review of current data. *J Clin Med Res.* 2017;9(6):451-456.

11. Emery LC, Cox SK, Souza MJ. Pharmacokinetics of nebulized terbinafine in Hispaniolan Amazon parrots (*Amazona ventralis*). *J Avian Med Surg.* 2012;26(3):161-166.

12. Carpenter JW, et al. Appendix 1. Table of Common Drugs and Approximate Doses. In: Speer B, Ed. *Current Therapy in Avian Medicine and Surgery.* St. Louis, MO: Elsevier Health Sciences; 2015.

Avian Cardiac Diagnostics

Maria-E. Krautwald-Junghanns, ME, DVM, Prof Dr med vet, and
Cornelia Konicek, Dr med vet

Session #3002

Affiliation: From Department for Birds and Reptiles, Veterinary Teaching Hospital, University of Leipzig, An den Tierkliniken 17, 04103 Leipzig Germany (Krautwald-Junghanns), Service for Birds and Reptiles, Clinical Unit of Internal Medicine Small Animals, Department for Companion Animals and Horses, University of Veterinary Medicine Vienna, Veterinärplatz 1, 1210 Vienna, Austria (Konicek).

Abstract: Limitations of diagnostic techniques in avian medicine make ante mortem diagnosis of cardiovascular diseases in birds challenging. Indications for examination of the cardiovascular system are often unspecific. A scoring system for criteria in anamnestic risk factors and clinical signs can facilitate the decision. Score assignment is discussed during the masterclass. Further, different pathologies are demonstrated in the individual diagnostic methods that are discussed.

Introduction

Ante mortem diagnosis of cardiovascular diseases in birds is challenging due to the limitations of several diagnostic techniques in avian medicine. Frequently described cardiovascular lesions in pet birds include atherosclerosis, pericardial effusions, pericarditis, myocarditis, dilatation or hypertrophy of the ventricles and valvular endocarditis. Right ventricular or biventricular heart failure seems to be more common than left ventricular heart failure. Some case reports are available on myxomatous degeneration or atrioventricular valve endocardiosis, bacterial endocarditis and arteriopathy of the right brachiocephalic artery. Indications for examination of the cardiovascular system are often unspecific. A scoring system for given criteria in anamnestic risk factors and clinical signs can facilitate the decision.

Diagnostic Approach

Anamnestic data

Relative predisposition to cardiovascular diseases is given by the species, breed, age, gender, lifestyle and diet. Examination parameters that might indicate an increased risk of cardiac disease and atherosclerosis especially may include the following (score 1 point for each):

- Predisposed species
- Female gender
- Age > 15-20 years
- Unbalanced, inappropriate feeding
- Decreased activity

- Inappropriate husbandry
- Pre-existing disease that may adversely affect the cardiovascular system

Cardiac disease is common in larger, elderly parrots living in captivity, due to inadequate housing conditions, malnutrition, and long-life expectancy. A high prevalence of atherosclerosis for example is reported in African grey parrots (*Psittacus erithacus*), Amazon parrots (*Amazona* spp.) and cockatiels (*Nymphicus hollandicus*), in contrast to cockatoos and macaws. A higher incidence of atherosclerosis was also noted in birds of 15- 20 years and older and in the female sex, especially in association with reproductive diseases. The predisposition to the development of atherosclerosis in female psittacine birds likely relates to the profound effects of estrogen in reproductively active birds.

In addition to age, female sex, and species, risk factors for development of atherosclerosis also include high-calorie and fat diets, dyslipidemia (e.g., hypercholesterolemia) and limited physical activity.

Clinical examination

Clinical signs

Although often unspecific, common examination findings that may indicate cardiac disease include the following (assign 1 point for each finding):

- Periorbital bluish discoloration/cyanosis
- Central nervous disorders/epileptiform seizures/ataxia
- Prolonged respiratory recovery with increased activity
- Exercise intolerance
- Disorientation or confusion
- Chronic untreatable skin changes
- Delayed absorption of administered subcutaneous medications

Clinical signs associated with atherosclerosis can have a subtle, insidious onset. Common findings are exercise intolerance, lethargy, dyspnea, cyanosis, seizures, as well as prolonged healing of dermal lesions.

Many birds do not show any clinical signs until they are stressed or the patient decompensates. As a result, atherosclerosis often is diagnosed postmortem. Severe lesions in the great vessels of the heart are often incidental findings at necropsy.

In general, owners of birds with cardiovascular disease mostly report unspecific clinical signs such as progressively declining activity level, and reduced appetite. Central nervous disorders, such as paresis or rigidity of one or more limb or seizure activity, as well as behavioral changes including disorientation or confusion, may be observed. Persistent exercise intolerance occurs regularly in birds with existing heart disease. Congestive heart failure can lead to a clinical presentation of coelomic effusion and hepatomegaly. Due to avian anatomy, the lack of a diaphragm, coelomic effusion reduces the patient's ability to ventilate. This may result in tachypnea, increased respiratory effort, open beak breathing, tail bobbing, and wide stance may be observed. The typical heart cough in mammals cannot be observed in birds in general, due to anatomic and physiologic particularities. Furthermore, cyanosis of the skin (or integument) may be observed periocular (or around the eyes) in avian species lacking feathers in this area. Other dermal indicators of cardiac disease can include recurring or chronic skin lesions such as axillary or wing web dermatitis and focal feather loss, which have been reported in patients with atherosclerotic lesions. Congestive heart failure and peripheral venous congestion may be recognized by distension of the jugular and cutaneous ulnar veins. Edema is rarely found, most often appearing as swelling of the periorbital region or of the hocks and feet. In hospitalized patients receiving subcutaneous fluids, the first indication of circulatory compromise may be delayed absorption (> 48 hours).

Additional examination methods

Standard examination methods for heart disease, as found in mammals, are partly lacking or hard to evaluate due to numerous physiologic peculiarities in birds. These include the following:

- arterial pulse,
- auscultation,
- hematology,
- biochemistry, and
- blood pressure measurement.

For instance, arterial pulses cannot always be palpated or thoroughly assessed. The relative warmth of each extremity can be subjectively assessed. Auscultation of the avian heart only plays a subordinate role due to many limitations; however, it can provide subjective information of the rate, sound quality and the rhythm. Due to rapid heart rates, individual heartbeats are difficult to note by the examiner, making muffled heart sounds, or heart murmurs very difficult to characterize. In the case of cardiogenic pulmonary edema, increased respiratory noise can be heard during auscultation in a calm bird, but characteristic crackles and wheezes are generally absent, except in severe cases.

Changes in the complete blood count have no correlation to cardiac disease; however, some alterations have been described in association with atherosclerosis. Plasma lipid levels may offer some observational data pertinent to population-level risk factors for cardiovascular diseases. Changes in plasma lipids, such as total cholesterol and total triglycerides, as well as HDL-C and LDL-C may be noticed in an individual bird, but single measurement results should be interpreted with caution. Until now, dyslipidemias, as well as lipoprotein ratio and non-HDL concentrations, have not been statistically linked to cardiovascular disorders. Similarly, atherosclerosis has also been shown in birds with normal plasma cholesterol levels, whereas hypercholesterolemia in the blood cannot always be associated with atherosclerosis. Furthermore, cholesterol may increase independently from the occurrence of atherosclerosis (e.g., females displaying vitellogenesis).

In human medicine, an increased level of triglycerides in the blood is a prominent risk factor for cardiovascular diseases due to the presence of atherogenic particles. To what extent this factor can be transferred to the class Aves is questionable. Since the triglyceride value in the blood changes rapidly after a change in the feed ration of a bird and is subject to more postprandial fluctuations, this value must also be assessed critically.

Unfortunately, there is poor agreement between invasive and non-invasive blood pressure measurements in birds. Using non-invasive techniques in psittacine birds, high variabilities in the readings from different cuff placement sites on the same individual have been reported. The accuracy of the method used must be therefore assessed critically. Because of these technical issues, the significantly higher arterial blood pressure of birds compared to mammals, and the lack of scientific data on hypertension in birds, blood pressure measurement plays a subordinate role for assessing heart diseases in birds; however, it may help to define hypertension.

Imaging techniques

Radiographs may show subjective changes in heart size and great vessel density (or radio-opacity); however, this is not an appropriate assessment of cardiac function. An echocardiogram is far more sensitive to assess cardiac function, while vascular health may require more advanced imaging. Because M-mode examination cannot be performed in birds, the spectral Doppler function is the only functional mode of evaluation of the heart in birds. It enables the investigator to visualize and measure the aortic root and take measurements of the aortic outflow. Reference intervals are available for the evaluation of the heart itself in 2D-mode as well as for aortic measurements in different species (Table 1). Other techniques such as fluoroscopic angiography and computed tomography may help to detect signs of conditions like atherosclerosis in birds. Electrocardiography (ECG) has been considered but is not commonly used in avian practice due to a lack of standardization, artifacts and problems with interpretation. In some instances, endoscopy, and rarely MRI, may be indicated as well.

Diagnostics of value for detection of cardiac disease include the following:

- Echocardiogram
- Doppler ultrasonography
- Radiologic native examinations (radiology, computed tomography)
- Angiocardiography
- ECG
- Endoscopy
- MRI

Echocardiogram

Echocardiography is currently the diagnostic method of choice for suspected cardiac disease, especially in larger Psittaciformes, Columbiformes and birds of prey. Echocardiograms in the awake avian patient is advisable in severely ill animals for which general anesthesia is associated with a higher risk. Sedatives, such as midazolam, may be warranted in severely stressed patients. Echocardiography is less invasive in comparison to other imaging methods and allows for the cardiovascular compromised patient to be evaluated in an upright position during the examination.

To assess the avian heart, two coupling sites are described: a) the ventromedian approach with two horizontal longitudinal views; and b) the parasternal approach with several short and long axis views, which may be helpful for examination of the valves and the vessels.

As in mammal medicine, echocardiography in birds allows for the visualization of cardiac structures in motion and detailed imaging of internal structures. Due to anatomic peculiarities, the established method of choice in birds is the B-mode method (2D echocardiography) and reference values are already available for various species (Table 1). Furthermore, echocardiography is well suited for determining the degree of success in treating hydropericardium or controlling the contractility of the ventricle once the patient has been administered medication. The use of echocardiography to image blood vessel walls in birds limited. At least the assessment of the aorta close to the heart is possible and clinical examination has shown that hyperechoic areas indicate the presence of calcification within the vessel wall; however, the examination of vascular disease using echocardiography is very subjective and requires an experienced examiner.

A routine echocardiogram exam assesses:

- Size of the ventricles
- Thickness of the interventricular septum
- Morphology and function of the valves
- Assessment of the aorta
- Contractility of the ventricles

The following parameters are important for evaluating the heart's morphology and function. The size of the ventricles and the thickness of the interventricular septum. Furthermore, the left atrioventricular valves, the aortic valves and the muscular right AV valve can be examined independent of the image quality. The diameter of the aortic root can be assessed, as well as the blood outflow velocity (see below).

The aortic diameter is not an accurate measurement to evaluate for the presence of atherosclerosis in birds. In addition to cardiac size, ventricular contractility (fractional shortening) of the transverse diameter of the left ventricle is a valuable parameter for the assessment of cardiac function. To obtain ventricular contractility, fractional shortening should be calculated using the following formula:

Fractional shortening (%) = (diastolic diameter – systolic diameter) × 100/diastolic value

Table 1. Reference values for selected bird species (mean ± standard deviation).

		Psittacus erithacus	*Amazona* spp.	*Cacatua* spp.	Diurnal birds of prey	*Columbiformes*
	Ventromedian approach					Parasternal approach
Body weight (g)		493 ± 55	353 ± 42	426 ± 162	720 ± 197	434 ± 52
Left ventricle	Systolic length [mm]	22.5 ± 1.9	21.1 ± 2.3	19.0 ± 1.3	14.7 ± 2.8	17.9 ± 1.0
	Diastolic length [mm]	24.0 ± 1.9	22.1 ± 2.2	19.9 ± 1.6	16.4 ± 2.7	20.1 ± 1.4
	Systolic width [mm]	6.8 ± 1.0	6.7 ± 1.2	6.4 ± 1.7	6.3 ± 1.1	5.2 ± 0.4
	Diastolic width [mm]	8.6 ± 1.0	8.4 ± 1.0	8.3 ± 1.5	7.7 ± 1.2	7.4 ± 0.6
	Fractional shortening [%]	22.6 ± 4.4	22.8 ± 4.2	25.6 ± 7.0	n/k[a]	27.2 ± 4.5
Right ventricle	Systolic length [mm]	9.2 ± 1.4	9.4 ± 1.8	10.3 ± 1.2	12.7 ± 2.7	n/k
	Diastolic length [mm]	11.5 ± 1.9	10.3 ± 1.3	11.3 ± 2.3	13.9 ± 2.5	9.9 ± 0.8
	Systolic width [mm]	2.8 ± 0.9	3.1 ± 0.7	2.3 ± 0.0	2.1 ± 0.6	n/k
	Diastolic width [mm]	4.8 ± 1.1	5.2 ± 1.3	3.5 ± 0.5	2.5 ± 0.8	4.0 ± 0.5
	Fractional shortening [%]	40.8 ± 11.9	34.1 ± 3.7	33.3 ± 10.3	n/k	n/k
Inter-ventricular septum	Systolic thickness [mm]	2.9 ± 0.5	2.2 ± 0.1	1.9 ± 0.3	1.9 ± 0.6	3.8 ± 0.1
	Diastolic thickness [mm]	2.5 ± 0.3	2.1 ± 0.4	1.7 ± 0.4	1.9 ± 0.5	3.3 ± 0.2

[a]Not known.

Doppler ultrasonography

The color Doppler function has been successfully utilized in various avian species and is considered a valuable tool in the diagnosis of aneurysms and valvular insufficiency; however, color Doppler function significantly reduces the frame rate. The ultrasound image is known as a frame. On the ultrasound monitor, we see multiple frames produced in rapid succession. The frame rate, then, is the speed at which those images are shown. A reduced frame rate is associated with a reduced temporal resolution, making the examination results strongly depending on the examiner`s experience. Therefore, its applicability in avian echocardiography is demanding.

Spectral Doppler ultrasonography is used for assessing the velocity of the blood flow. For this measurement the ventromedian coupling side should be utilized. Spectral Doppler measures the rate of blood flow (inflow, outflow) can be seen in a 2D curve against time. In most species, it is possible to visualize the diastolic blood flow at the left and the right atrioventricular valves and therefore obtain inflow rates to the right and the left ventricles, as well as aortic systolic outflow. For these measurements, reference values are available for some psittacine and raptor species. Due to stress induced bias (increased heart rate and cardiac outflow), the examination is recommended to be done under general anesthesia or at least, mild sedation.

In birds, spectral Doppler, specifically pulsed wave Doppler (PW Doppler), is preferred because the examiner can select a gate (the sample site from which the signal is obtained with pulsed Doppler) for the examination. This gate is positioned directly distal to the aortic valve in the aortic root. With the help of the angle correction on the ultrasonic device, the direction of measurement of the blood flow velocity in the aorta can be aligned manually. Afterwards the aortic flow is shown in a graphic representation on the ultrasonic device. By tracing the aortic flow with the trackball the velocity time integral (VTI) and the heart rate can be calculated.

In the author's experience, measurement of the blood velocity in the aortic root can provide useful information on possible atherosclerotic changes. This is because wall alterations result in a smaller vessel diameter and therefore increase the aortic outflow. Values above 1-1.5 m/s are often documented in birds with suspected atherosclerotic vessels.

Another spectral Doppler, the continuous wave Doppler (CW Doppler) is rarely used in birds, as measurements the blood flow velocities are required from an inaccessible point. Recently tissue Doppler imaging was used, to quantify myocardial velocities by measuring the longitudinal peak velocities in the systole and diastole in awake racing pigeons.

Radiographic diagnostics (conventional CR and CT)

In native lateral and ventrodorsal radiographs the position, size and shape of the heart silhouette as well as the radiopacity of the major vessels can be assessed. A discernible radiographic image of the of the apex of the heart is visible in few avian species (e.g., cockatoos). In other species, a clear view of the apex of the heart indicates the presence of an air sac rupture as a result of trauma, as the air surrounding the heart works as a contrast medium. In such cases, echocardiographic imaging is limited because ultrasound waves striking a tissue-air interface are mostly reflected, limiting further tissue penetration.

To determine the size of the avian heart, measurements are described for various species, using conventional radiographs. The maximal width of the cardiac silhouette and the thoracic width should be measured in ventrodorsal (VD) projection. The length of the sternum and the cardiac length is measured in lateral projection. Ratios are then calculated and compared to the published data.

A common radiographic finding is cardiomegaly, noted by an enlarged cardiac silhouette. Cardiomegaly indicates multiple etiologies such as cardiac hypertrophy, cardiac dilatation, pericardial effusions, aneurysms, inflammation or neoplasms; however, the differentiation of these etiologies can only be determined by further diagnostics means. Additional radiographic findings, such as hepatomegaly, coelomic effusion, respiratory disease may be diagnosed secondary to cardiac diseases. A rare radiographic finding in birds is microcardia. Microcardia is most likely associated with severe dehydration.

By the means of radiography, pathologies of the great vessels, such as atherosclerotic lesions, are only visible if they are very pronounced. Subjectively, soft tissue density or increased radio-opacity of the aorta can be assigned to atherosclerotic lesions. A massively increased radiographic density, which is comparable to that of the grit stones, can be rated as calcification. Such a detectable mineralization of the aortic wall, which is described as stage VII of atherosclerosis, is considered highly suspicious for the presence of atherosclerosis. Often these animals also have very high blood flow velocities in the aortic root.

Compared to radiography, computed tomography enables better identification of the cardiovascular system without interference from the overlaying structures. Also, CT examination provides an excellent assessment of major arteries particularly with the use of intravascular (IV) contrast media. Likewise, the four distinct chambers of the heart can be visualized using IV contrast media. The limits to imaging structures are dependent on the spatial resolution of the CT scanner, differences in density between the object of interest, surrounding structures, and size of the object. Motion artifact may be noted due to the rapid heart rate in birds, which may decrease the accuracy of the diameter measurements. Severe limitations also exist for small bird species when attempting to visualize the smaller arteries within the patient.

Angiography

Currently, there are only a few case reports on the use of angiocardiography in avian medicine. The examination can be performed either by real time fluoroscopy in conventional radiography or by computed tomography. Due to the high heart rate in birds, an examination of cardiac contractility is challenging. Angiocardiography, allows the brachiocephalic trunk, the aorta, pulmonary arteries and veins and the caudal vena cava to be visualized. Thus, angiocardiography can provide useful information especially for diagnosing changes in the heart vessels such as stenoses, dilatations and aneurysms.

The angiocardiographic examination should include the following:

- sedation/anesthesia
- warmed iodinated contrast agent
- iodine content 380 ml iodine/ml
- dosage of 2-4 ml/kg BW IV

Angiographic techniques can be utilized to diagnose arterial luminal stenosis and atheromatous plaques. Angiograpy is likely to become the diagnostic test of choice for atherosclerosis with continued improvement of resolution artifacts through technologic advances; however, disagreements of the luminal arterial diameters were found when comparing three different CT angiocardiography protocols, limiting the diagnostic power with the existing protocols and current available technical possibilities. Further studies of CT angiocardiography and protocols are warranted to become diagnostically useful in cases of cardiovascular disease. Again, the availability and the high cost of the examination must be considered in birds.

Electrocardiography (ECG)

Clinicians must be aware of the different characteristics of the avian ECG compared to small animals. Due to the patient's high susceptibility to stress, the result of the electrocardiography in the awake bird may be falsified and should therefore be classified as an exercise ECG. The application of ECG in awake birds is difficult due to restraint and patient stress. Problems may occur with the connection of the leads to the skin and, as previously mentioned, stress may contribute to alterations of the recorded ECG. Therefore, electrocardiography, for detection of heart disease in birds, is still infrequently used in comparison to mammals.

For recording an electrocardiogram in birds, both the use of needle electrodes and the use of spring electrodes have been described in the literature. As in human medicine, the position of the leads is based on the scheme of Einthoven. A recording speed of 100 to 200 mm/s is necessary. Reference values for various bird species have been established in studies focused on electrocardiograms in selected species.

In principle, an ECG may be useful to evaluate heart rate and rhythm, but one should remember that electrical activity does not ensure mechanical (pumping) activity. The ECG can actually remain relatively normal in birds, with severe cardiopulmonary compromise or even cardiac arrest from anesthetic overdose. It is also important to keep in mind, that arrhythmias can be explained by extra-cardiac and cardiac causes, as well as changes in the autonomic nervous system. It is therefore important to rule out metabolic abnormalities. Changes in the P waves and/or QRS complex can suggest enlargement of the cardiac chambers and a low voltage ECG may indicate the presence of pericardial effusion, ascites or obesity.

Endoscopy

Endoscopic examination and guided sampling can be especially valuable in the diagnosis of masses associated with the heart or the major vessels. Endoscopy allows direct visualization and detection of gross abnormalities, such as cardiomegaly, pericardial effusion, pericardial thickening, or exudate, arterial discoloration, gross structural changes, neighboring masses or granulomas. Standard lateral approaches can be used to visualize the abdominal organs and arteries. The more uncommon interclavicular approach allows direct visualization of the great arteries and the base of the heart but is more invasive and may impaired by fat deposits in overweight birds.

Magnetic resonance imaging (MRI)

MRI is relatively rarely used in clinical avian medicine, due to high noise level and the long scanning time, which is a significant concern for the anesthesiologist who has limited access to the patient. The other reason is the small size of the structures to be imaged and motion artifacts due to rapid respiratory and heart rates, which makes it difficult to achieve images of diagnostic quality. Motion artifacts prevent adequate imaging of the avian heart using MRI and it is also not recommended for the vascular system because of the rate of contrast circulation. In addition, there is little empirical data available for avian MRI cardiac image assessment.

References

1. Krautwald-Junghanns M-E, Schulz U, Pees M. Evaluation of diagnostic criteria in grey parrots (*Psittacus erithacus erithacus*) with suspected atherosclerosis. *Tierärztliche Praxis K.* 2022;epub.

2. Krautwald-Junghanns M-E, Pees M, Reese S, Tully T. *Diagnostic Imaging of Exotic Pets.* Avian Chapters: Radiology, Ultrasonography, Computed tomography, Heart and large vessels. Hannover: Schlütersche Verlagsgesell.; 2011, ISBN 978-3-89993-049-8.

3. Konicek C, Krautwald-Junghanns M-E. Heart disease in pet birds - diagnostic options. *Vet Clin North Am Exot Anim Pract.* 2022;25(2):409-433. doi: 10.1016/j.cvex.2022.01.004. PMID: 35422260.

Treatment of Cardiovascular Diseases in Companion Birds

Brenna Fitzgerald, DVM, DABVP (Avian Practice)

Session #3003

Affiliation: From Colorado Exotic Animal Hospital, 7120 E. Hampden Ave, Denver, CO 80224, USA.

Abstract: Cardiovascular disease, including atherosclerosis, congestive heart failure, and pericardial disease, commonly occurs in companion birds. Treatment approaches are largely empirical and extrapolated from small animal and human medicine; there is a paucity of pharmacokinetic and pharmacodynamic data and no clinical trials in birds for cardiovascular therapeutic agents used in mammals. Nevertheless, there are numerous medications, including diuretics, vasodilators, positive and negative inotropes, and antiarrhythmic agents, that can be employed in the treatment of cardiovascular disease in avian species. Although long-term prognosis in most cases is guarded to poor, the treatment options and management strategies outlined in this review have the potential to maintain quality of life and extend survival time in these patients.

Introduction

Cardiovascular disease is frequently encountered in avian practice, predominantly in psittacine birds, and poses a serious threat to the quality of life and longevity of these and many other avian species. Successful intervention requires a foundational understanding of relevant anatomy and physiology, heightened awareness of risk factors and clinical disease states, accurate and timely diagnosis, and innovative treatment approaches.

At the present time, therapeutic interventions for cardiovascular disease in birds are largely empirical and extrapolated, where possible, from small animal and human medicine. Treatment has also been guided by a growing number of case reports in which treatment was attempted. These include cases of vascular disease, such as atherosclerosis and its associated pathologies of stroke and intermittent claudication, cardiac disease, arrhythmias, and congestive heart failure (CHF), and pericardial disease and effusion.

The long-term prognosis for most cardiovascular diseases is considered guarded to poor, given that treatment is limited to management for a finite period, rather than resolution of disease in most cases. Prognosis is partly contingent on timely diagnosis, which proves challenging given the absence or subtlety of clinical signs and the limited sensitivity of available diagnostic modalities before disease has become advanced. Primary goals are to identify and control risk factors, where possible, and following diagnosis of cardiovascular disease, to maintain quality of life and extend survival time. The following sections review treatment options that show promise for management of recognized disease states in birds, including atherosclerotic disease, CHF and related conditions, and pericardial disease and effusion.

Atherosclerosis

Atherosclerosis is a chronic inflammatory and degenerative disease of the arterial wall wherein the lumen narrows by progressive accumulation of fibro-fatty atheromatous plaques within the intima. Advanced lesions are characterized by severe arterial stenosis and occlusion, and by calcification and osseous metaplasia. Atherosclerosis is likely an underlying factor in the majority of non-infectious cardiovascular diseases in pet birds. Lesions are most frequently recognized in the ascending aorta, brachiocephalic trunks, and pulmonary arteries, but occur in the peripheral vasculature as well (notably the coronary and carotid arteries, descending aorta, subclavian arteries, coeliac artery, and ischiatic arteries). Prevalence is highest in grey parrots (*Psittacus* spp.), Amazon parrots (*Amazona* spp.), and cockatiels (*Nymphicus hollandicus*). Other risk factors include increasing age, female sex and female reproductive activity, high-calorie, fat, and cholesterol diets, dyslipidemia (e.g., hypercholesterolemia, hypertriglyceridemia), and limited physical activity.

Clinical signs are attributed to advanced lesions, whereas early and intermediate lesions are generally silent and subclinical. Unlike humans, recognizable clinical disease is primarily the product of progressive, flow-limiting arterial stenosis rather than thromboembolism and acute arterial obstruction. Clinical signs vary depending on the vessels affected, severity of atherosclerotic lesions, and presence of concurrent disease, including cardiac disease and CHF. Patients often present for falling or collapse, frequently accompanied by transient or persistent weakness and dysfunction of one or more limbs. Exercise-induced intermittent weakness and pain in the pelvic limbs, resolving with rest, is termed intermittent claudication. There may be persistent neurologic abnormalities identified upon physical examination, including reduced mentation, blindness, anisocoria, seizures, vestibular signs, paresis of one or both pelvic limbs, and ataxia. These signs are considered most consistent with stroke, but rarely is this confirmed diagnostically.

Treatment of atherosclerosis involves both controlling risk factors and managing sequelae, including peripheral hypoperfusion, intermittent claudication, ischemic stroke, and CHF. Atherosclerotic lesions cannot be resolved, but diet, husbandry, and lifestyle changes may help to prevent, slow progression, or decrease the size of lesions. Medical management focuses primarily on improving peripheral perfusion. Peripheral vasodilators, used either singly or in combination, may have symptomatic benefit by decreasing vascular resistance and reducing afterload.

Vasodilators

Isoxsuprine: Isoxsuprine is a peripheral vasodilator that causes vascular smooth muscle relaxation predominately through α-adrenoceptor blockade. To a much lesser degree, it is a β-adrenoreceptor agonist and as such may have positive chronotropic and inotropic effects (via β-1 adrenoreceptors) and further vasodilatory effects (via β-2 adrenoreceptors). In addition, it is known to increase erythrocyte deformability in humans. In veterinary medicine, isoxsuprine is used to increase peripheral blood flow in horses with vascular disorders of the lower limb and to address trauma-induced wing tip edema in raptors. In a published report, a 35-year old yellow-naped Amazon parrot with presumptive atherosclerosis was treated with isoxsuprine (10 mg/kg PO q24h). Clinical signs of lethargy, weakness, hyporexia, weak grip, and ataxia resolved with treatment, and repeatedly recurred when the drug was discontinued, again resolving once it was reinstituted. The author has observed similar, apparent symptomatic improvement when using isoxsuprine (10-15 mg/kg PO q12h) in numerous cases of clinical, presumed atherosclerosis (some of which were later confirmed at necropsy and by histopathology). Many of these patients have been treated and followed for several months to several years, over which time frequency and severity of stroke-like events, intermittent claudication episodes, and other clinical signs appeared to decrease.

Angiotensin-converting enzyme inhibitors (ACEI): ACEI, including enalapril, result in vasodilation by blocking the formation of angiotensin II (AII). AII promotes vasoconstriction and venoconstriction by mediating release of catecholamines, which act on vascular smooth muscle via α-adrenergic receptors. By blocking formation of AII, ACEI reduces both total peripheral resistance and pulmonary vascular resistance. Although the relative vasodilatory effect of an ACEI compared to isoxsuprine in birds is not known, it is conceivable that the two used in combination would have synergistic effects: an ACEI by limiting α-adrenoreceptor stimulation and isoxsuprine by α-adrenoreceptor antagonism. The author has treated patients with clinical (presumed) atherosclerotic disease using enalapril at a dose and frequency of 1.25-5 mg/kg PO q8-12h. In more severe cases this is paired with isoxsuprine.

Sildenafil: Sildenafil is a selective pulmonary vasodilator that reduces pulmonary vascular resistance and right ventricular afterload. It has predominantly been used for treatment of pulmonary hypertension (PH) in humans and dogs. In birds, the drug may have merit in cases with suspected atherosclerosis of the pulmonary arteries, potentially in combination with other vasodilators.

Other medical managements

Pentoxifylline: Pentoxifylline has been used in human medicine for treatment of peripheral vascular and cerebrovascular disease by improving microcirculatory blood flow. In small mammal models, the drug was found to increase tissue perfusion, mitigate inflammation, and attenuate atherosclerotic plaque formation. Based on these findings, pentoxifylline may have value in improving peripheral perfusion in birds with atherosclerotic disease, including those with concurrent polycythemia. The author has used it for these indications in numerous psittacine patients (15-25 mg/kg PO q8-12 h).

Statins: Statins are a group of lipid-lowering drugs used extensively in human medicine for their anti-atherosclerotic effects through inhibition of cholesterol synthesis and other mechanisms. Several products are commercially available for human use, including atorvastatin (Lipitor®) and rosuvastatin (Crestor®). Statins have been employed empirically in psittacine birds, but their use is controversial because target lipid levels that would reduce atherosclerosis risk are unestablished, and because their efficacy is not supported by available pharmacodynamic and pharmacokinetic data. Consequently, the author does not recommend use of statins in psittacines considered either to have or to be at risk for atherosclerotic disease. Instead, vasodilatory therapy (for symptomatic patients) and dietary and lifestyle changes to prevent and address dyslipidemia are more appropriate.

Supportive care and husbandry considerations

In addition to medical treatment, patients with clinical atherosclerosis and associated disease can benefit greatly from the following measures:

Supportive care: Patients with signs of stroke may have marked neurologic deficits to include reduced mentation, limb paresis, and ataxia that prevent normal eating and drinking and impair mobility. They may experience seizures or suffer injuries from falls. Supportive care measures to consider for these patients include fluid and nutritional support, analgesia, anticonvulsant therapy when needed, and management of secondary conditions such as trauma and aspiration pneumonia. Benzodiazepines diazepam (0.5-2 mg/kg IM) or midazolam (1-2 mg/kg IM) can be used for emergency control of seizures. Options for longer-term anticonvulsant therapy are levetiracetam (50-150 mg/kg PO q8-12h), zonisamide (20 mg/kg PO q8-12h), and gabapentin (15-20 mg/kg PO q8-12h). The author has used levetiracetam alone or in combination with zonisamide and gabapentin to control seizure activity in psittacine birds with severe atherosclerotic disease.

Exercise restriction versus promotion of exercise: For patients with advanced, clinical atherosclerotic disease, exercise restriction should be part of the longer-term treatment plan, as well as appropriate housing modifications to accommodate and protect birds with persistent deficits. In contrast, increasing opportunities for exercise (especially flight) from early in life may have preventive value. Physical activity can be promoted through training and by designing captive environments to facilitate locomotion and foraging behaviors.

Dietary management: Dietary management to avoid dyslipidemia may have both therapeutic and preventive value, particularly for at-risk species. Such measures include:

- Moderation of dietary calories and fat and prevention and resolution of obesity
- Provision of formulated (rather than seed-based) diets, supplemented with fresh vegetables and fruits
- Avoidance of dietary sources of cholesterol (animal-based products)
- Supplementation with omega-3 fatty acids, particularly α-linolenic acid (found in flaxseed oil), has been shown to improve lipid metabolism, reduce inflammation, and minimize development (or slow progression) of atherosclerosis in several avian species.

Other:

- Along with dietary changes, control of female reproductive activity may help prevent atherosclerotic disease.
- For patients receiving medications long-term, training to allow low-stress administration is indicated.

Congestive Heart Failure

CHF occurs when the heart is unable to empty the venous reservoirs, manifested by vascular congestion and transudation of fluid within tissues and body cavities (congestive signs). In the case of right-sided CHF, peripheral venous congestion, hepatic congestion, ascites, and pericardial effusion are often present. Pulmonary edema and congestion of the pulmonary veins occur with left-sided CHF, and a combination of signs may be seen with biventricular failure. Heart failure can further be characterized as systolic (inadequate ventricular ejection), diastolic (inadequate ventricular filling), or a combination of the two. In either scenario, stroke volume and cardiac output decrease.

CHF is not a primary disease in itself, but an ultimate consequence of structural or functional abnormalities of the cardiovascular or pulmonary systems, compounded by the chronic effects of compensatory mechanisms. Not all cardiovascular disease necessarily leads to CHF, but it is a frequent clinical end-point encountered in avian species. Heart failure can result from primary myocardial failure, including dilated cardiomyopathy (DCM), ventricular pressure overload (as occurs with outflow obstruction and systemic or pulmonary hypertension) or volume overload (as occurs with valvular insufficiency), conduction disturbances, or diastolic dysfunction (as occurs with myocardial hypertrophy and cardiac tamponade).

In a patient presenting with acute or decompensated heart failure, initial treatment aims to stabilize the patient, followed by design of a longer-term management strategy. In the author's experience, this has met with highly variable success, but some birds have been maintained in stable condition for up to 9 years.

The mainstays of treatment of CHF in small animal medicine, namely diuretics, vasodilators (including ACEI), and positive inotropes, can be applied to treatment of the condition in birds. Beta-blockers (BB), considered part of standard treatment in humans, may also have potential application. In addition, known or hypothesized underlying cause(s) should be addressed, if possible. Some etiologies, including bacterial and fungal infections, parasitic disease, and certain toxic insults, carry a better prognosis for recovery.

Diuretics

Marked reduction of excessive circulating plasma volume (hypervolemia), edema, and effusion is an immediate treatment priority in any avian patient with CHF. Alleviation of this relative fluid overload can be accomplished through the use of diuretics, principally furosemide. Diuretics should not be used alone long-term as they further activate the renin-angiotensin-aldosterone system (RAAS). Of the various diuretics available, furosemide is most commonly used in companion birds.

Furosemide: Furosemide is a potent loop diuretic that inhibits the sodium, potassium, and chloride cotransporter in the ascending limb of the loop of Henle, thereby promoting diuresis and excretion of sodium and chloride. Parenteral administration of furosemide is an essential step in management of the acute crisis in avian patients as this route allows for a rapid onset of action. A relatively high initial dose is often required. Following stabilization of the patient, a maintenance dose of furosemide can be used to enable the long-term management of chronic CHF.

Furosemide has been found to be efficacious with a rapid onset of action in birds, despite the presence of only 10-30% of looped nephrons in the avian kidney. It has been used successfully for treatment of pericardial effusion, CHF, pulmonary edema, and ascites. Route of administration may significantly influence bioavailability, as suggested by the results of a study examining the diuretic effects of furosemide in chickens (adult laying hens). In this study, it was found that urine output significantly increased following parenteral administration (2.5 mg/kg), but did not in birds receiving the drug orally, even at twice the parenteral dose (5 mg/kg).

Doses that have been used vary widely. Similar to dogs and cats, the dose and administration interval is best determined by clinical response (namely normalization of respiratory rate and effort) balanced with maintenance of acceptable hydration. The degree of clinical response can be of diagnostic and prognostic value; a positive response supports the working diagnosis of CHF and suggests a more favorable prognosis. Alternatively, a poor response warrants reevaluation of the working diagnosis or, if the diagnosis is sound, suggests a poorer prognosis. In the author's experience, a dose range of 1-5 mg/kg intramuscularly (IM) has been efficacious when stabilizing most psittacine patients. An initial frequency of 2 hours is usually followed by a shift to 6- to 12-hour intervals. There have been isolated cases with severely decompensated disease in which higher doses (7-11 mg/kg IM q12h) were required to gain control of congestive signs. In patients whose disease is less severe and advanced, a lower initial dose and frequency (1-5 mg/kg IM q12h) may be adequate. Once the patient has been stabilized, furosemide can be administered orally, with the dose increased at least 2-fold (presuming oral bioavailability is 60-75% as in humans) given q8-12h. Ultimately, the goal for long-term management is to identify the lowest dose and frequency that controls congestive signs; adjustments will be needed over time as the disease condition changes or progresses. Patients with more severe, advanced disease may require a higher dose and frequency (30-70 mg/kg PO q6-12h) to maintain control of congestive signs.

Spironolactone: Spironolactone, an aldosterone antagonist that is classified as a potassium-sparing diuretic, may have merit as part of the treatment regime in cases where congestive signs cannot be controlled with furosemide and an ACEI alone. Its combined use with furosemide may also be considered to offset potassium loss. Aside from its diuretic effects, spironolactone is thought to prevent or decrease myocardial fibrosis in humans, and counteract the myriad deleterious effects of aldosterone in CHF; however, thus far, reports on its use or efficacy in avian species are scarce. In the author's experience, doses between 1-13 mg/kg PO q8-12h have been well tolerated in numerous CHF patients.

Vasodilators

ACEI: ACEI are typically combined with diuretics and positive or negative inotropes and comprise an essential component of long-term medical management of CHF by blunting the effects of the RAAS. By interfering with AII formation and limiting aldosterone production, ACEI promote vasodilation and reduce sodium and water retention, thereby decreasing total peripheral resistance and pulmonary vascular resistance (afterload) and circulating volume (preload), allowing an increase in cardiac output. Enalapril is thought to attenuate myocardial remodeling in humans and dogs. In both species, ACEI have been found beneficial to increase survival times.

Of the different ACEI that are available, enalapril has been the most commonly used in birds, with empirical evidence suggesting it is both safe and efficacious. Enalapril has been used, both alone and in combination with furosemide, to treat CHF with reduction in pericardial effusion, ascites, and hepatic congestion documented by echocardiography, and reportedly increased quality of life and longevity in birds with severe cardiac pathology. The author has found that when edema and effusion are mild, they can in some cases be resolved with enalapril alone without concurrent use of a diuretic. Pharmacokinetic data supports a dose of at least 1.25 mg/kg PO q8-12h in pigeons and Amazon parrots, although the half-life was shorter and lower maximum plasma concentrations were reached in the latter species. As a result, the author has used higher doses (4-11 mg/kg PO q8-12h) in psittacine birds with appreciable symptomatic benefit and no apparent adverse effects.

Sildenafil: Compared to mammals, birds are thought to have a greater propensity for developing PH and right-sided, rather than left-sided CHF. The inclusion of sildenafil should be considered in treatment of CHF patients with cor pulmonale. The author has used sildenafil for this indication in psittacine birds at dose and frequency of 1-11 mg/kg PO q8-12h.

Positive inotropes

Positive inotropes such as digoxin and pimobendan have been used or proposed for treatment of heart failure in avian species, but pharmacodynamic data is lacking and pharmacokinetic data and information as to their efficacy and margin of safety is extremely limited. Both drugs are used to enhance myocardial contractility and are appropriate for treatment of heart failure due to systolic dysfunction. They are contraindicated in cases of hypertrophic cardiomyopathy (HCM), where diastolic dysfunction is the primary problem, and of outflow obstruction (such as aortic stenosis). Likewise, it should be questioned whether their use is appropriate in birds with heart failure secondary to atherosclerotic disease and luminal stenosis of major arteries. In those cases with severe systolic dysfunction; however, it is fair to consider their inclusion in the treatment regime in an effort to stabilize the patient, if not also to manage CHF over the longer term. Once the patient has been stabilized, the clinician can also consider either alternative or concurrent use of a BB (see following section). Digoxin may have value in certain heart failure cases, particularly when supraventricular tachyarrhythmia is a feature.

Pimobendan: Pimobendan, a calcium sensitizer and phosphodiesterase inhibitor, is a positive inotrope and vasodilator (inodilator), as well as a positive lusitrope. It enhances myocardial contractility, primarily through calcium sensitization of cardiac myofibrils and by phosphodiesterase III inhibition, without increasing myocardial oxygen consumption. Inhibition of phosphodiesterase III and V promotes systemic and pulmonary arterial and venous dilation, thereby reducing afterload and preload, respectively. Pimobendan is commonly used in small animal cardiology and has been shown to increase both survival time and quality of life in dogs with dilated cardiomyopathy and heart failure secondary to mitral valve disease.

A pharmacokinetic study in Hispaniolan Amazon parrots (*Amazona ventralis*) demonstrated that a single oral dose of 10 mg/kg is required to achieve a peak plasma concentration of 8.26 ng/ml, comparable to levels considered therapeutic in dogs and humans. Extrapolation of the dose used in this study to other species groups should be done with caution. Plasma concentrations in a Harris hawk (*Parabuteo unicinctus*) with CHF receiving 10 mg/kg pimobendan PO peaked at 25,196 ng/ml; however, there was no indication of toxicosis.

Bio-availability of pimobendan can be affected by pharmacologic composition. An oral suspension formulated from commercially-available tablets (crushed and combined with a suspending vehicle) produced 6 times greater plasma concentrations than a suspension made from the bulk chemical (powder). The authors proposed that the difference might be attributed to the citric acid excipient present in the commercially-available tablets, which facilitated oral absorption.

The author has incorporated pimobendan (6-20 mg/kg PO q8-12h) into the treatment regime for management of CHF in many companion birds.

Negative inotropes

Negative inotropes, including BB and calcium channel blockers (CCB), may have merit as adjunctive treatment for CHF, particularly when ventricular hypertrophy and tachyarrhythmia are contributing factors.

Beta blockers: BB are sympatholytic agents that block the binding of endogenous catecholamines to β-adrenoreceptors. They are negative inotropes, chronotropes, and lusitropes that also slow AV nodal conduction:

- First-generation, non-selective BB (e.g., propanolol, carteolol, sotalol) block both β_1- and β_2-adrenoreceptors.
- Second-generation BB (e.g., atenolol, metoprolol) are relatively β_1-selective.
- Carvedilol, a third-generation BB, is both a non-selective β-adrenoreceptor antagonist and selective α_1-adrenoreceptor antagonist, such that it also has vasodilatory action to reduce afterload.

BB are considered part of the core therapy for heart failure in humans, both in early and advanced stages. They counter the increased sympathetic tone and RAAS activation characteristic of the neuroendocrine system that is central to the pathogenesis and progression of CHF. By retarding its deleterious effects, β-blockade is beneficial in treatment of heart failure, ultimately improving systolic function in spite of the attendant negative inotropic effect. Longer-term actions of BB include regression of myocardial hypertrophy, reversal of remodeling, and normalization of ventricular geometry. In addition, the antioxidant properties of some BB, including carvedilol, may also contribute to their beneficial effects. In human cardiovascular medicine, second- and third-generation BB are utilized in combination with diuretics and ACEI for treatment of chronic, stable CHF related to left ventricular systolic dysfunction (including DCM).

In dogs, these drugs have also garnered interest as adjunctive treatment for CHF, though their efficacy has not yet been established. BB and ACEI are also used in humans to prevent heart failure in cases with structural cardiac abnormalities (such as left ventricular hypertrophy and valvular disease) and systolic dysfunction that are as yet asymptomatic. Similarly, BB have been suggested as part of treatment for preclinical chronic degenerative AV valve disease (CVD) and preclinical DCM in dogs, as they might delay progression to heart failure.

The beneficial effects of BB have been documented in avian models of heart failure. The author suggests that BB may have application in avian cardiology as an adjunctive treatment of CHF, provided the condition is first stabilized using conventional management strategies. The author has used carvedilol (1-9 mg/kg PO q12-24h), in conjunction with diuretic(s) and an ACEI, as part of treatment for CHF in psittacines, specifically when concentric ventricular hypertrophy, tachycardia, and diastolic dysfunction were features. Concurrent administration of a positive inotrope, such as pimobendan, during the period of BB uptitration may allow patients to tolerate initiation of treatment.

Calcium channel blockers: CCBs, including nifedipine, verapamil, diltiazem and amlodipine, inhibit influx of extracellular calcium ions across cardiomyocyte and vascular smooth muscle cell membranes, thereby inhibiting contraction. Effects are negative inotropy and chronotropy, slowing of the sinus rate and AV nodal conduction, and vasodilation. CCB are classified by their relative selectivity for the vasculature or for the myocardium; those that predominantly promote peripheral vasodilation and reduce total peripheral resistance (nifedipine and amlodipine) are indicated primarily for treatment of hypertension, and those that influence conduction (verapamil and diltiazem) indicated for treatment of supraventricular tachyarrhythmias. Adverse effects of CCB include bradycardia, hypotension and reflex tachycardia, and AV block. Contraindications include SA nodal dysfunction, second- or third-degree AV block, and decompensated heart failure.

In human heart failure patients with systolic dysfunction, CCB confer no survival benefit and may exacerbate the condition. Nevertheless, CCBs warrant mention because their cardioprotective effects have been investigated in avian models of heart failure and found to rival those of BB. Both nifedipine and verapamil are cardioprotective (nifedipine > verapamil), but the mechanisms by which these drugs confer their benefits are unknown.

Treatment of Related Conditions

CHF may be accompanied by disease processes such as HCM, systemic and/or pulmonary hypertension, and arrhythmias. These conditions thus deserve special consideration.

Hypertrophic cardiomyopathy

HCM comprises hyperplasia or thickening of individual muscle fibers of the heart. HCM in pet birds is usually secondary to pressure overload states (cardiomyopathy of overload) including arterial luminal stenosis due to advanced atherosclerosis, and PH. As in mammals, the affected ventricle(s) ultimately undergoes concentric hypertrophy, in which the ventricular wall thickens with a corresponding decrease in chamber volume. Eventual ischemia of the hypertrophied myocardium results in fibrosis and increased collagen content, impairing both systolic and diastolic function. Analogous conditions in human and small animal medicine are ventricular outflow obstructions such as subvalvular aortic stenosis (SAS) and primary HCM to include the obstructive type characterized by systolic anterior motion (SAM) of the mitral valve. BB are utilized in treatment of these conditions; their potential benefit is based on reduction of heart rate and myocardial oxygen consumption, as well as improved diastolic filling and coronary artery flow; however, in cats, evidence of their efficacy is scarce, with little to no indication that they improve survival in patients with CHF.

Anecdotally, BB can have merit for treatment of HCM in birds. The author has utilized carvedilol, with or without an ACEI inhibitor, in a small number of psittacines with suspected atherosclerotic disease and concentric ventricular hypertrophy that had not yet progressed to failure. In each of these cases, there has been symptomatic improvement in the form of increased energy and activity level, appetite, and body weight. One of these cases was that of a 34-year-old, female yellow-naped Amazon parrot (*Amazona auropalliata*) that was treated for biventricular CHF secondary to advanced atherosclerotic disease. There was marked concentric hypertrophy of both ventricles, diastolic dysfunction, and sinus tachycardia. Following initial management with standard therapy (furosemide, spironolactone, enalapril, sildenafil, and pimobendan) clinical status deteriorated and carvedilol was added to the treatment regime; pimobendan was discontinued. Following this change in the treatment protocol, rapid and marked clinical improvement was seen, after which the patient remained stable for an additional 7 months.

Systemic hypertension

Systemic hypertension occurs in poultry species and can result in severe ventricular hypertrophy and CHF. Although not defined in psittacines, systemic hypertension probably is a disease entity in this group as well. In human and small animal medicine, BB are often used in conjunction with ACEI, CCB, and/or diuretics to reduce arterial blood pressure (BP). In humans, such treatment markedly reduces the risk of developing heart failure, and combined therapy with a BB and ACEI is recommended in cases of diastolic heart failure with concurrent hypertension. In cats, the CCB amlodipine is the treatment of choice for systemic hypertension; it has been shown to significantly reduce BP, and in one study, resolved secondary ventricular hypertrophy in 50% of subjects.

In companion birds, application of BB, CCB, or other antihypertensive drugs for treatment of systemic hypertension is complicated by the fact that there is no practical means to obtain accurate and repeatable arterial BP measurements in the clinical setting and no established definition of hypertension exists. Direct BP measurement, while accurate, requires invasive techniques; indirect BP measurements do not seem to correlate well with direct arterial BPs in birds; however, if in select cases, BP measurements are obtained and support a diagnosis of hypertension (direct systolic BP $>$ 240 mmHg; indirect systolic BP consistently $\geq$ 270 mmHg), combined treatment of an ACEI with a BB or a CCB may be considered.

Pulmonary hypertension

Compared to mammals, birds are thought to have a greater propensity for developing PH and right-sided, rather than left-sided CHF owing to the morphology of the right AV valve, less deformable nucleated erythrocytes, and the rigid, non-distensible lungs which limit the ability of the blood capillaries to expand and accommodate greater blood flow. PH can result from pulmonary vascular disease (to include atherosclerosis), chronic pulmonary disease and/or hypoxia, congenital left-to-right shunts, or left-sided heart disease and failure. Secondary polycythemia, which develops as a consequence of chronic hypoxemia, can be a complicating factor in a number of disease conditions, including PH. Increased blood viscosity and larger and less deformable erythrocytes result in increased resistance to blood flow in the lung and other tissues.

Certainly the known or hypothesized underlying cause(s) of PH should be addressed, if possible, but pulmonary vasodilators (sildenafil), CCB, pentoxifylline, and periodic phlebotomy may have symptomatic benefit in birds with a presumptive or definitive diagnosis of PH. Similarly, the inclusion of these drugs should be considered in treatment of CHF patients with cor pulmonale.

Pulmonary vasodilators: Sildenafil is a selective pulmonary vasodilator that acts by specifically inhibiting phosphodiesterase V (PDE5), an enzyme that degrades cyclic guanosine monophosphate (cGMP) in pulmonary vascular smooth muscle cells. The resulting increase in cGMP enhances nitric oxide (NO)-mediated pulmonary vasodilation, thereby reducing pulmonary vascular resistance and right ventricular afterload. The drug does not typically lower systemic arterial BP or alter heart rate. Its hypotensive effects may be increased upon concurrent use of α-adrenoceptor blockers, amlodipine, or other hypotensive drugs. In addition, its metabolism may be reduced by administration of azole antifungals. In humans and dogs, sildenafil is an effective treatment for PH, and has been shown to improve quality of life and mitigate secondary polycythemia in dogs.

In birds, prolonged (>15 months) use of sildenafil (2.5-3 mg/kg PO q8h) has been described in a 25-year-old male mealy Amazon parrot (*Amazona farinosa*) with suspected PH (based on clinical signs, secondary, absolute polycythemia, and consistent echocardiographic findings), in addition to therapeutic phlebotomy, and supplemental oxygen. In this bird, resolution and reappearance of the clinical signs (lethargy, anorexia, ataxia, tachypnea/dyspnea, and oxygen dependence) were found to coincide with changes to the dosing frequency (from q8-12h). In addition, the author has treated an unknown-age, female rose-breasted cockatoo (*Eolophus roseicapilla*) with suspected PH, in which the clinical signs (lethargy and exercise intolerance) rapidly resolved with administration of sildenafil (2 mg/kg PO q12h) and enalapril (3 mg/kg PO q12h). Over a period of 1 year, doses of both drugs were gradually increased to 5 mg/kg PO q12h. Right ventricular hypertrophy decreased 6 months into treatment.

Other treatment options: Patients with suspected PH may also benefit from an ACEI, pimobendan, CCB, pentoxifylline, and therapeutic phlebotomy. Pentoxifylline, a methylxanthine derivative, increases flexibility and deformability of erythrocytes and leukocytes and decreases blood viscosity in mammals. It thereby promotes blood flow through damaged or occluded microvasculature, and thus may have merit in avian patients with PH and secondary polycythemia and hyperviscosity syndrome.

Arrhythmias

Cardiac arrhythmias range from clinically insignificant to life-threatening, and/or terminal events. Clinical signs may be absent or include weakness, syncope, or sudden death. Arrhythmias rarely constitute a primary disease process; they can develop secondary to cardiac chamber dilatation, myocarditis, or cardiomyopathy of any cause, as well as toxicoses, nutritional deficiencies, electrolyte imbalances, and various anesthetic agents. They can be potentiated by catecholamine release as occurs with handling stress and painful conditions. Clinically significant cardiac arrhythmias likely represent the minority, but those that are symptomatic, causing hemodynamic instability, and complicating or precipitating heart failure warrant characterization and appropriate treatment. To date, reports of antiarrhythmic therapy in birds are extremely few.

Treatment of tachyarrhythmias: Aside from digoxin, BB and CCB are utilized in small animal cardiology for treatment of supraventricular tachyarrhythmias, including atrial fibrillation, in order to slow the ventricular response rate. Their use is generally contraindicated in patients with acute or decompensated CHF unless the arrhythmia is contributing to the condition such that conventional treatment alone has failed to stabilize the patient; however, even in this scenario, these drugs must be used with great caution, beginning at low dosages, with the aim of decreasing ventricular rate only marginally.

Sotalol (1-2 mg/kg PO q12h) was employed to control supraventricular tachycardia in a 19-year-old golden eagle (*Aquila chrysaetos*) with marked left atrial and ventricular dilatation, marked left ventricular systolic dysfunction, and stenosis of the right brachiocephalic trunk (most consistent with an atherosclerotic lesion). Sotalol was used in conjunction with isoxsuprine and pimobendan, resulting in resolution of the arrhythmia and restoration of normal chamber sizes and systolic function. The patient had remained in stable condition for 15 months at the time of case report publication.

Lidocaine is a parenteral antiarrhythmic agent used to control life-threatening ventricular tachyarrhythmias. At usual doses, there is minimal effect on the cardiac conduction system or on myocardial contractility. To the author's knowledge, there are no reports describing its use for treatment of tachyarrhythmias in avian species, but its pharmacokinetics and cardiovascular effects have been evaluated during experimental studies.

Treatment of bradyarrhythmias: Bradyarrhythmias, and in particular, second- and third-degree AV blocks, have been noted in birds with cardiac disease. Escape beats and escape rhythms may be seen with severe bradyarrhythmias; they are of ectopic origin and perform an essential salvage function by preventing asystole at low heart rates. QRS morphology of ventricular escape beats may be abnormal (wide and bizarre), but these should not be confused with premature beats (additional beats added to already normal or rapid heart rate), because in the case of escape beats, antiarrhythmic therapy is contraindicated.

Antimuscarinic agents, including atropine and glycopyrrolate, can be used for both the diagnosis and treatment of bradycardias related to increased vagal tone, including sinoatrial (SA) arrest and first- and second-degree AV block. They are also used as antidotes for organophosphate and carbamate intoxication. These drugs are contraindicated in patients with tachycardia or tachyarrhythmias and must be used cautiously in patients with heart failure. Since the activity of atropine and glycopyrrolate is short-lived, they are mainly used during treatment of bradycardia during anesthesia. For long-term use, propantheline, an oral antimuscarinic agent, has been used to normalize heart rate and rhythm in a 30-year-old Moluccan cockatoo (*Cacatua moluccensis*) with second-degree AV block (Mobitz type 2).

Supportive care and husbandry considerations

Along with medical management of CHF, it is often necessary to address concurrent problems and meet specific supportive needs. Patients may be presented in a severely debilitated state, with cachexia, dehydration, secondary renal dysfunction, and injuries. In-patient supportive care measures may include oxygen supplementation, nutritional support, fluid therapy (though parenteral fluid administration must be carefully questioned), and analgesia. Once the patient has been stabilized and discharged, longer-term management strategies should also incorporate exercise restriction, housing modifications, and dietary and lifestyle changes.

Supportive care: General supportive principles apply to the avian heart failure patient, namely rest, minimizing stress with judicious, limited handling and restraint, and provision of supplemental oxygen when appropriate. Patients are typically dyspneic due to pulmonary edema, intracoelomic air sac compression by ascitic fluid, and/or pericardial effusion. Oxygen supplementation is indicated for patients with pulmonary edema, whereas physical fluid removal is more efficacious to stabilize those with air sac compression and cardiac tamponade.

Coelomocentesis will rapidly relieve air sac compression in case of ascites. It is justified as a short-term stabilization strategy in patients with severe respiratory compromise. When large fluid volumes are present, the procedure can be performed relatively safely with (ideally) or without ultrasound guidance. Small fluid volumes will usually resolve with medical management alone without the need for physical fluid removal. Although coelomocentesis can be performed periodically over the longer-term, it should not be substituted for pharmacologic management.

Fluid therapy challenges: Maintaining the delicate balance between management of hypervolemia and concurrent dehydration and renal dysfunction is a profound clinical challenge, requiring close monitoring for changes in clinical status and adjustment of the treatment plan accordingly. Fluid therapy is generally not indicated in the treatment of CHF where a primary, immediate goal is reduction of fluid overload. Initial treatment of acute and decompensated CHF frequently results in some degree of dehydration and pre-renal azotemia, but these abnormalities can resolve over a few days without fluid therapy once food and water intake have normalized.

In cases with persistent, severe azotemia, it may be necessary to administer small volumes of parenteral fluids, but patients for which hemodynamic stability cannot be achieved without severe renal compromise have a poor prognosis. Subcutaneous fluid absorption may fail in cases of right-sided CHF.

Exercise restriction and housing: Weakness and ataxia may limit patient mobility and access to food and water, and predispose to falls. This necessitates housing modifications designed to limit exertion and allow easy, immediate access to food and water. As even debilitated birds will often persist in trying to climb or perch, enclosures that prevent or limit this activity are ideal. Incubators, aquarium set-ups, or plastic or acrylic bins with a soft substrate serve this purpose well and permit visual monitoring. Depending on patient strength and stability, a low, secure and stable perch can be provided (such as a rope perch or rolled towel situated on the bottom of the enclosure). Once the patient regains strength and exercise tolerance, housing can be adjusted to allow greater activity (as in a small cage or crate, furnished with low perches and readily accessible dishes); some patients will ultimately be able to return to a traditional cage with no specific exercise restrictions.

Diet and lifestyle: Longer-term husbandry considerations, including dietary changes (entailing optimum nutrition and sodium restriction) and lifestyle changes, mirror those discussed for atherosclerosis; however, dietary changes should not be made until the patient is well-stabilized. Given that CHF often arises secondary to atherosclerotic disease in psittacine birds, preventive considerations may be applicable to both conditions.

Other: Considering that most patients will require long-term (if not life-long) medical management, operant learning methods and food vehicles should be employed to facilitate low-stress medication administration.

Pericardial Effusion and Cardiac Tamponade

Pericardial effusion is characterized by an inappropriate accumulation of fluid within the pericardial sac. Several causes have been reported, though in some cases an underlying cause cannot be identified and the condition is ruled idiopathic. Severe pericardial effusion or restrictive pericarditis can compress the heart, resulting in impaired ventricular filling (diastolic dysfunction), and subsequent decreases in stroke volume and cardiac output. This condition, which can become life-threatening, is referred to as cardiac tamponade. Since the intramural pressure of the thinner-walled right ventricle is overcome more rapidly than that of the left, cardiac tamponade results more quickly in right-sided CHF than in left-sided failure.

Treatment for pericardial effusion and cardiac tamponade is based first on removal of the fluid, and second on treatment of the underlying etiology of fluid accumulation. Fluid removal can be accomplished either by ultrasound-guided or endoscopic pericardiocentesis, or by endoscopic or surgical fenestration of the pericardium.

Although diuretics are generally contraindicated in cardiac tamponade because they decrease the preload necessary to maintain ventricular filling pressure and cardiac output, furosemide with or without an ACEI may be useful to reduce pericardial fluid of low volume related to CHF in birds.

References

1. Beaufrere H, Nevarez J, Gaschen L, et al. Diagnosis of presumed acute ischemic stroke and associated seizure management in a Congo African grey parrot. *J Am Vet Med Assoc.* 2011;239(1):122-128.

2. Beaufrere H, Holder KA, Bauer R, et al. Intermittent claudication-like syndrome secondary to atherosclerosis in a yellow-naped Amazon parrot (*Amazona ochrocephala auropalliata*). *J Avian Med Surg*. 2011;25(4): 266-276.

3. Beaufrere H. Avian atherosclerosis: parrots and beyond. *J Exotic Pet Med.* 2013;22(4):336-347.

4. Brady SM, Burgdorf-Moisuk A, Silverman S, et al. Successful treatment of suspected pulmonary arterial hypertension in a mealy Amazon parrot (*Amazona farinosa*). *J Avian Med Surg.* 2016;30:368-373.

5. Fitzgerald BC, Beaufrere H. Cardiology. In: Speer BL, ed. *Current Therapy in Avian Medicine and Surgery*. St. Louis, MO: Elsevier; 2016:252-328.

6. Fitzgerald BC, Dias S, and Martorell J. Cardiovascular drugs in avian, small mammal, and reptile medicine. *Vet Clin North Am Exotic Anim Pract.* 2018;21(2):399-442.

7. Guzman DS, Beaufrere H, KuKanich B, et al. Pharmacokinetics of single oral dose of pimobendan in Hispaniolan Amazon parrots (*Amazona ventralis*). *J Avian Med Surg.* 2014;28(2):95-101.

8. Guzman DSM. Advances in avian clinical therapeutics. *J Exot Pet Med.* 2014;23:6-20.

9. Scagnelli A. Therapeutic review: pentoxifylline. *J Exotic Pet Med.* 2017;26:238-240.

10. Straub J, Pees M, Enders F, Krautwald-Junghanns ME. Pericardiocentesis and the use of enalapril in a Fischer's lovebird (*Agapornis fischeri*). *Vet Rec.* 2003;152:24-26.

Pododermatitis in Birds of Prey

Thomas Dutton, BVM&S Cert AVP (ZooMed) DipECZM (Avian), MRCVS

Session #3004

Affiliation: From Great Western Exotics, UK.

Abstract: Pododermatitis is the most frequently encountered foot disease of captive raptors and it is a significant welfare concern. It is also the most common husbandry-related disease. Pododermatitis is rarely seen in free-living birds and consequently, management of cases frequently requires identification and correction of husbandry and environmental deficiencies that have causal roles in the condition. Pododermatitis can be simply defined as inflammation of the skin of the foot. In avian medicine it is considered a degenerative, inflammatory condition of the plantar surface of the foot, most commonly involving the plantar metatarsal pad or plantar digital pads. It is best considered a disease syndrome where disruption of the plantar epithelium allows microbial invasion and secondary infection. If not identified quickly and managed effectively, pododermatitis can progress and affect deeper tissues including tenosynovitis, osteomyelitis and septic arthropathy. This masterclass describes causes, investigation, and treatment of pododermatitis in birds of prey.

Introduction

Falconiformes and Strigiformes represent a group of birds commonly referred to as birds of prey, or raptors. The Order of the Falconiformes comprises five families: Cathartidae, Pandionidae, Accipitridae, Sagittariidae, and Falconidae.

Pododermatitis is a condition commonly encountered in captive birds of prey, particularly the larger species.[1-3] Different etiologic factors have been proposed as responsible for the development of pododermatitis in birds of prey including increased body weight, inadequate perches, lack of exercise, sudden cessation of exercise, bruising of the plantar surface of the feet, unequal weight-bearing on one foot, overgrown talons, and inadequate diets.[1-4] Members of the Falconiformes, but in particular the gyr (*Falco rusticolus*) falcon and saker (*Falco cherrug*) falcon are more likely to develop pododermatitis. Owls are the next most presented with larger species such as snowy owls (*Bubo scandiacus*) particularly susceptible. Hawks and eagles have a significantly lower prevalence, and typically only develop significant pododermatitis due to concurrent injury or significant husbandry limitations.

Clinical Examination

Initial diagnosis is made based on clinical examination and clinical history. Body condition score, general condition, and weight are important to note when assessing underlying causes of pododermatitis. It is worth also examining other birds from the collection to determine if the case is localized or a more general problem in the collection.

Differential diagnoses for pododermatitis include the following:

- trauma (soft tissue, fracture, dislocations);
- gout; and
- septic arthropathy (hematologic spread).

With the majority of these conditions, there will not be disruption to the plantar skin, facilitating a diagnosis. These will need to be ruled out based on signalment, blood work, imaging and cytology.

Diagnostic tests

To determine whether infection has spread to underlying tissues and to differentiate the higher grades of pododermatitis (3 to 5), imaging often will be required. Radiography or computed tomography (CT) imaging are best used to evaluate skeletal involvement (e.g., osteomyelitis). Imaging also is required to assess associated disease that could be contributing to uneven weight distribution and be an underlying cause of the pododermatitis (e.g., degenerative joint disease, previous misaligned fracture healing, metabolic bone disease, intra-abdominal disease-causing sciatic nerve compression). Ultrasound, on the other hand, can be used to assess soft tissue structures such as tendon sheaths and tendons, particularly in larger birds, although the scales may hinder proper visualization of the structures. CT or magnetic resonance imaging (MRI) also may be used to detect more subtle bone and soft tissue pathologies.

The author regularly collects clinical pathology samples from the lesions for cytology and/or culture. Cytologic samples should be stained with Gram's stain as this can help with the selection of first-line antimicrobial therapy while awaiting culture results. Culture and sensitivity performed on a deep portion of the lesion or a piece of infected tissue will generally lead to the best results. Frequently, *Staphylococcus aureus* is isolated from lesions, although other bacteria, such as *E. coli*, *Corynebacterium* spp., *Pseudomonas* spp., and yeasts are regularly cultured from the lesions.

A complete blood cell count (CBC) and biochemistry panel can be performed to identify concurrent diseases or predisposing factors such as elevated uric acid with gout. In advanced stage of disease, there is frequently an elevated white blood cell count (WBC) with heterophilia. Fibrinogen also is elevated to boost clotting (hyperfibrinogenemia). Changes in the protein electrophoresis fractions may be seen with severe cases.

Grading Pododermatitis

Pododermatitis grades by identification characteristics, treatment options, and prognosis are presented in Table 1. Representative photographs of pododermatitis grades 1-5 are presented in Figures 1-6.

Table 1. Pododermatitis grades, treatment options and prognosis.

Grade	Identification	Treatment options	Prognosis
1	Erythema, smoothing skin Flattened papillae Discoloration (darkening, pink erythema)	Environmental and husbandry changes Pressure relieving dressings/casts Topical treatments (emollients, foot bathes, soaks) No indication for antibiotics	Excellent, if underlying cause can be identified and corrected
2	Hyperkeratosis, thickening plantar skin Localized inflammation Early necrosis Crusts	As above for grade 1 disease Systemic anti-inflammatories Topical antibiotics Systemic antibiotics (rare cases where secondary bacterial colonization likely Surgery considered in refractory cases	Good
3	Progressed hyperkeratosis Diffuse swelling feet Mild to moderate pain Necrotic and hyperkeratotic core Abscess formation	As for grade 1 and 2 Surgical debridement necrotic/avascular tissues Remove caseous material and pyogenic membranes Systemic antibiotics based on culture Topical antibiotics (absorbable calcium sulphate impregnated beads) Topical dressings promoting granulation	Fair
4	Spread of infection into tendon sheaths, joints and adjacent tissues	Flush tendon sheaths and surgical debridement as required Advanced imaging foot can help with surgical planning Local, regional antibiotics particularly beneficial Regional limb perfusion antibiotic administration Long courses of treatment required where bone and synovial structures affected	Guarded
5	Involvement of skeletal structures Osteomyelitis and/or septic arthropathies Marked swelling with moderate to severe pain is often observed	Viability and ethics of treating needs to be assessed Even if infection is controlled long-term secondary degenerative joint disease will affect welfare Euthanasia may need to be considered	Poor

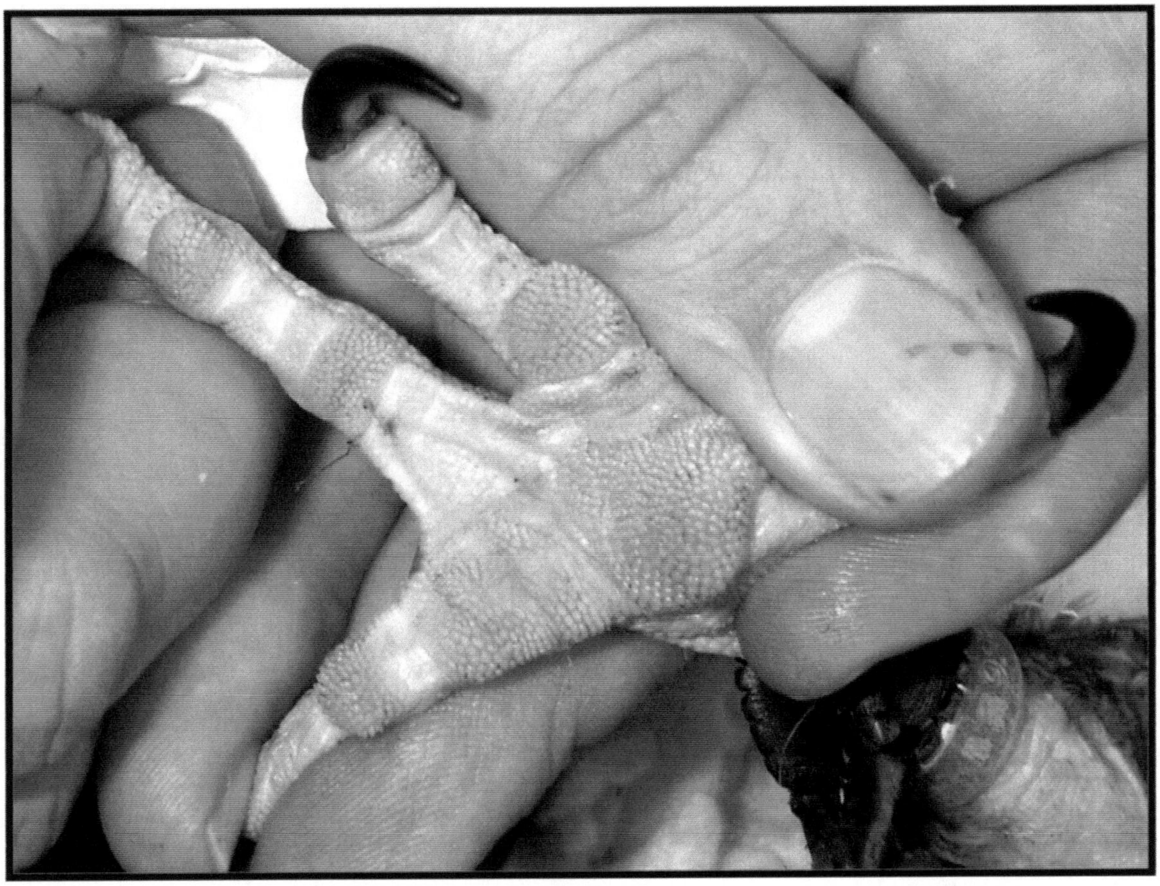

Figure 1. Grade 1 pododermatitis.

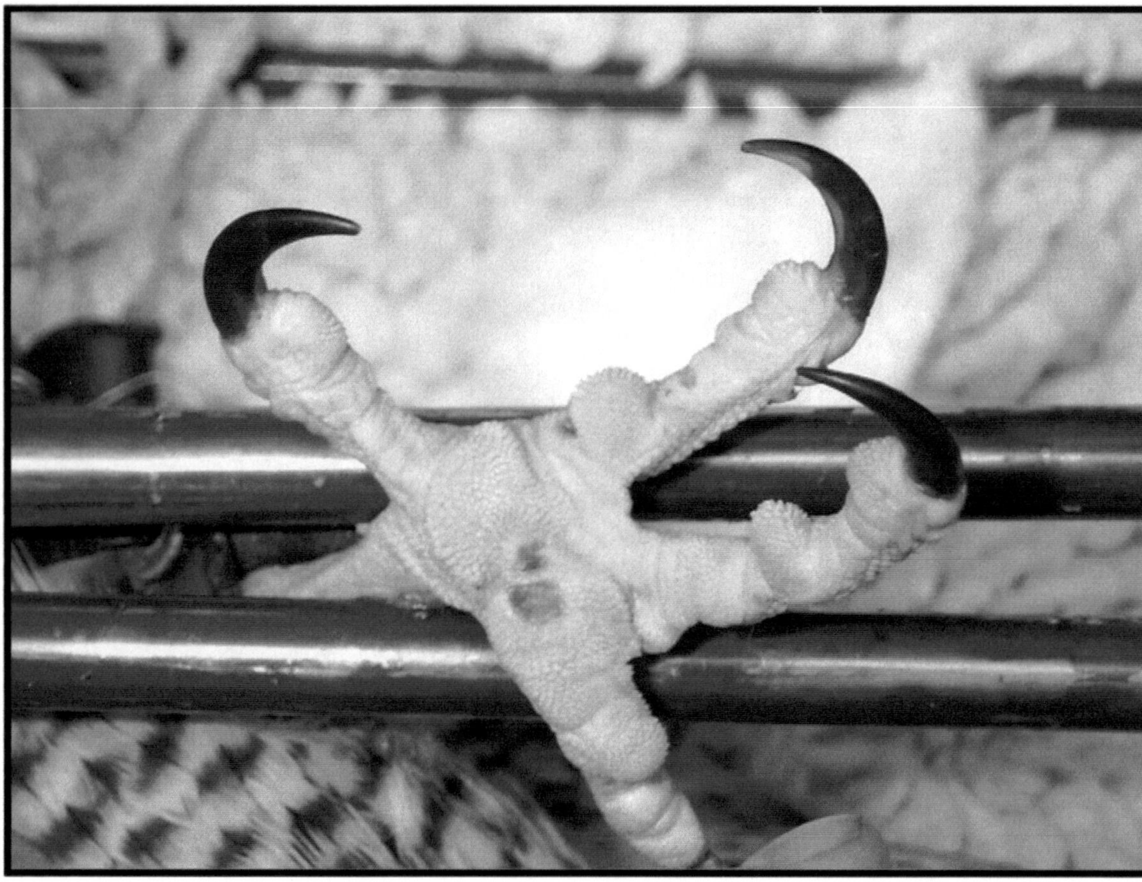

Figure 2. Grade 2 pododermatitis.

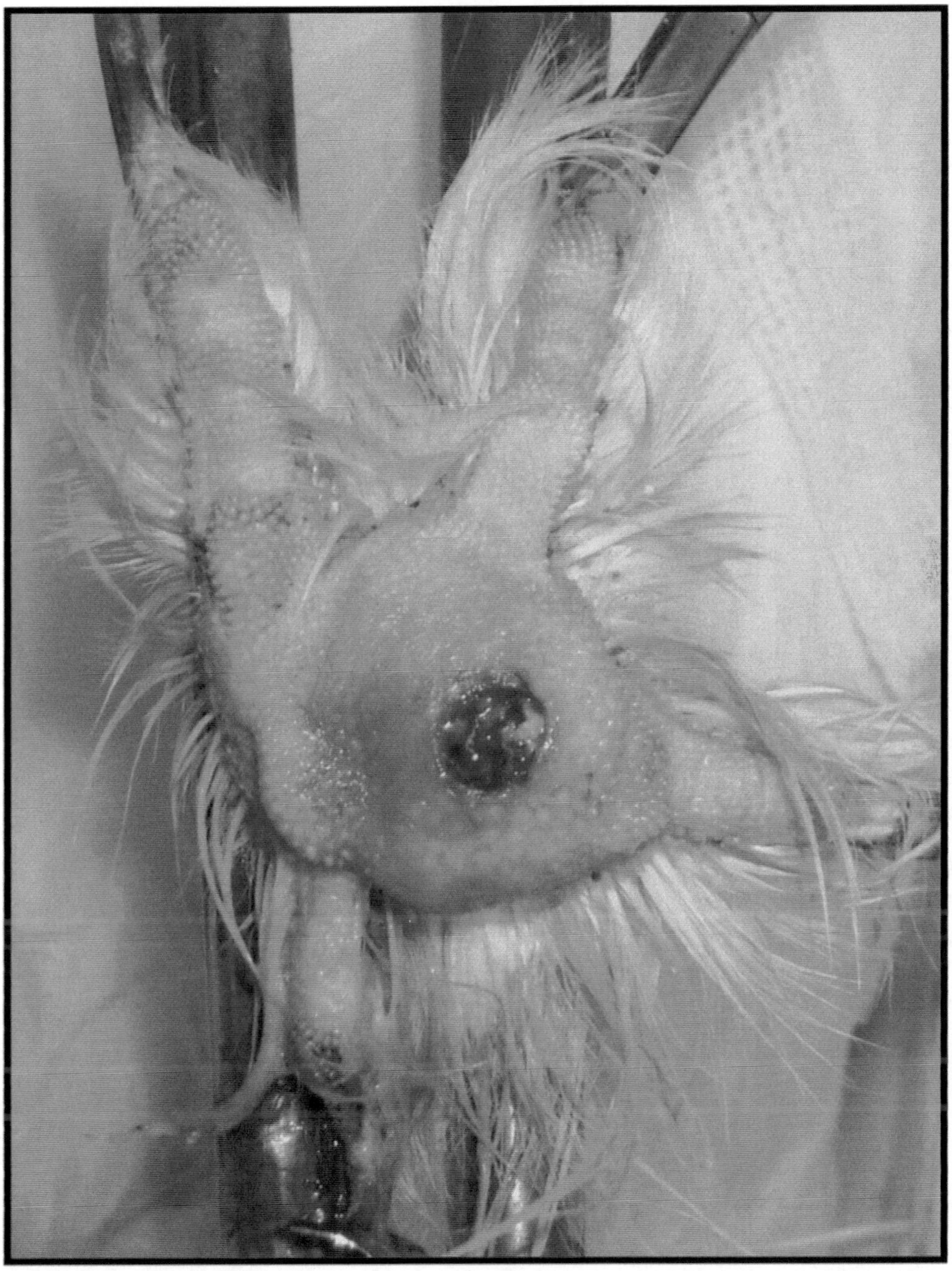

Figure 3. Grade 3 pododermatitis.

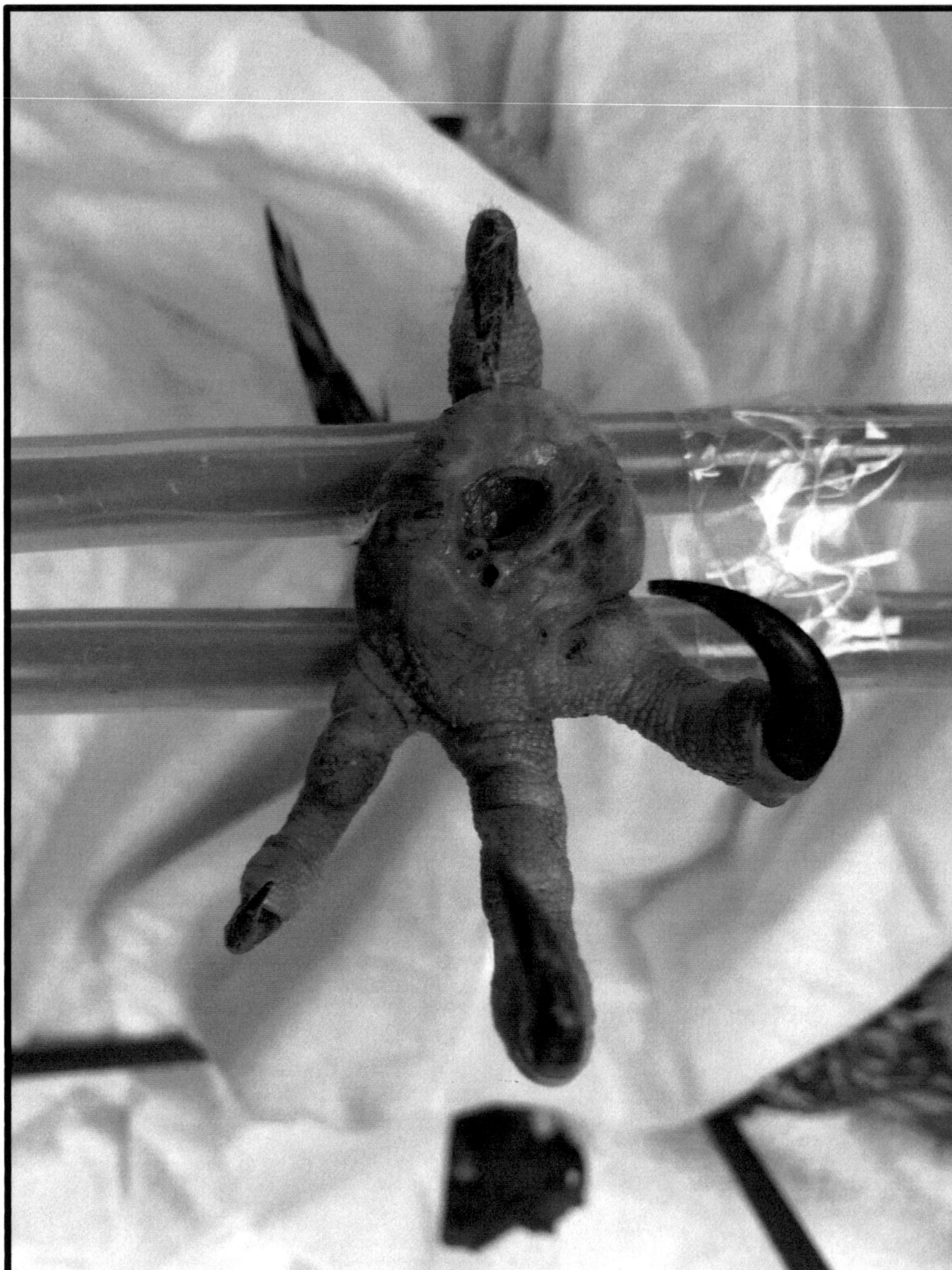

Figure 4. Grade 4 pododermatitis.

Figure 5. Grade 5 pododermatitis.

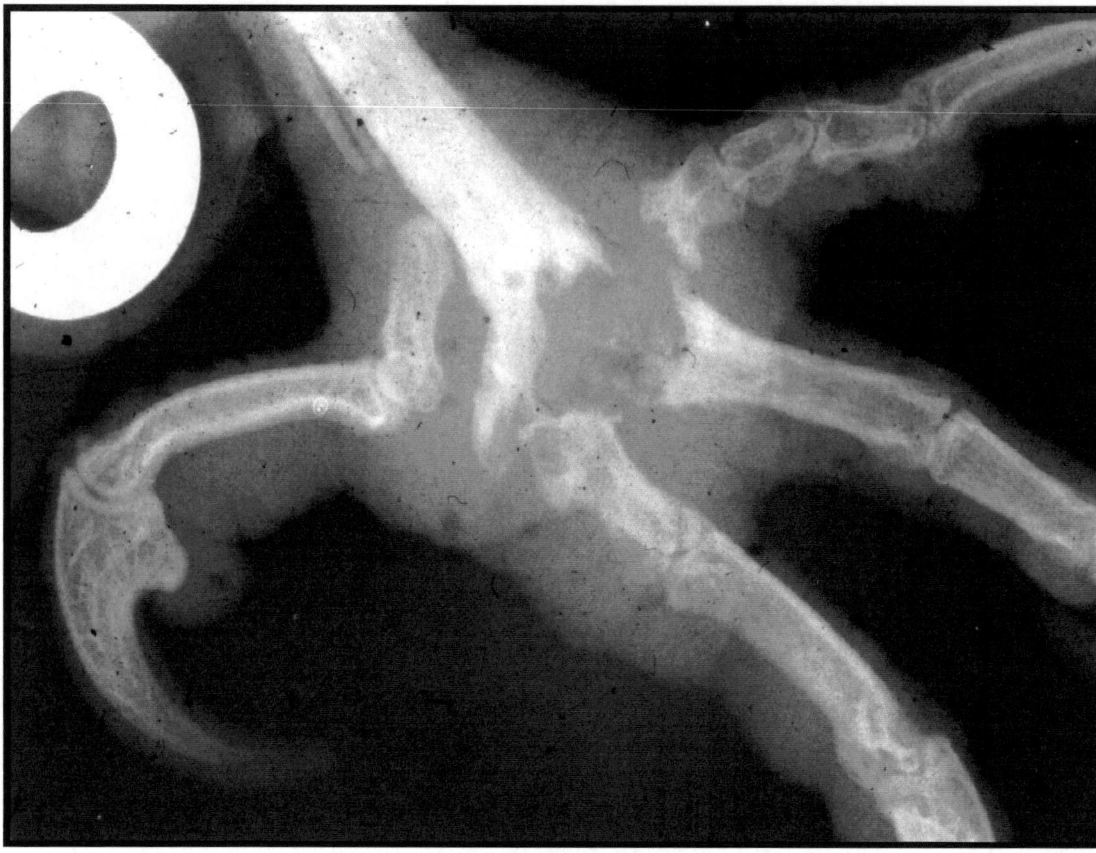

Figure 6. Radiographic changes in grade 5 pododermatitis.

Treatment

Low grade (grade 1 or 2) pododermatitis is frequently managed with husbandry changes and topical therapies, with or without the addition of pressure relieving dressings. Overall aims of this treatment are to:

- provide analgesia,
- control swelling and inflammation,
- prevent formation of (further) necrotic tissue,
- prevent progression of pododermatitis to higher grade disease,
- prevent colonization of pathogens, and
- promote granulation and healing (thus avoiding cytotoxic therapies).

Higher grade pododermatitis (i.e., grade 3-5) frequently requires more extensive therapy to be successful. Surgical intervention is often required and aims to:

- debride necrotic and avascular tissues, and remove caseous material and pyogenic membranes;
- minimize disruption to the normal foot structure;
- allow primary closure, where possible; many cases will require healing by granulation and secondary intention;
- expedite healing and allow the limb to return to full function as soon as possible; and
- prevent spread of infection into bone, tendons, tendon sheaths or joints.

References

1. Cooper J. *Veterinary Aspects of Captive Birds of Prey.* London, UK: Standfast Press; 1978.
2. Degernes L, Talbot B, Mueller L. Raptor foot care. *J Assoc Avian Vet.* 1990;4(2):93. doi:10.2307/30136889
3. Cooper J. Foot conditions. In: Cooper JE, ed. *Birds of Prey: Health and Disease.* 1st Ed. Oxford, UK: Blackwell Science; 2002:121-131.
4. Zsivanovits P, Monks D. Bumblefoot (pododermatitis). In: Samour J, ed. *Avian Medicine.* 3rd Ed. St Louis, MO: Elsevier; 2016:260-264.
5. Samour J Wernick M, Zsivanovits P. Therapeutic management of pododermatitis in falcon medicine: historical and modern perspective. *Arch Vet Anim Sci.* 2021;3(1):1-11.

The Long-Lost Gem of Avian Ophthalmology

Bret A. Moore, DVM, PhD, DACVO

Session #3005

Affiliation: From University of Florida, Department of Small Animal Clinical Sciences, 2015 SW 16th St. Gainesville, FL 32608, USA.

Abstract: Avian ophthalmology was founded on the interest of comparative anatomists, physiologists, and visual ecologists. With the advancement of medicine and the study of ophthalmic disease, clinical medicine deviated from the critical study of first the animal, their visual system, and how the animals use that system to interact visually with their environment. This lecture discusses how to approach an avian patient based on the individual species at hand, even prior to the examination, by following three simple steps: 1) birds, 2) birds' eyes, and 3) birds + birds' eyes. Using this approach is essential to understanding species differences in treatment and prognosis, as well as welfare and potentially release of free-living individuals.

Introduction

It was in the late 1800s to the early 1900s when a spark of fascination for the comparative anatomy and physiology of animal eyes ignited, as did our search to understand how animals see our shared world. As stated in 1942 by an early visual ecologist:

"If the comparative ophthalmologists of the world should ever hold a convention, the first resolution they would pass would say: 'Everything in the vertebrate eye means something.' Except for the brain, there is no other organ in the body of which that can be said... Man can make optical instruments only from such materials as brass and glass. Nature has succeeded with only such things as leather and water and jelly; but the resulting instrument is so delicately balanced that it will tolerate no tampering."[1]

This statement laid a foundation under which the minds of the founders of the American College of Veterinary Ophthalmologists began to grow. Over the following 70 years, our knowledge of comparative ophthalmology expanded at a near exponential rate; however, during that expansion, we nearly lost our early ways as comparative visual ecologists along with our wonder for the beauty of the enormous diversity that is the vertebrate eye. Today, a major gap in our field remains the true comparative nature of how our profession began. Despite much scientific progress, most of our knowledge for wild and exotic animal ophthalmology has depended upon extrapolation from domestic animal ophthalmology. By reaching back to our roots and studying an animal in their natural environment, studying the multidimensional components of their visual system, and then understanding how they use their visual system to interact with their environment, only then are we able to approach comparative ophthalmology more accurately. The purpose of the session is to discuss these often-forgotten principles of the ophthalmic examination in the avian patient.

Birds

The first step is familiarizing oneself with the visual ecology of a given species. It includes:

- habitat,
- activity pattern,
- prey and foraging behaviors, and
- predators and anti-predator behaviors.

Most birds are recognized as having superior visual performance;[2-6] however, most birds are highly visual in very different ways. Despite our increasing knowledge of avian visual systems, we know very little about their comparative differences. The first step toward understanding the visual system of a given bird species, and thus the first step in our approach to an avian ophthalmic patient, is to study why there are such differences between species. That is, to familiarize oneself with the visual ecology of the species at hand.

A bird's visual system is tuned for a specific set of uses as it interacts with its environment. The more highly visual the species, the more specific the tuning of the visual system. Gordon Wall's (1942) statement, "Everything in the vertebrate eye means something" is true and exists because everything in the environment also means something, and it is the function of the eye to interact with a specific environment with precision.[1] Such diversity of visual systems is because the diversity of possible environmental interactions is even greater! One species could live in a barren desert with very little visual complexity other than the vastness of depth and the homogenous contrast of sand and those organisms camouflaging within it. Another species may live within the dense, tropical rainforest with infinite layers of spatial complexity. But of course, spanning from the forest floor to the canopy you encounter numerous other ecosystems that offer different visual interactions and thus visual requirements. The possibilities are nearly endless.

The diversity only increases (as does our understanding of how a species sees) when we consider at what time period they are most active visually. That is, their activity pattern. Some species are active only in bright daylight, others in the dark of night, and finally others spend time in both night and day or in the between periods of time when the sun is rising or setting. Each of these activity patterns provides an important component of the story that is being told about how they use their eyes.

Finally, we must consider how a species visually interacts with its prey and predators. The methods used to capture prey, from pecking motionless seeds in a high-contrast substrate to actively chasing near-microscopic insects in flight, require very different visual capabilities. And the trade-off? Simultaneously needing to maintain watch for predators in the peripheral visual fields. The combination of foraging modes and methods of monitoring for predators often shape key components of a bird's visual system (namely the visual field and positions of retinal centers of specialization) as discussed in the next section.

Birds' Eyes

The second step is understanding the function of the visual system of a given species. It includes:

- orbit convergence and visual field configuration,
- spatial resolution,
- temporal resolution,
- color vision, and
- retinal processing and beyond.

Despite relative conservation of the avian eye, the species differences discussed above result in thousands of iterations of the avian eye. These differences go way beyond the face-value of iris color, eye size, and various adnexal structures. Such differences as visual field width and the degree of binocular overlap, visual acuity, and color vision can vary dramatically across species.[7,8] Avian visual systems are multidimensional, and the key to understanding the magnitude of their interspecific differences is to assess each difference and how they function together, something that has traditionally been lacking in vision studies. In addition, we tend to view avian vision through the lens of our own visual experience, thus providing misguided ideas of how birds may perceive their environment.

Specific regions across the retina are specialized for high spatial resolution (visual acuity),[7-9] high temporal resolution (speed of sight),[10] and color vision.[10,11] Retinal specializations for visual acuity include the area centralis, visual streak, and the fovea. Specializations for high temporal resolution or color vision can be contained with these regions, overlap with them, or be fully separate (i.e., one part of the retina may function for high visual acuity whereas another may contain a proportionally high number of red cones for color vision). Where these retinal specializations are positioned in space depends on the orbit convergence and the configuration of the visual field. Species with high binocularity and corresponding small visual fields (i.e., a greater degree of blind space) tend to be higher predatory species, whereas wider visual fields are a trade-off for smaller binocular space common in prey species as they have a greater need to keep watch for predators.[8] Multiple retinal specializations (e.g., the bifoveate retina of some predatory birds) can be present in a single eye, and the position of specializations can vary, meaning that different retions of a visual field can have difference degrees of visual importance or are used for completely different visual tasks.[12] These areas, and what a species uses them for, are important to note when clinically assessing the potential not only for vision, but for acute and highly functional vision, as we will discuss below. Finally, processing of retinal signals and the corresponding areas in the visual cortex are very poorly understood. Together, these different components of the visual system work together to help a species obtain the visual information they need to survive based upon their specific ecology.

Birds + Birds' Eyes

The third step is understanding how birds visually sample and interact with their ecology. It includes:

- head movements,
- eye movements, and
- other visual behaviors.

Finally, by taking the ecology of a given species, and applying what we know about their visual system, we can begin to understand how they visually sample and interreact with their ecology. For example, the combination of the type and location of retinal specializations, along with the orientation of the visual field, are key determinants driving head and eye movement behavior.[8] For example, songbirds tend to have very wide visual fields, forming small binocular but also small blind areas. While this orientation maximizes predator detection, they need to position their single, centrotemporal region of high acuity (fovea) onto an object of interest (e.g., a predator), and thus perform rapid head and eye movements to increased high acuity visual sampling. Similarly, while foraging, the fovea in each eye projects laterally and does not overlap in their narrow binocular field, resulting in side-to-side head movements to visualize food objects with high acuity. Other species have relatively immobile eyes but extreme ranges of head movement (e.g., owls), or high degrees of eye movements in a specific plane to enable them to remain motionless while hunting below their bill (e.g., American bittern).

Clinically, we can deploy our knowledge of a bird's visual ecology combined with their eyes to help make important decisions regarding treatment, prognosis, and even release of an injured or ill free-living individual. It is critical to assess the visual potential of a patient in a specific way to make appropriate decisions, which unfortunately can be a matter of life and death for the patient. The above discussion provides a groundwork for the following guidelines toward assessing the visual potential for a bird in its natural habitat.[12]

Extent and location of the lesion

Extensive lesions are understandably concerning (e.g., complete cataract, rhegmatogenous retinal detachment); however, small focal lesions can be detrimental to visual success depending on their location within the eye of a given species. For example, lesions less than 2 mm in diameter at either fovea in the eye (even in just a single eye of a diurnal raptor) should deem the bird a poor prognosis for release given the dependence on high acuity vision for foraging and flight. Owls, on the other hand, can even experience complete retinal damage in one eye and still be considered for release because of their dependence on other sensory systems (e.g., hearing). The same consideration should exist for iatrogenic visual compromise. For example, cataract surgery without the placement of an intraocular lens renders a bird severely myopic. Some species tend to severely fibrose their lens capsule following phacoemulsification thus limiting their vision to that of the capsulorrhexis, and if a planned posterior capsulorrhexis is not performed, can completely obscure vision. Corneal scars following ulcer debridement, or surgical repair/grafting, can be substantial and lethal to a bird in the wild depending on the lesion size and location.

Chronicity of the lesion

If a generally well bird is diagnosed with chronic ocular lesions (e.g., chorioretinal scars and pigmentary changes, cataracts), even if advanced, it is reasonable to suggest that the bird has had success for some time despite the visual disturbance. It is difficult to know whether the ocular lesion was the cause of their presentation, but their thriftiness suggests they had been doing well up to that point. Evidence of trauma on the same side as the chronic ocular lesions should raise suspicion for the chronic lesion being the cause for the clinical presentation. Additionally, vision testing can be challenging in these patients because they often have adapted well to reduced vision.

Age of the patient

Age typically brings wisdom and increased survivability. Young birds many have not learned all the skills necessary to survive in the wild, let alone with visual compromise. Considering that most raptor young do not survive past 1 year of age without visual compromise, and ocular lesion that reduces their visual potential should be raise concern. Adults often tolerate a greater degree of visual impairment.

Ecological impact of the lesion

Even closely related species can have markedly different visual ecologies, including hunting style, habitats, the importance of binocular vision, the activity pattern during foraging, and even human environmental interferences (e.g., windmills, powerlines, traffic, etc.). Ultimately, the impact of a lesion needs to be projected into their visual space to properly determine how it may impact the visual ecology of that species.

Success in flight and feeding trials

The final step of evaluating vision if a patient is deemed potentially releasable is observing successful flight and feeding behaviors. Although future trauma cannot always be predicted despite successful flight and foraging tests, some certainty of ability for successful flight and visual navigation can be obtained through behavioral experiments.

References

1. Walls GL. *The Vertebrate Eye and Its Adaptive Radiation.* Bloomfield Hills, MI: Cranbrook Institute of Science; 1942.

2. Martin GR. The subtlety of simple eyes: the tuning of visual fields to perceptual challenges in birds. *Philos Trans R Soc B.* 2014;369:SI1-12.

3. Moore BA, Paul-Murphy JR, Tennyson AJD, et al. Blind free-living kiwi offer a unique window into the evolution of vertebrate vision. *BMC Biol.* 2017;15:85.

4. Moore BA, Tyrrell LP, Kamilar JM, et al. Structure and function of regional specializations in the vertebrate retina. In: Kaas JH, ed. *Evolution of Nervous Systems.* Oxford, UK: Elsevier; 2017:351-372.

5. Moore BA, Maggs DJ, Kim S, et al. Clinical findings and normative ocular data for free-living Anna's (*Calypte anna*) and black-chinned (*Archilochus alexandri*) hummingbirds. *Vet Ophthalmol.* 2019;22:13-23.

6. Moore BA, Hawkins M, Fernandez-Juricic E, et al. Introduction to avian ophthalmology. In: Montiani-Ferreira F, Moore BA, Ben-Shlomo G, eds. *Exotic and Wild Animal Ophthalmology.* London, UK: Springer Nature; 2022:321-348.

7. Moore BA, Doppler M, Young JE, et al. Interspecific differences in the visual system and scanning behavior of three forest passerines that form heterospecific flocks. *J Comp Physiol A.* 2013;199:263-277.

8. Moore BA, Tyrrell LP, Pita D, et al. Does retinal configuration make the head and eyes of foveate birds move? *Nat Sci Rep.* 2017; doi:10.1038/srep38406.

9. Fernandez-Juricic E, Moore BA, Doppler M, et al. Testing the terrain hypothesis: Canada geese see their world laterally and obliquely. *Brain Behav Evol.* 2011;77:147-158.

10. Tyrrell LP, Teixeira LBC, Dubielzig RR, et al. A novel cellular structure in the retina of insectivorous birds. *Nat Sci Rep.* 2019;9:15230.

11. Moore BA, Baumhardt P, Doppler M, et al. Oblique color vision in an open-habitat bird: spectral sensitivity, photoreceptor distribution, and behavioral implications. *J Exp Biol.* 2012;215:3442-3452.

12. Montiani-Ferreira F, Moore BA. Ophthalmology of raptors. Introduction to avian ophthalmology. In: Montiani-Ferreira F, Moore BA, Ben-Shlomo G, eds. *Exotic and Wild Animal Ophthalmology.* London, UK: Springer Nature; 2022:429-504.

Advances in Maximizing Behavioral Welfare for the Avian Patient

Barbara Heidenreich

Session #3006

Affiliation: From Barbara's Force Free Animal Training, Austin, TX 78715, USA.

Abstract: The topic of maximizing behavior welfare for the avian patient has been explored with increased interest in recent years. As the conversation continues, more information sharing amongst scientific disciplines has resulted in further advances that facilitate maximizing benefits and minimizing harms for animals requiring medical attention. Practitioners can build upon the foundation of information that has proven useful and incorporate additional evidence-based findings to better overcome difficult challenges. This presentation reviews behavior science and practical application that remains effective, efficient, and optimal, as well as introduce emerging practices that are improving welfare with alternative approaches for addressing distress.

Introduction

Contemporary avian veterinary practitioners have been migrating away from traditional practices that can potentially evoke distress such as handling and restraint upon initial intake. Depending on the condition of the patient, strategies can include a wide range of options including combinations of the following.[1]

- Observe first without handling or restraint
- Give the patient time to habituate to the exam room
- Associate appetitives such as food, touch, toys, with procedures
- Moving slowly around the patient
- Carefully introducing novel stimuli
- Conducting procedures in small increments of time and providing "breaks"
- Arranging the environment so that capture for restraint is easily accomplished (thus minimizing distress)
- Arranging the environment so that the patient can emit desired responses voluntarily (such as stand on a scale, or enter a transport container)
- Learn and respond to body language to increase patient's comfort
- Utilize conscious sedation
- Train desired responses using reinforcement in the clinic

- Teach clients to train desired responses using reinforcement prior to coming to the clinic
- Provide clients with instructions prior to arriving at the clinic to facilitate avian examinations

While these practices have improved animal welfare, it is advantageous to continue to explore emerging information to fine-tune the understanding of behavior science and its application to strive for optimal welfare for avian patients. After a brief review of how animals learn, the following sections will focus on discussions that have been leading to important advances.

How Animals Learn

Animal behavior draws upon several scientific disciplines. Some of these include ethology, phylogeny, and ontogeny. Ethology is the scientific study of animal behavior, especially as it occurs in a natural environment.[2] Phylogeny is the natural selection of traits due to genetic inheritance over generations.[3,4] Practitioners are better able to predict some potential behavioral responses based on knowledge of ethology and phylogeny of the patient; however, ontogeny often can supersede all. Ontogeny is the selection of behavior by consequences during the lifetime of the individual organism.[3,4] An individual's learning history over its lifetime has the potential to tremendously alter responses.

There are some basic principles or laws of nature that demonstrate an organism responds to changes in the environment in systematic ways. Understanding these provides a foundation upon which all animal behavior change is accomplished.

The first of these is that behavior is selected by the environment.[3,4] Behavior is often misunderstood to be a phenomenon that comes from a drive or motivation within the animal; however, the behaviorist approach is like that of evolution. Environmental conditions select behavior. This then leads to the relationship between the behavior and the resulting consequences, which can be described as a contingency.[5]

A contingency is the relationship between two events, one being contingent or a consequence of the other event. Consequences can occur on a variety of schedules from continuous to almost none and still maintain behavior. The range in between is referred to as intermittent.[3] It is also important to remember that contingencies occur under specific conditions.[5] This is how behavior change is produced.

Behavior is what the organism is doing. Overt behavior is observable and measurable. There is also covert behavior such as thoughts and emotions.[4,6]

Consequences are the outcome for doing the behavior.[3-5] This can increase or decrease the future probability of a behavior being emitted. Some possible consequences of behavior include the following.[7]

- Escape/avoidance
- Tangible items (e.g., food, enrichment)
- Attention/social affiliations
- Sensory or automatically reinforcing experiences (such as self-stimulatory behaviors)
- Preferred activities

Reinforcement is a learning process in which consequences serve to increase the future probability of a behavior being emitted under certain conditions. Punishment is a learning process in which consequences serve to decrease the future probability of behavior being emitted under certain conditions.[3,4]

These basic underpinnings of animal learning and behavior have served to provide a foundation upon which interventions that maximize benefits and minimize harms are built. Understandably, this section serves only as a brief introduction to how animals learn. Explaining the science of behavior change is beyond the scope of this discussion. It is recommended practitioners review the references for additional materials to continue their journey.

Emerging Advances

Using functional reinforcers

Functional reinforcers are the reinforcers already existing in the environment that have been identified via assessment as maintaining the behavior.[5] It can be particularly helpful to look for the functional reinforcer when evaluating undesired behavior and the consequence that maintains it. For example, an animal that lunges or moves away when a veterinary professional approaches, may want distance as a reinforcer. This is information that can be used to develop an appropriate behavior intervention plan.

Using the functional reinforcer results in the animal getting the outcome it desires. It also avoids delays to appropriate interventions in the case of problem behavior. This can also potentially help address continued reinforcement and rehearsal of undesired responses.

Animal caregivers and practitioners often overlook that animals emit behaviors such as aggressive responses, escape and avoidance to gain distance as a reinforcer from an aversive stimulus. This stimulus could be a person, object, place, sound, or other stimulus. When this is observed, it is an indicator that a negative reinforcement contingency is likely maintaining the behavioral response.

Negative reinforcement

Rosales-Ruiz pointed out that fear responses and aggressive behaviors have generally been regarded as respondent behaviors.[8] Traditionally, practitioners have been advised to use only respondent conditioning techniques, such as desensitization or counterconditioning. The focus of these procedures is on changing the animal's emotional state towards the aversive stimulus; however, these can many times be difficult to successfully apply in veterinary settings with animals presenting such extreme responses that no appetitive will overcome the strength of the aversive stimulus. Often no significant changes in behavior are observed, or the response is that of an obviously uncomfortable animal attempting to collect the desired item while forced to tolerate the presence of the aversive stimulus. This might be described as the animal behaving as if conflicted. Additionally, the respondent behavior may have occurred in the initial interactions with the stimulus. Subsequent responses are likely operant in which behaviors emitted by the animal in the presence of the stimulus result in desired outcomes, such as increased distance as described previously.[9]

Other common procedures to address fear responses and aggressive behavior include trying to directly shape a calm response using food or other appetitives. Again, if the animal is unreceptive to appetitives there is often little option for success unless the practitioner is willing to consider deprivation tactics that may have the potential to compromise animal welfare. In such an example this has the potential to make positive reinforcement coercive if the animal has only one way to access the appetitive.

As has been discussed, another option is to recognize that fear responses and aggressive behaviors can be seen as successful behaviors that are maintained by negative reinforcement. Engaging in these behaviors leads to giving the animal a desired outcome, increased distance from the aversive stimulus. For example, a parrot learns that when it lunges, hands will go away and leave the bird alone.[8] This view offers an alternative approach to addressing these problem behaviors. When applied properly, distance can be used as a reinforcer to shape calm behaviors that replace fear responses or aggressive behaviors. Using this approach has the potential to expose the animal to a much less potent aversive stimulus than using counter conditioning, in which the pressure for the animal to tolerate the presence of the aversive stimulus may be greater.

Practitioners (and caregivers) have been led to believe negative reinforcement is to be avoided at all costs; however, this requires further clarification. Negative reinforcement is a learning principle. How it is applied in a procedure can look vastly different. Often in the veterinary environment, the aversive stimulus is already present. It is not introduced or contrived to coerce behavior. This distinction is important. This changes the objective to creating conditions in which the aversive stimulus can be far enough away the animal can emit desired responses that can be reinforced by the removal of the aversive stimulus. This benign application of negative reinforcement is giving the animal the outcome it desires under conditions in which it can be successful. Applied in this manner, the procedure can be used to introduce novel people, medical instruments, and to teach birds laying calmly in a towel will result in looser restraint. See the reference list for additional resources with details on practical application.[10-13]

Emotions as contingency descriptors

Emotions have often been removed from discussions of animal behavior due to their inaccessibility. They have been described as internal, mentalistic, indescribable, requiring verbal communication, and anthropomorphic; however, an alternative approach is to consider emotions as covert, and emotional behavior as overt. Emotions may not be observable but emotional behavior is observable and can include physical movements, vocalizations, feather positions, changes in eye position, among others that the bird emits in response to the environment. Emotional behavior can alert practitioners to aversive stimuli impacting behavior.

For example, if a bird is leaning away from a towel, giving darting looks, searching for an escape route, and crouching in preparation to fly, this can indicate a negative reinforcement contingency is in place. The emotional behaviors observed can be equated with a fear response, but they also give information about the conditions that are giving rise to this response. This overt behavior gives information about the contingencies that are impacting the animal that result in the emotional behavior. The contingencies can be changed which results in a change in the emotional behavior.[6] Therefore, the practitioner can change the contingencies to address the emotional response. In this example the functional reinforcer is distance from the aversive stimulus. Therefore, one potential option is to start with the towel far enough away that the bird can emit calm responses and then remove the towel. This distance is gradually decreased and the removal process for calm behavior is repeated, until the towel is close, and the bird remains calm. Usually at this stage, the bird is calm enough to be receptive to appetitives in the presence of the towel. A shaping procedure with positive reinforcement can then be initiated.

Viewing emotions and emotional behavior as the outcome of contingencies gives practitioners starting points towards addressing distress using systematic interventions that create new desired behavior change. Traditional approaches have focused on stopping the emotion of fear. Building a new repertoire of behavior ultimately teaches the animal how to respond in the presence of the stimulus, which can help change the response to the stimulus from aversive to appetitive.

Nonlinear contingency analysis

Nonlinear contingency analysis can help reveal the multiple contingencies that are operating on the same target behavior. This can be viewed as a package of contingencies (which includes the contingency that results from the direct linear analysis). Each of these contingencies involves a cost or benefit. If one of these contingencies involves a critical consequence (something important to the survival of the animal), it can override or suppress the impact of the other contingencies.[5,14] This type of analysis can help reveal when training strategies may be less than optimal. For example, if an animal is hesitant to go into a crate and allow the door to be closed, a strategy that might be considered is to only offer the animal's diet in the crate. While this may on the surface appear to be a positive reinforcement-based procedure (offering food for moving into the enclosed space), there are also other contingencies to consider. Hesitancy to move into the crate could indicate a negative reinforcement contingency is also maintaining behavior. The animal is moving to avoid or gain distance from the crate. The animal may also emit this response when closing the door is attempted, which would indicate another negative reinforcement contingency. Therefore, for the positive reinforcement procedure to be non-coercive and more effective, it would be beneficial to address the negative reinforcement contingencies first that are maintaining escape and avoidance behaviors.[15] This would remove the high cost involved in approaching and entering the crate and increase the benefit in approaching to access the food, instead of waiting for hunger to override the fear response. A nonlinear contingency analysis is used in programs such as the constructional approach.

For most practitioners and caregivers, it is usually easy to identify positive reinforcement contingencies. This has been promoted and used frequently in contemporary animal training. Positive reinforcement may also have been contrived, meaning the reinforcer selected is often food or some other convenient appetitive; however, as described above, functional reinforcers are those that already are maintaining the behavior and may be the preferred outcome for the animal. And as nonlinear contingency analysis demonstrates, they are not the only factor influencing behavior. The environment can also be impacting the behavior in other ways (e.g., loud places, unstable perching, slippery surfaces) Further investigation can reveal the impact of relevant prior learning history and the natural history of the species (ethology and phylogeny). This can help practitioners gain more understanding as to why a preferred food item may not always be successful in reaching desired behavior goals.

Consider behavioral freedom

Behavioral freedom is the possibility of genuine alternative choices that result in equally reinforcing consequences.[16]

Using the behavioral freedom model provides a more accurate way of measuring choice rather than using cultural or colloquial based assumptions about choice (and control). It also allows the word control to remain aligned with its already understood relationships to behavioral control (e.g., control by contingent consequences, stimulus control).[17] It causes practitioners to ask the important questions of what is the outcome for the animal when it doesn't participate? If the result is "nothing" then true choice has not been provided. Often in traditional animal training there is only one option for gaining desired outcomes. This results in zero degrees of freedom. Increasing degrees of freedom gives us information to consider about why our animal does or does not participate in training.

For example, a parrot may be trained to touch the tip of a syringe to access a preferred food item. If that same preferred food item is offered freely nearby and the bird actively avoids the syringe to access the freely available food, this can provide information that distance from the syringe is a desired outcome. This indicates a negative reinforcement contingency exists that must be addressed. By providing one degree of freedom, the practitioner can discover when positive reinforcement is coercive and implement procedures that better address superimposition of contingencies. Superimposition occurs when contingencies are competing. In this example, adding one degree of freedom reveals the negative reinforcement contingency is still maintaining escape and avoidance responses and that positive reinforcement for touching the syringe with zero degrees of freedom was coercive.

Practitioners can regularly evaluate items and experiences provided as consequences to determine if access, availability, and conditions have created critical consequences. Practitioners should regularly evaluate the degrees of freedom tied to each critical consequence to determine the behavioral freedom. The smaller the number of degrees of freedom, the greater the coercion.[16]

Genuine choice exists when animals have many equally available opportunities to earn equally potent consequences.[18] This approach helps practitioners understand that coercion cannot be reduced to the oversimplified interpretation of aversive control.[16] It helps illustrate when there is power over resources (critical consequences) and access to those resources, it is possible to be coercive.

Assent

Assent means to agree with an opinion, approval, or permission. In other words, the participant agrees with what is happening. This agreement can be non-verbal and expressed by one's actions or conduct.[19] In animal training this means assessing body language and often providing freedom of movement to escape or avoid participation if desired. Assent usually signals stronger agreement than consent which suggests acquiescence and little more. For example, a human can sign their consent to a procedure but may not agree with it. Practitioners are really looking for assent in animal training although the word consent is often conflated with assent. Animal caregivers are also monitoring for assent throughout the training process. Assent also requires that animals have multiple ways to access desired outcomes as described in the behavioral freedom section. Limiting options to only one, especially for critical consequences such as food, can be coercive.

Assent can also be evaluated by looking for responses such as affiliative behaviors (e.g., soliciting play, allogrooming, allopreening), offering trained behaviors, approaching the practitioner, engaging with the practitioner, accepting food items or other appetitives. It is desired to see animals showing relaxed and calm responses or what might be described as eager to participate type responses.

Assent withdrawal can be evaluated by looking for responses such as escape or avoidance behaviors, apathy/ ignoring the practitioner, engaging in other more appetitive activities, fear responses, aggressive behavior, conflicted responses, among others.

Start buttons and assent

A start button is a trained behavior with a great deal of reinforcement history tied to a specific behavior chain. This term has become popularized recently, but the practice is not new. Prior to naming this phenomenon many caregivers had learned to look for body language that preceded an animal engaging in a desired trained response. The start button strategy teaches people to either look for or train this specific behavior and add it to the chain; however, this can only be true assent if two or more start buttons are available at the same time that result in equally reinforcing consequences.

Resources and Tools

Behavior intervention guidelines

Whenever it is determined that a change of behavior is essential, animals deserve the most effective, efficient, and optimal interventions and behavior change plans.[7] Fortunately, there are resources available to help guide practitioners through this process, whether it is addressing an undesired behavior or working towards a new goal.

Functional behavior assessment is a process used for gathering information used to maximize the effectiveness and efficiency of behavioral support.[20] It includes providing an operationalized definition of the behavior, identification of events that are functionally related to the behavior, identification of consequences that maintain the behavior, hypothesis about the function of the behavior, and direct observation to confirm/support hypothesis, as well as careful monitoring of interventions once implemented. This gathering of information helps caregivers speculate what might be the reasons behind the behaviors observed.[21,22] Many practitioners may recognize this as the A-B-C or three term contingency analysis of a target behavior. This is a helpful starting point, especially when trying to determine the functional reinforcer. This can be used in conjunction with nonlinear contingency analysis for further understanding of what maintains a behavior.

The free behavior intervention guidelines at BIGforAnimals.com provide step by step instructions on how to conduct a functional behavior assessment.[7]

The constructional approach to animal welfare and training

Many traditional behavior-change programs focus on reducing or eliminating undesired responses. This can result in undesired outcomes that are occasioned from using extinction and punishment-based procedures. The constructional approach developed by Goldiamond[23] focuses exclusively on building repertoires of behaviors.

The constructional approach is a program that includes five elements for attaining desired behavioral outcomes. The five elements include the following: 1) Identify a specific behavior goal 2) Identify existing behaviors (a starting point from which behavior can be built) 3) Determine appropriate change procedures based upon 4) the consequences (linear and nonlinear) that maintain the behavior including functional reinforcers, and 5) implement a means of monitoring progress. This can be as simple as taking video recordings of sessions.[5]

This summary does not do justice to the nuances that differentiate this program from others; however, resources such as Nonlinear Contingency Analysis,[5] The Constructional Approach to Animal Welfare and Training,[24] Exotic Animal Training. A Constructional Approach to Addressing Fear Responses and Aggressive Behavior,[13] and various publications on the Organization for Reinforcement Contingencies with Animals[25] can provide details on application that can lead to improved animal welfare.

Conclusions

Many fields are benefitting from the emerging trends in behavior analysis that are improving animal welfare. This includes a focus on identifying and using functional reinforcers. This has led to recognizing that negative reinforcement has been the maintaining contingency for many behaviors. This in turn has resulted in the development of shaping procedures using negative reinforcement that reinforce calm repertoires of behaviors under conditions in which animals can be successful. Furthermore, it is without question that identifying the many contingencies that are impacting a target behavior is essential to fully understand what maintains behavior. Behavioral freedom provides clarity to the questions of choice, control, and assent. These often-misunderstood topics are far more complex than what procedures were used, or whether an animal can opt in or opt out. As practitioners seek to add more structure and systematic approaches to their behavioral interventions, resources such as BIG for Animals and the constructional approach can facilitate continued growth towards evidence-based behavioral practices.

References

1. McLaughlin A, Heidenreich B, Strunk A, Clark P. Fear Free Veterinary Certification Program – Avian. Available at: https://fearfreepets.com/fear-free-certification-overview-avian/. Accessed August 19, 2022.

2. Fisher W. *Handbook of Applied Behavior Analysis.* New York, NY: Guilford Publications; 2011.

3. Cooper JO, Heron TE, Heward WL. *Applied Behavior Analysis.* 3rd Ed. London, UK: Pearson Education; 2019.

4. Pierce WD, Cheney CD. *Behavior Analysis and Learning.* 4th Ed. New York, NY: Psychology Press; 2017.

5. Layng TVJ, Andronis PT, III RTC, Abdel-Jalil A. *Nonlinear Contingency Analysis.* London, UK: Taylor & Francis; 2022.

6. Layng TVJ. Private emotions as contingency descriptors: emotions, emotional behavior, and their evolution. *Eur J Behav Anal.* 2017;18(2):168-179.

7. Heidenreich B, Farhoody P, Hetts S, et al. Behavior Intervention Guidelines for Animals. Available at: www.BigForAnimals.com. Accessed August 19, 2022.

8. Rosales-Ruiz J. From Fierce or Fearful to Friendly: A Shaping Program for Dogs, Cats, Horses and Beyond. Available at: https://www.caawt.com/cat. Accessed August 19, 2022.

9. Snider K. CAT with Kellie Snider. The Animal Training Fundamentals Podcast with Barbara Heidenreich. Available at: https://animaltrainingfundamentals.com/podcasts/c-a-t-with-kellie-snider/. Accessed August 19, 2022.

10. Heidenreich B, McLaughlin A. Addressing Fear and Aggression in Avian Patients. A Constructional Approach. Available at: https://fearfreepets.com/courses/addressing-fear-aggression-in-avian-patients-a-constructional-approach/. Accessed August 19, 2022.

11. Fernandez EJ. Training petting zoo sheep to act like petting zoo sheep: an empirical evaluation of response-independent schedules and shaping with negative reinforcement. *Animals.* 2020;10(7):1122.

12. Katz M, Rosales-Ruiz J. Constructional fear treatment: teaching fearful shelter dogs to approach and interact with a novel person. *J Exp Anal Behav.* 2022;118(2):278-291.

13. Heidenreich B. Exotic Animal Training. A Constructional Approach to Addressing Fear Responses and Aggressive Behavior. Available at: https://animaltrainingfundamentals.com/courses/the-constructional-approach-to-addressing-fear-and-aggressive-behavior/. Accessed August 19, 2022.

14. Layng TVJ. The search for an effective clinical behavior analysis: the nonlinear thinking of Israel Goldiamond. *Behav Anal.* 2009;32:163-184.

15. Layng TVJ. Consequences superimposition, coercion, and their nonlinear alternatives. *Proc Anim Behav Manag Alliance.* Available at: https://www.animalprofessional.com/abma-2022-tuesday.html. Accessed August 19, 2022.

16. de Fernandes RC, Dittrich A. Expanding the behavior-analytic meanings of "freedom": the contributions of Israel Goldiamond. *J Behav Soc Sci.* 2018;27:4-19.

17. Farhoody P. Animal training revisited. Operants. Double Issue II and III. 2021.

18. Layng TVJ. Coercion without aversive stimuli. *Proc Art Sci Anim Training.* 2020.

19. Heidenreich B. Assent or Consent? Which one do you think is most relevant to animal training? Available at: https://youtu.be/OBnoRh_9Bl8. Accessed August 19, 2022.

20. O'Neill R, Horner R, Albin R, et al. *Functional Assessment and Programme Development for Problem Behaviour: A Practical Handbook.* Pacific Grove, CA: Brooks/Cole Publishing; 1997.

21. Ingram K, Lewis-Palmer T, Sugai G. Function-based intervention planning: comparing the effectiveness of fba function-based and non-function-based intervention plans. *J Posit Behav Interv.* 2005;7(4):224-236.

22. Missouri Department of Elementary and Secondary Education. Functional Behavioral Assessment. Tim Lewis. Available at: https://youtu.be/C_AKrr_mCJ8. Accessed August 19, 2022.

23. Goldiamond I. Toward a constructional approach to social problems: ethical and constitutional issues raised by applied behavior analysis. *Behav Soc Sci.* 2002;11(2):108-197.

24. The Constructional Approach to Animal Welfare and Training. Available at: https://www.caawt.com/. Accessed August 19, 2022.

25. University of North Texas. The Organization for Reinforcement Contingencies with Animals. Available at: https://orca.unt.edu/publications/. Accessed August 19, 2022.

Strategic Use of Enrichment for Feather Damaging Behavior

Yvonne van Zeeland, DVM, MVR, PhD, DipECZM (Avian), DipECZM (Small Mammal), CPBC

Session #3007

Affiliation: From the Division of Zoological Medicine, Department of Clinical Sciences, Faculty of Veterinary Medicine, Utrecht University, Yalelaan 108, 3584 CM Utrecht, The Netherlands.

Abstract: Feather damaging behavior is frequently displayed by parrots kept in captivity and can be a complex and frustrating problem for veterinarians to deal with due to the multitude of medical, neurobiologic and socio-environmental factors that can underlie the condition. While management of the condition is presumably most effective when targeting any and all involved risk factors, the strategic use of enrichment in its many forms can be beneficial in many cases, even those with a medical background. When designing and implementing an enrichment plan, a systematic and goal-oriented approach, which takes the species' biology, caregiver's possibilities, as well as the individual's abilities, preferences, temperament and (early) life experiences into account, combined with proper documentation and regular follow-up are key to ensuring the end result is tailored to the needs of the individual bird.

Introduction

Feather damaging behavior (FDB), also referred to as feather destructive behavior, feather plucking, feather picking or pterotillomania, is common in captive psittacine birds, with a prevalence of up to 40% reported in cockatoos (*Cacatua* spp.) and grey parrots (*Psittacus erithacus*). Feather damaging behavior includes the chewing, plucking, biting or pulling of feathers that are accessible to the beak, and usually represents a multifactorial problem, in which a multitude of medical, genetic, neurobiologic and socio-environmental factors can play a role.[1] As such, a thorough medical and behavioral work-up, which includes a detailed history, complete physical examination, and – dependent on the findings – additional diagnostic tests, is pertinent to help identify the various factors that could be underlying the behavior.

A targeted approach, which addresses any and all factors that have been identified, is highly recommended when trying to manage a bird with FDB. Should any medical issues be encountered, these should be treated appropriately. In absence of (signs that point towards) an underlying medical cause, a primary behavioral origin for FDB becomes more likely. In such patients, the behavior is considered to originate from either an attempt of the bird to deal with a suboptimal living environment (i.e., maladaptive behavior) or from an altered neurochemistry (i.e., malfunctional behavior). While maladaptive behavior will respond favorably to changes that optimize the living environment to the bird's needs, this will no longer be the case in birds with malfunctional behavior, as the behavior has become ritualized (similar to stereotypic and other abnormal repetitive behaviors). Therefore, it will be displayed regardless of any inciting stimuli and generally require the use of psychoactive drugs to alter brain neurochemistry and make the bird more receptive to (changes in) its living environment. As maladaptive behavior may develop into malfunctional behavior following prolonged and continued exposure to a suboptimal living environment, hence hindering easy distinction of the two conditions, the use of environmental enrichment is commonly used as the first go-to treatment as this is the least intrusive and most beneficial to the (welfare of the) patient; however, even in patients in which medical conditions have not been ruled out, or in those in which the inciting and/or reinforcing stimuli have not been clearly identified, environmental enrichment will promote a more stimulating environment that can help to alleviate FDB by offering the bird a distraction or redirect the underlying 'trigger(s)' or 'motivation' for FDB. Thus, regardless of the cause, use of enrichment should be considered an important cornerstone in the management of any patient with FDB.

Environmental Enrichment

According to Newberry (1995),[2] environmental enrichment comprises "an improvement in the biological functioning of captive animals resulting from modifications to their environment." Offering enrichment to a parrot will allow it to perform species-typical behaviors, provide mental and physical stimulation, and enable the bird to exert control over its environment, which are all beneficial to its well being. Five categories of enrichment can be distinguished, which include:

1) Social enrichment: interaction with conspecifics, humans, other birds or animals through direct or indirect (vocal, visual, physical) contact;

2) Occupational enrichment: enrichment that provides a mental challenge (cognitive enrichment such as puzzles, training, novelty) or promotes physical exercise (e.g., flying, climbing, walking);

3) Physical enrichment: changes to the bird's habitat in terms of position, type, and size of the enclosure, as well as the climate (sunlight, temperature, humidity) and furnishings within the enclosure (e.g., perches, substrate, nest boxes, bathing opportunities, gnawing/chewing materials, swings, ropes and other toys);

4) Sensory enrichment: enrichment targeting one of the five senses: vision, audition, touch, taste and olfaction;

5) Nutritional and foraging enrichment: enrichment aimed at stimulating food-related activities.

Please note that these categories are partially overlapping one another and that items are grouped according to their most important function.

Enrichment for Behavior Modification

Enrichment is frequently used in the treatment of problem behaviors. In the context of a behavior modification strategy, enrichment can be implemented as part of antecedent arrangement to help reduce the likelihood of the behavior occurring or as a reinforcer (reward) to stimulate alternative, incompatible or other behavior. In humans, a review of 71 publications showed 265 uses of environmental enrichment, either as a standalone treatment or in combination with behavior manipulations used to either prompt or reinforce alternative behaviors or to block or restrain problem behavior.[3] When used alone, environmental enrichment was effective in 41% of the cases, hence indicating its value in the treatment of problem behavior; however, when combined with manipulations of alternate or problem behaviors, the treatment success further increased, suggesting that treatment should preferably include additional behavior modification methods.

Environmental enrichment has also been proven to be effective in the management of parrots with feather damaging behavior. In particular, stimulation of foraging activity is recommended as FDB has been inversely correlated to foraging effort.[4,5] As foraging behavior (i.e., the act of searching for and consuming food) is highly conserved throughout evolution, many species (including parrots) are motivated to work for food even if the same food is freely available (i.e., contrafreeloading). Compared to their wild counterparts, companion parrots are found to spend significantly less time on foraging (i.e., 30-60 minutes per day versus 4-8 hours per day) as food is often readily available in abundance. In absence of appropriate stimuli that promote captive foraging, the motivation to forage may be redirected or displaced to the feathers, leading to FDB.[6] Provision of foraging enrichment will enable the bird to perform its species-typical foraging activities and fulfill this foraging motivation. In the management of FDB, stimulating foraging activity is considered especially beneficial as it is incompatible with FDB (i.e., both activities require use of the beak and cannot be performed simultaneously). Moreover, the food obtained through foraging may act as a reinforcer that can result in the bird electing to forage over destroying its feathers if it has equal or greater value than the (extrinsic or intrinsic) reinforcement that is obtained from the feather destruction. Because of food being a primary reinforcer, foraging enrichment also tends to maintain its stimulating value for longer periods of time. It furthermore has the added benefits of increasing activity, providing mental stimulation and alleviating stress and boredom.

Foraging enrichment

In essence, four basic strategies can be distinguished to promote foraging in parrots. They include:

1) Increasing the spatial or temporal variability of feeding (e.g., through provision of smaller, more frequent meals in multiple bowls and/or locations)

2) Increasing search time for food (e.g., by scattering or hiding food in the enclosure or mixing it with inedible items)

3) Increasing extraction time (e.g., through use of foraging devices or puzzle feeders)

4) Increasing the time needed to process and ingest food (e.g., by providing whole food items, larger food particles or frozen ice blocks with food inside)

For birds that are new to foraging, starting with easy foraging tasks (e.g., multiple bowls, scatter feeding, mixing with inedible items) is recommended, while tasks can gradually be made more complex (e.g., through use of puzzle feeders or a combination of foraging strategies) for more experienced birds. Aside from offering foraging opportunities, the source of nutrition or the food itself can also function as enrichment (e.g., by presenting different types of or novel foods, varying the type of food offered over time or by using seasonal produce, or limiting provision of favorite foods to training or foraging activities).

Other types of enrichment

While foraging enrichment is seemingly highly effective in reducing FDB, promoting a more stimulating environment by means of social contact, physical exercise, mental stimulation, chewing toys, sensory input and other forms of environmental enrichment should also be considered in the therapeutic plan. As previously mentioned, causes for FDB can be multifactorial and complex. In essence, any environmental factor that results in a negative affective state (e.g., stress, fear, anxiety, boredom, frustration) can lead to displaced or redirected FDB that develops in an attempt of the bird to cope with a suboptimal living environment. Both absence of appropriate stimuli, which limit the ability to perform species-typical behaviors (e.g., solitary housing, lack of gnawing opportunities, flight deprivation) and exert control over the environment, as well as presence of potential aversive stimuli (e.g., presence of a predator, exposure to novelty, sudden changes in routine, family structure, etc.) can contribute to FDB and should be addressed appropriately when trying to manage such cases.

Social enrichment

Parrots in particular are known to be highly social, and lack of social contact can result in problem behaviors such as excessive screaming to attract attention, anxiety, and FDB. Social enrichment can be achieved through vocal, visual and/or physical contact with conspecifics; however, for parrots that have been hand-reared without conspecifics and are heavily imprinted on humans, other birds may be a stressor. As such, the parrot's rearing history needs to be taken into account to help determine whether it will favor a human, conspecific or other bird or animal as its social companion. In human-bonded birds, the type of interaction with the caregiver is important as well, as too much physical contact may stimulate reproductive-related behaviors (e.g., regurgitation, nesting behavior) that in turn can lead to 'sexual frustration' and associated FDB.

Occupational enrichment

Occupational enrichment will generally also be beneficial to consider in the treatment plan, particularly those activities that promote beak exercise (e.g., exploration of novel objects, chewing opportunities), as these are incompatible with FDB (similar to foraging). Cognitive enrichment in particular is important for highly intelligent animals such as parrots, allowing them to be mentally stimulated and (re)gain a certain control over their environment. For example, training can be used to teach a parrot specific husbandry behaviors such as entering a carrier or acceptance of a towel, thereby offering the bird an opportunity to cooperate rather than being forced to undergo these activities. In many cases, such training will require the use of so-called desensitization and counterconditioning techniques, as birds may initially be fearful and therefore hesitant or unwilling to participate in the training.

Physical enrichment

Physical enrichment encompasses the various physical features of the bird's living environment, including bathing opportunities (which are important for stimulating preening activity and maintaining optimal plumage condition) and toys or other objects that stimulate independent play. Preference should be given to items that are favored by the bird and elicit its curiosity. Many parrots will favor items that are easily destroyed (e.g., objects made of paper, cardboard or soft wood). Aside from the type of material, characteristics such as size, color or texture can also influence the parrot's interest or aversion to a certain object. For example, many of the larger parrot species will prefer items that can be held in the foot while manipulating it with the beak (so-called podomandibulation). Several studies have shown preferences for certain characteristics within a species that can be helpful as a starting point;[7-10] however, it should be remembered that preferences can vary greatly among individuals, in part dependent on their previous life experiences, thereby emphasizing the need to carefully evaluate the parrot's responses to the items and objects presented to it and adjusting the type of environment, as well as the frequency with which it is rotated, renewed or replaced, to their needs. Positive reinforcement, desensitization, counter conditioning and other behavior modification techniques may again be helpful to reward the bird's interest and stimulate interaction with the (novel) items provided.

Sensory enrichment

For parrots, sight, smell and touch comprise the main senses, thereby emphasizing the need for sensory enrichment by visual, tactile and auditory stimulation, including offering of different colored items, mirrors, moving or noise-making toys, radio (music) or television sounds; offering of variable substrates, perching surfaces or toys of different textures; or a change of scenery (novel surroundings). Similar to physical enrichment, preferences can be highly dependent on the individual.

Designing an Enrichment Program

When creating an enrichment program for a FDB bird, it is highly recommended to use the so-called SPIDER model.[11] The model consists of the following:

S: Setting goals

The goals of the enrichment program should be specified and be as concrete and measurable as possible, clearly identifying which behavior(s) the enrichment tends to encourage or discourage and to what extent. Larger, complex goals should ideally be broken into smaller, achievable goals that contribute to achieving the larger goal (e.g., FDB may be reduced by an increase in foraging time, an increase in bathing time and initiation of training activities).

Gain in-depth knowledge of the individual (e.g., rearing history, socialization, early life experiences) and natural history (e.g., natural habitat, dietary preferences, social structure, daily activity patterns) of the species involved, which will help determine its biological, social and cognitive needs.

P: Planning

An enrichment plan needs to be created which specifies the type of behaviors (intended) to be encouraged with the enrichment, the resources needed to create the enrichment initiative, and the location, time frame and circumstances during which the enrichment will be offered. The available space, resources, and time available by the caregiver should be carefully weighed as this will help to ensure feasibility of the approach. In addition, the species' biology, individual's temperament and early life experiences should be carefully considered, as these will all influence the value that the offered stimuli will have to the bird. Generally speaking, enrichments will have higher value if they are biologically relevant for the species, thereby emphasizing the importance of gaining more in-depth knowledge of the species' natural habitat, dietary preferences, social structure, and daily activity pattern. In addition, level of novelty offered should be considered, as neophobia and exploration have been tightly linked to species' ecology, personality traits, and early life experiences.[12-15] Equally important is to ensure that the plan is properly aligned with the capabilities of the individual bird (e.g., offering complex foraging puzzles may not be helpful in a bird that has never foraged in its life), allows the bird to maintain a free choice and exert control over its living environment, and addresses any possible safety concerns that warrant additional monitoring (e.g., a bird may accidentally ingest or get entangled in a toy, offering of enrichment may increase risk for aggression in group-housed birds).

I: Implementation

Implementation should be carried out in a logical and gradual way. Consider timing of and location where the enrichment is offered, as these can be vital to the overall success of the enrichment plan. Also ensure gradual introduction of the enrichment, as too much stimulation and novelty at once can cause undue stress in certain individuals. Challenges should be graduated so that they remain achievable yet allow the bird to continue to learn and evolve, which is important to help prevent frustration of both the bird and its caregiver. For anxious, nervous or fearful birds, the use of desensitization and counterconditioning techniques may be required.

D: Documentation

Ideally, a detailed record should be kept on when, how and what enrichment was presented; how the bird responded to it (e.g., type, frequency and duration of interactions); and on its behavior in general. Video recording may be useful to document enrichment use and should be encouraged where possible. This will help to determine which components of the enrichment program are the most and least useful. Unfortunately, observations can be easily skewed by wishful thinking, available time for observation, and misinterpretation, especially in cases of FDB, as this often occurs at night or when the owner is not present.

E: Evaluation

Periodic evaluation of the enrichment plan will help to determine whether the enrichment has the effects as originally intended on FDB and whether there are any safety or other concerns with its implementation. To accurately assess the value of the enrichment plan, it is essential for the clinician to receive uninterpreted, objective and accurate observations from the caretaker or to be able to review raw video footage. As an alternative or adjunct method, the use of serial photographs or assessment of the plumage condition, which serves as an indirect measurement of (changes in) FDB, can be considered.

R: Re-adjustment

Modify the enrichment plan by adjusting the enrichment or the way it is offered to optimize its effects on the behavior until the achieved goal is reached.

Following evaluation, modifications may be made to optimize the enrichment plan. Modifications may include removal or modification of poorly used elements (which can be revisited at a later time), addition of more of the well-used elements, or addition of new elements of enrichment. Sometimes changing the way in which an enrichment is offered (e.g., moving it to a new location) can spike the interest of the bird, despite the item being previously ignored. For fearful birds, repeated exposure may lead to habituation, thereby allowing them to become more accepting of objects they initially feared. Complexity, difficulty level and/or amount of food offered via foraging enrichment can be increased as the bird becomes more experienced. Similarly, physical exercise can and should progress as the bird develops better physical stamina and condition. Training can begin simple and gradually include more (complex) behaviors. Sensory enrichment can be adjusted to the preferences of the bird.

Frequent consultations will generally be required as the process is fluid and requires continuous evaluation and readjustment. Nevertheless, the SPIDER method is helpful and promising to gradually adjust an enrichment plan until it is tailored to meet the needs of the individual bird.

References

1. Van Zeeland YRA, Friedman SG, Bergmann L. Behavior. In: Speer B, ed. *Current Therapy in Avian Medicine and Surgery.* St. Louis, MO: Elsevier Health Sciences; 2016:177-251.

2. Newberry RC. Environmental enrichment: increasing the biological relevance of captive environments. *Appl Anim Behav Sci.* 1995;44:229-243.

3. Gover HC, Fahmie TA, McKeown CA. A review of environmental enrichment as treatment for problem behavior maintained by automatic reinforcement. *J Appl Behav Anal.* 2019;52:299-314.

4. Lumeij JT, Hommers CJ. Foraging 'enrichment' as treatment for pterotillomania. *Appl Anim Behav Sci.* 2008;111:85-94.

5. Mellor EL, McDonald Kinkaid HK, Mendl MT, et al. Nature calls: intelligence and natural foraging style predict poor welfare in captive parrots. *Proc Royal Soc B: Biol Sci.* 2021;288.

6. van Zeeland YRA, Spruit BM, Rodenburg TB, et al. Feather damaging behaviour in parrots: a review with consideration of comparative aspects. *Appl Anim Behav Sci.* 2009;121:75-95.

7. Kim LC, Garner JP, Millam JR. Preferences of orange-winged Amazon parrots (*Amazona amazonica*) for cage enrichment devices. *Appl Anim Behav Sci.* 2009;120:216-223.

8. Webb NV, Famula TR, Millam JR. The effect of rope color, size and fray on environmental enrichment device interaction in male and female orange-winged Amazon parrots (*Amazona amazonica*). *Appl Anim Behav Sci.* 2010;124:149-156.

9. Rozek JC, Millam JR. Preference and motivation for different diet forms and their effect on motivation for a foraging enrichment in captive orange-winged Amazon parrots (*Amazona amazonica*). *Appl Anim Behav Sci.* 2011;129:153-161.

10. Gupfinger R, Kaltenbrunner M. The design of musical instruments for grey parrots: an artistic contribution toward auditory enrichment in the context of ACI. *Multimodal Technol Interact.* 2020;4:16.

11. Alligood CA, Dorey NR, Mehrkam LR, et al. Applying behavior-analytic methodology to the science and practice of environmental enrichment in zoos and aquariums. *Zoo Biol.* 2017;36:175-185.

12. Mettke-Hofmann C, Winkler H, Leisler B. The significance of ecological factors for exploration and neophobia in parrots. *Ethol.* 2002;108:249-272.

13. Van Zeeland YRA, van der Aa MMJA, Vinke CM, et al. Behavioural testing to determine differences between coping styles in grey parrots (*Psittacus erithacus erithacus*) with and without feather damaging behaviour. *Appl Anim Behav Sci.* 2013;148:218-231.

14. Cussen VA, Mench JA. The relationship between personality dimensions and resiliency to environmental stress in orange-winged Amazon parrots (*Amazona amazonica*), as indicated by the development of abnormal behaviors. *PLoS ONE.* 2015;10.

15. Fox RA, Millam JR. Novelty and individual differences influence neophobia in orange-winged Amazon parrots (*Amazona amazonica*). *Appl Anim Behav Sci.* 2007;104:107-115.

Association of Avian Veterinarians
PO Box 9
Teaneck, NJ 07666
USA

ISBN 978-1-7138-7085-2

9 781713 870852

CURRAN ASSOCIATES INC.
proceedings.com

19th Industrial Conference on Advances in Data Mining (ICDM 2019)

New York, New York, USA
17-21 July 2019

ibai - publishing